36.99

PLUMBING
INSTANT ANSWERS

PLUMBING INSTANT ANSWERS

R. Dodge Woodson

McGRAW-HILL

New York   Chicago   San Francisco   Lisbon   London
Madrid   Mexico City   Milan   New Delhi   San Juan
Seoul   Singapore   Sydney   Toronto

McGraw-Hill

A Division of *The* **McGraw·Hill** *Companies*

1 2 3 4 5 6 7 8 9 0   DOC/DOC   0 9 8 7 6 5 4 3 2

ISBN 0-07-137957-6

The sponsoring editor for this book was Larry S. Hager and the production supervisor was Pamela Pelton. It was set in Stone Sans by Lone Wolf Enterprises, Ltd.

Printed and bound by R. R. Donnelley & Sons.

This book is printed on recycled, acid-free paper containing a minimum of 50% recycled, de-inked fiber.

McGraw-Hill books are available at special quantity discounts to use as premiums and sales promotions, or for use in corporate training programs. For more information, please write to the Director of Special Sales, McGraw-Hill, Professional Publishing, Two Penn Plaza, New York, NY 10121-2298. Or contact your local bookstore.

For Afton, Adam, and Victoria
my best friends.

CONTENTS

INTRODUCTION

Woodson has done it again! He has created yet another book that is an indispensable tool for anyone working within the plumbing industry. If you are looking for a unique resource that provides easy access to instant answers from a seasoned, hands-on expert, this is it. The graphic design of this book makes finding fast answers to your questions simple. Whether in the office or in the field, you can't beat this detailed guide to nearly every plumbing situation.

Hundreds upon hundreds of tables and figures give you visual reference to countless plumbing scenarios. There are conversion tables, quick-reference tables, graphics, fast facts, and hard-hitting, easy-to-find details to make your work go faster, smoother, and more profitably. Who else but America's plumber could have thought of and developed such a complete plumbing reference that is so fast and easy to use?

R. Dodge Woodson is the owner of Advanced Plumbing and a licensed master plumber and licensed master gasfitter with over 25 years of field experience. He has written many best-selling books for McGraw-Hill and has served as adjunct faculty for Central Maine Technical College as an instructor of both plumbing code classes and plumbing apprentice classes. Woodson has been called America's plumber and is well known internationally for his expertise in the trade. His reputation in the trade is undisputed.

What will you gain from owning this book? More than you can imagine. Take a few minutes to review the Contents. Thumb through the pages. Look at how easy it is to find the data you need. Where else have you ever seen so much information offered in such a fast, accessible, easy-to-understand format? There is no other plumber's reference like this one. No plumbing contractor or plumber should be without this valuable tool. Woodson has packed it full of over 25 years of real-world experience that is priceless. This book is as essential to a plumber as a right-angle drill!

FAST-TRACK CODE FACTS

The plumbing code is complex. This chapter is not a replacement for the code, but it will give you a lot of pertinent information that you may use daily in a concise, accessible manner. A majority of the tables provided here were generously provided by the International Code Council, Inc. and the *International Plumbing Code 2000*. The visual nature of this chapter will allow you to answer many of your code questions by simply reviewing the numerous tables.

In most cases, the tables will speak for themselves. While there may be some confusion, I will provide some insight in the use of a table. For the most part, this is a reference chapter that will not require heavy reading. Consider this your fast track to code facts.

WATER DISTRIBUTION PIPE

MATERIAL	STANDARD
Brass pipe	ASTM B 43
Chlorinated polyvinyl chloride (CPVC) plastic pipe and tubing	ASTM D 2846; ASTM F 441; ASTM F 442; CSA B137.6
Copper or copper-alloy pipe	ASTM B 42; ASTM B 302
Copper or copper-alloy tubing (Type K, WK, L, WL, M or WM)	ASTM B 75; ASTM B 88; ASTM B 251; ASTM B 447
Cross-linked polyethylene (PEX) plastic tubing	ASTM F 877; CSA CAN/CSA-B137.5
Cross-linked polyethylene/aluminum/cross-linked polyethylene (PEX-AL-PEX) pipe	ASTM F 1281; CSA CAN/CSA-B137.10
Galvanized steel pipe	ASTM A 53
Polybutylene (PB) plastic pipe and tubing	ASTM D 3309; CSA CAN3-B137.8

FIGURE 1.1 Approved materials for water distribution. (*Courtesy of International Code Council, Inc. and the International Plumbing Code 2000*)

2

WATER SERVICE PIPE

MATERIAL	STANDARD
Acrylonitrile butadiene styrene (ABS) plastic pipe	ASTM D 1527; ASTM D 2282
Asbestos-cement pipe	ASTM C 296
Brass pipe	ASTM B 43
Copper or copper-alloy pipe	ASTM B 42; ASTM B 302
Copper or copper-alloy tubing (Type K, WK, L, WL, M or WM)	ASTM B 75; ASTM B 88; ASTM B 251; ASTM B 447
Chlorinated polyvinyl chloride (CPVC) plastic pipe	ASTM D 2846; ASTM F 441; ASTM F 442; CSA B137.6
Ductile iron water pipe	AWWA C151; AWWA C115
Galvanized steel pipe	ASTM A 53
Polybutylene (PB) plastic pipe and tubing	ASTM D 2662; ASTM D 2666; ASTM D 3309; CSA B137.8
Polyethylene (PE) plastic pipe	ASTM D 2239; CSA CAN/CSA-B137.1
Polyethylene (PE) plastic tubing	ASTM D 2737; CSA B137.1
Cross-linked polyethylene (PEX) plastic tubing	ASTM F 876; ASTM F 877; CSA CAN/CSA-B137.5
Cross-linked polyethylene/ aluminum/cross-linked polyethylene (PEX-AL-PEX) pipe	ASTM F 1281; CSA CAN/CSA B137.10
Polyethylene/aluminum/ polyethylene (PE-AL-PE) pipe	ASTM F 1282; CSA CAN/CSA-B137.9
Polyvinyl chloride (PVC) plastic pipe	ASTM D 1785; ASTM D 2241; ASTM D 2672; CSA CAN/CSA-B137.3

FIGURE 1.2 Approved materials for water service piping. (*Courtesy of International Code Council, Inc. and the International Plumbing Code 2000*)

BUILDING SEWER PIPE

MATERIAL	STANDARD
Acrylonitrile butadiene styrene (ABS) plastic pipe	ASTM D 2661; ASTM D 2751; ASTM F 628
Asbestos-cement pipe	ASTM C 428
Cast-iron pipe	ASTM A 74; ASTM A 888; CISPI 301
Coextruded composite ABS DWV sch 40 IPS pipe (solid)	ASTM F 1488
Coextruded composite ABS DWV sch 40 IPS pipe (cellular core)	ASTM F 1488
Coextruded composite PVC DWV sch 40 IPS pipe (solid)	ASTM F 1488
Coextruded composite PVC DWV sch 40 IPS pipe (cellular core)	ASTM F 1488
Coextruded composite PVC IPS DR - PS DWV, PS140, PS200	ASTM F 1488
Coextruded composite ABS sewer and drain DR - PS in PS35, PS50, PS100, PS140, PS200	ASTM F 1488
Coextruded composite PVC sewer and rain DR - PS in PS35, PS50, PS100, PS140, PS200	ASTM F 1488
Concrete pipe	ASTM C 14; ASTM C 76; CSA A257.1; CSA CAN/CSA A257.2
Copper or copper-alloy tubing (Type K or L)	ASTM B 75; ASTM B 88; ASTM B 251
Polyvinyl chloride (PVC) plastic pipe (Type DWV, SDR26, SDR35, SDR41, PS50 or PS100)	ASTM D 2665; ASTM D 2949; ASTM D 3034; ASTM F 891; CSA B182.2; CSA CAN/CSA-B182.4
Stainless steel drainage systems, Type 316L	ASME/ANSI A112.3.1
Vitrified clay pipe	ASTM C 4; ASTM C 700

FIGURE 1.3   Approved materials for building sewer piping. *(Courtesy of International Code Council, Inc. and the International Plumbing Code 2000)*

UNDERGROUND BUILDING DRAINAGE AND VENT PIPE

MATERIAL	STANDARD
Acrylonitrile butadiene styrene (ABS) plastic pipe	ASTM D 2661; ASTM F 628; CSA B181.1
Asbestos-cement pipe	ASTM C 428
Cast-iron pipe	ASTM A 74; CISPI 301; ASTM A 888
Coextruded composite ABS DWV sch 40 IPS pipe (solid)	ASTM F 1488
Coextruded composite ABS DWV sch 40 IPS pipe (cellular core)	ASTM F 1488
Coextruded composite PVC DWV sch 40 IPS pipe (solid)	ASTM F 1488
Coextruded composite PVC DWV sch 40 IPS pipe (cellular core)	ASTM F 1488
Coextruded composite PVC IPS - DR, PS140, PS200 DWV	ASTM F 1488
Copper or copper alloy tubing (Type K, L, M or DWV)	ASTM B 75; ASTM B 88; ASTM B 251; ASTM B 306
Polyolefin pipe	CSA CAN/CSA-B181.2
Polyvinyl chloride (PVC) plastic pipe (Type DWV)	ASTM D 2665; ASTM D 2949; ASTM F 891; CSA CAN/CSA-B181.2
Stainless steel drainage systems, Type 316L	ASME/ANSI A112.3.1

FIGURE 1.4 Approved materials for underground building drainage and vent pipe. *(Courtesy of International Code Council, Inc. and the International Plumbing Code 2000)*

PIPE FITTINGS

MATERIAL	STANDARD
Acrylonitrile butadiene styrene (ABS) plastic	ASTM D 3311; CSA B181.1; ASTM D 2661
Cast iron	ASME B16.4; ASME B16.12; ASTM A 74; ASTM A 888; CISPI 301
Coextruded composite ABS DWV sch 40 IPS pipe (solid or cellular core)	ASTM D 2661; ASTM D 3311; ASTM F 628
Coextruded composite PVC DWV sch 40 pipe IPS-DR, PS140, PS200 (solid or cellular core)	ASTM D 2665; ASTM D 3311; ASTM F 891
Coextruded composite ABS sewer and drain DR-PS in PS35, PS50, PS100, PS140, PS200	ASTM D 2751
Coextruded composite PVC sewer and drain DR-PS in PS35, PS50, PS100, PS140, PS200	ASTM D 3034
Copper or copper alloy	ASME B16.15; ASME B16.18; ASME B16.22; ASME B16.23; ASME B16.26; ASME B16.29; ASME B16.32
Glass	ASTM C 1053
Gray iron and ductile iron	AWWA C110
Malleable iron	ASME B16.3
Polyvinyl chloride (PVC) plastic	ASTM D 3311; ASTM D 2665
Stainless steel drainage systems, Types 304 and 316L	ASME/ANSI A112.3.1
Steel	ASME B16.9; ASME B16.11; ASME B16.28

FIGURE 1.5 Approved materials for pipe fittings. *(Courtesy of International Code Council, Inc. and the International Plumbing Code 2000)*

ABOVE-GROUND DRAINAGE AND VENT PIPE

MATERIAL	STANDARD
Acrylonitrile butadiene styrene (ABS) plastic pipe	ASTM D 2661; ASTM F 628; CSA B181.1
Brass pipe	ASTM B 43
Cast-iron pipe	ASTM A 74; CISPI 301; ASTM A 888
Coextruded composite ABS DWV sch 40 IPS pipe (solid)	ASTM F 1488
Coextruded composite ABS DWV sch 40 IPS pipe (cellular core)	ASTM F 1488
Coextruded composite PVC DWV sch 40 IPS pipe (solid)	ASTM F 1488
Coextruded composite PVC DWV sch 40 IPS pipe (cellular core)	ASTM F 1488
Coextruded composite PVC IPS - DR, PS140, PS200 DWV	ASTM F 1488
Copper or copper-alloy pipe	ASTM B 42; ASTM B 302
Copper or copper-alloy tubing (Type K, L, M or DWV)	ASTM B 75; ASTM B 88; ASTM B 251; ASTM B 306
Galvanized steel pipe	ASTM A 53
Glass pipe	ASTM C 1053
Polyolefin pipe	CSA CAN/CSA-B181.2
Polyvinyl chloride (PVC) plastic pipe (Type DWV)	ASTM D 2665; ASTM D 2949; ASTM F 891; CSA CAN/CSA-B181.2; ASTM F 1488
Stainless steel drainage systems, types 304 and 316L	ASME/ANSI A112.3.1

FIGURE 1.6 Approved materials for above-ground drainage and vent pipe. *(Courtesy of International Code Council, Inc. and the International Plumbing Code 2000)*

SIZE OF PIPE IDENTIFICATION

PIPE DIAMETER (Inches)	LENGTH OF BACKGROUND COLOR FIELD (Inches)	SIZE OF LETTERS (Inches)
$3/4$ to $1^1/4$	8	0.5
$1^1/2$ to 2	8	0.75
$2^1/2$ to 6	12	1.25
8 to 10	24	2.5
over 10	32	3.5

For SI: 1 inch = 25.4 mm.

FIGURE 1.7 Requirements of pipe identification. *(Courtesy of International Code Council, Inc. and the International Plumbing Code 2000)*

Caulking Ferrules

Pipe size (inches)	Inside diameter (inches)	Length (inches)	Minimum weight each Lb. Oz.	
2	2-1/4	4-1/2	1	0
3	3-1/4	4-1/2	1	12
4	4-1/4	4-1/2	2	8

Caulking Ferrules (Metric)

Pipe size (mm)	Inside diameter (mm)	Length (mm)	Minimum weight each (kg)
50	57	114	0.454
80	83	114	0.790
100	108	114	1.132

Soldering Bushings

Pipe size (inches)	Minimum weight each Lb. Oz.		Pipe size (inches)	Minimum weight each Lb. Oz.	
1-1/4	0	6	2-1/2	1	6
1-1/2	0	8	3	2	0
2	0	14	4	3	8

Soldering Bushings (Metric)

Pipe size (mm)	Minimum weight each (kg)	Pipe size (mm)	Minimum weight each (kg)
32	0.168	65	0.622
40	0.224	80	0.908
50	0.392	100	1.586

FIGURE 1.8 Requirements for ferrules and bushings. *(Courtesy of International Code Council, Inc. and the International Plumbing Code 2000)*

MINIMUM NUMBER OF PLUMBING FACILITIES*

OCCUPANCY		WATER CLOSETS (Urinals, see Section 419.2) Male	Female	LAVATORIES	BATHTUBS/ SHOWERS	DRINKING FOUNTAINS (see Section 410.1)	OTHERS
A S S E M B L Y	Nightclubs	1 per 40	1 per 40	1 per 75	—	1 per 500	1 service sink
	Restaurants	1 per 75	1 per 75	1 per 200	—	1 per 500	1 service sink
	Theaters, halls, museums, etc.	1 per 125	1 per 65	1 per 200	—	1 per 500	1 service sink
	Coliseums, arenas (less than 3,000 seats)	1 per 75	1 per 40	1 per 150	—	1 per 1,000	1 service sink
	Coliseums, arenas (3,000 seats or greater)	1 per 120	1 per 60	Male 1 per 200 Female 1 per 150	—	1 per 1,000	1 service sink
	Churches[b]	1 per 150	1 per 75	1 per 200	—	1 per 1,000	1 service sink
	Stadiums (less than 3,000 seats), pools, etc.	1 per 100	1 per 50	1 per 150	—	1 per 1,000	1 service sink
	Stadiums (3,000 seats or greater)	1 per 150	1 per 75	Male 1 per 200 Female 1 per 150	—	1 per 1,000	1 service sink
	Business (see Sections 403.2, 403.4 and 403.5)	1 per 50		1 per 80	—	1 per 100	1 service sink
	Educational	1 per 50		1 per 50	—	1 per 100	1 service sink
	Factory and industrial	1 per 100		1 per 100	(see Section 411)	1 per 400	1 service sink
	Passenger terminals and transportation facilities	1 per 500		1 per 750	—	1 per 1,000	1 service sink
I N S T I T U T I O N A L	Residential care	1 per 10		1 per 10	1 per 8	1 per 100	1 service sink
	Hospitals, ambulatory nursing home patients[c]	1 per room[d]		1 per room[d]	1 per 15	1 per 100	1 service sink per floor
	Day nurseries, sanitariums, nonambulatory nursing home patients, etc.[c]	1 per 15		1 per 15	1 per 15[e]	1 per 100	1 service sink
	Employees, other than residential care[c]	1 per 25		1 per 35	—	1 per 100	—
	Visitors, other than residential care	1 per 75		1 per 100	—	1 per 500	—
	Prisons[c]	1 per cell		1 per cell	1 per 15	1 per 100	1 service sink
	Asylums, reformatories, etc.[c]	1 per 15		1 per 15	1 per 15	1 per 100	1 service sink
	Mercantile (see Sections 403.2, 403.4 and 403.5)	1 per 500		1 per 750	—	1 per 1,000	1 service sink
R E S I D E N T I A L	Hotels, motels	1 per guestroom		1 per guestroom	1 per guestroom	—	1 service sink
	Lodges	1 per 10		1 per 10	1 per 8	1 per 100	1 service sink
	Multiple family	1 per dwelling unit		1 per dwelling unit	1 per dwelling unit	—	1 kitchen sink per dwelling unit; 1 automatic clothes washer connection per 20 dwelling units
	Dormitories	1 per 10		1 per 10	1 per 8	1 per 100	1 service sink
	One- and two-family dwellings	1 per dwelling unit		1 per dwelling unit	1 per dwelling unit	—	1 kitchen sink per dwelling unit; 1 automatic clothes washer connection per dwelling unit[f]
	Storage (see Sections 403.2 and 403.4)	1 per 100		1 per 100	(see Section 411)	1 per 1,000	1 service sink

a. The fixtures shown are based on one fixture being the minimum required for the number of persons indicated or any fraction of the number of persons indicated. The number of occupants shall be determined by the *International Building Code*.
b. Fixtures located in adjacent buildings under the ownership or control of the church shall be made available during periods the church is occupied.
c. Toilet facilities for employees shall be separate from facilities for inmates or patients.
d. A single-occupant toilet room with one water closet and one lavatory serving not more than two adjacent patient rooms shall be permitted where such room is provided with direct access from each patient room and with provisions for privacy.
e. For day nurseries, a maximum of one bathtub shall be required.
f. For attached one- and two-family dwellings, one automatic clothes washer connection shall be required per 20 dwelling units.

FIGURE 1.9 Minimum number of plumbing facilities. *(Courtesy of International Code Council, Inc. and the International Plumbing Code 2000)*

MINIMUM SIZES OF FIXTURE WATER SUPPLY PIPES

FIXTURE	MINIMUM PIPE SIZE (inch)
Bathtubs (60″ × 32″ and smaller)[a]	$1/2$
Bathtubs (larger than 60″ × 32″)	$1/2$
Bidet	$3/8$
Combination sink and tray	$1/2$
Dishwasher, domestic[a]	$1/2$
Drinking fountain	$3/8$
Hose bibbs	$1/2$
Kitchen sink[a]	$1/2$
Laundry, 1, 2 or 3 compartments[a]	$1/2$
Lavatory	$3/8$
Shower, single head[a]	$1/2$
Sinks, flushing rim	$3/4$
Sinks, service	$1/2$
Urinal, flush tank	$1/2$
Urinal, flush valve	$3/4$
Wall hydrant	$1/2$
Water closet, flush tank	$3/8$
Water closet, flush valve	1
Water closet, flushometer tank	$3/8$
Water closet, one piece[a]	$1/2$

For SI: 1 inch = 25.4 mm, 1 foot = 304.8 mm, 1 psi = 6.895 kPa.

a. Where the developed length of the distribution line is 60 feet or less, and the available pressure at the meter is a minimum of 35 psi, the minimum size of an individual distribution line supplied from a manifold and installed as part of a parallel water distribution system shall be one nominal tube size smaller than the sizes indicated.

FIGURE 1.10 Minimum sizes of fixture water supply pipes. *(Courtesy of International Code Council, Inc. and the International Plumbing Code 2000)*

**MAXIMUM FLOW RATES AND CONSUMPTION
FOR PLUMBING FIXTURES AND FIXTURE FITTINGS**

PLUMBING FIXTURE OR FIXTURE FITTING	MAXIMUM FLOW RATE OR QUANTITY[b]
Water closet	1.6 gallons per flushing cycle
Urinal	1.0 gallon per flushing cycle
Shower head[a]	2.5 gpm at 60 psi
Lavatory, private	2.2 gpm at 60 psi
Lavatory (other than metering), public	0.5 gpm at 60 psi
Lavatory, public (metering)	0.25 gallon per metering cycle
Sink faucet	2.2 gpm at 60 psi

For SI: 1 gallon = 3.785 L, 1 gallon per minute = 3.785 L/m, 1 psi = 6.895 kPa.

a. A hand-held shower spray is a shower head.

b. Consumption tolerances shall be determined from referenced standards.

FIGURE 1.11   Maximum flow rates and consumption for plumbing fixtures and fixture fittings. *(Courtesy of International Code Council, Inc. and the International Plumbing Code 2000)*

MANIFOLD SIZING

NOMINAL SIZE INTERNAL DIAMETER (inches)	MAXIMUM DEMAND (gpm)	
	Velocity at 4 feet per second	Velocity at 8 feet per second
$1/2$	2	5
$3/4$	6	11
1	10	20
$1 1/4$	15	31
$1 1/2$	22	44

For SI: 1 inch = 25.4 mm, 1 gallon per minute = 3.785 L/m, 1 foot per second = 0.305 m/s.

FIGURE 1.12   Manifold sizing. *(Courtesy of International Code Council, Inc. and the International Plumbing Code 2000)*

WATER DISTRIBUTION SYSTEM DESIGN CRITERIA
REQUIRED CAPACITIES AT FIXTURE SUPPLY PIPE OUTLETS

FIXTURE SUPPLY OUTLET SERVING	FLOW RATE[a] (gpm)	FLOW PRESSURE (psi)
Bathtub	4	8
Bidet	2	4
Combination fixture	4	8
Dishwasher, residential	2.75	8
Drinking fountain	0.75	8
Laundry tray	4	8
Lavatory	2	8
Shower	3	8
Shower, temperature controlled	3	20
Sillcock, hose bibb	5	8
Sink, residential	2.5	8
Sink, service	3	8
Urinal, valve	15	15
Water closet, blow out, flushometer valve	35	25
Water closet, flushometer tank	1.6	15
Water closet, siphonic, flushometer valve	25	15
Water closet, tank, close coupled	3	8
Water closet, tank, one piece	6	20

For SI: 1 psi = 6.895 kPa, 1 gallon per minute (gpm) = 3.785 L/m.

a. For additional requirements for flow rates and quantities, see Section 604.4.

FIGURE 1.13 Water distribution system design criteria: required capacities at fixture supply pipe outlets. *(Courtesy of International Code Council, Inc. and the International Plumbing Code 2000)*

HORIZONTAL FIXTURE BRANCHES AND STACKS[a]

DIAMETER OF PIPE (inches)	Total for a horizontal branch	MAXIMUM NUMBER OF DRAINAGE FIXTURE UNITS (dfu) Stacks[b]		
		Total discharge into one branch interval	Total for stack of three branch intervals or less	Total for stack greater than three branch intervals
1½	3	2	4	8
2	6	6	10	24
2½	12	9	20	42
3	20	20	48	72
4	160	90	240	500
5	360	200	540	1,100
6	620	350	960	1,900
8	1,400	600	2,200	3,600
10	2,500	1,000	3,800	5,600
12	2,900	1,500	6,000	8,400
15	7,000	Footnote c	Footnote c	Footnote c

For SI: 1 inch = 25.4 mm.

a. Does not include branches of the building drain. Refer to Table 710.1(1).

b. Stacks shall be sized based on the total accumulated connected load at each story or branch interval. As the total accumulated connected load decreases, stacks are permitted to be reduced in size. Stack diameters shall not be reduced to less than one-half of the diameter of the largest stack size required.

c. Sizing load based on design criteria.

FIGURE 1.14 Drainage fixture units allowed on horizontal fixture branches and stacks. (*Courtesy of International Code Council, Inc. and the International Plumbing Code 2000*)

BUILDING DRAINS AND SEWERS

DIAMETER OF PIPE (inches)	MAXIMUM NUMBER OF DRAINAGE FIXTURE UNITS CONNECTED TO ANY PORTION OF THE BUILDING DRAIN OR THE BUILDING SEWER, INCLUDING BRANCHES OF THE BUILDING DRAIN[a]				
	Slope per foot				
	1/16 inch	1/8 inch	1/4 inch	1/2 inch	
1 1/4	—	—	1	1	
1 1/2	—	—	3	3	
2	—	—	21	26	
2 1/2	—	—	24	31	
3	—	36	42	50	
4	—	180	216	250	
5	—	390	480	575	
6	—	700	840	1,000	
8	1,400	1,600	1,920	2,300	
10	2,500	2,900	3,500	4,200	
12	3,900	4,600	5,600	6,700	
15	7,000	8,300	10,000	12,000	

For SI: 1 inch = 25.4 mm, 1 inch per foot = 0.0833 mm/m.

a. The minimum size of any building drain serving a water closet shall be 3 inches.

FIGURE 1.15 Drainage fixture units allowed for building drains and sewers. *(Courtesy of International Code Council, Inc. and the International Plumbing Code 2000)*

FIGURE 1.16 Maximum unit loading and maximum length of drainage and vent piping.
(Courtesy of International Code Council, Inc. and the International Plumbing Code 2000)

Maximum Unit Loading and Maximum Length of Drainage and Vent Piping

Size of Pipe, inches (mm)	1-1/4 (32)	1-1/2 (40)	2 (50)	2-1/2 (65)	3 (80)	4 (100)	5 (125)	6 (150)	8 (200)	10 (250)	12 (300)
Maximum Units											
Drainage Piping[1]											
Vertical	1	2[2]	16[3]	32[3]	48[4]	256	600	1380	3600	5600	8400
Horizontal	1	1	8[3]	14[3]	35[4]	216[5]	428[5]	720[5]	2640[5]	4680[5]	8200[5]
Maximum Length											
Drainage Piping											
Vertical, feet (m)	45 (14)	65 (20)	85 (26)	148 (45)	212 (65)	300 (91)	390 (119)	510 (155)	750 (228)		
Horizontal (Unlimited)											
Vent Piping (See note)											
Horizontal and Vertical											
Maximum Units	1	8[3]	24	48	84	256	600	1380	3600		
Maximum Lengths, feet (m)	45 (14)	60 (18)	120 (37)	180 (55)	212 (65)	300 (91)	390 (119)	510 (155)	750 (228)		

1 Excluding trap arm.
2 Except sinks, urinals and dishwashers.
3 Except six-unit traps or water closets.
4 Only four (4) water closets or six-unit traps allowed on any vertical pipe or stack; and not to exceed three (3) water closets or six-unit traps on any horizontal branch or drain.
5 Based on one-fourth (1/4) inch per foot (20.9 mm/m) slope. For one-eighth (1/8) inch per foot (10.4 mm/m) slope, multiply horizontal fixture units by a factor of 0.8.

Note: The diameter of an individual vent shall not be less than one and one-fourth (1-1/4) inches (31.8 mm) nor less than one-half (1/2) the diameter of the drain to which it is connected. Fixture unit load values for drainage and vent piping shall be computed from Tables 7-3 and 7-4. Not to exceed one-third (1/3) of the total permitted length of any vent may be installed in a horizontal position. When vents are increased one (1) pipe size for their entire length, the maximum length limitations specified in this table do not apply.

15

SIZE OF COMBINATION DRAIN AND VENT PIPE

DIAMETER PIPE (inches)	MAXIMUM NUMBER OF DRAINAGE FIXTURE UNITS (dfu)	
	Connecting to a horizontal branch or stack	Connecting to a building drain or building subdrain
2	3	4
$2^{1}/_2$	6	26
3	12	31
4	20	50
5	160	250
6	360	575

For SI: 1 inch = 25.4 mm.

FIGURE 1.17   Size of combination drain and vent pipe. *(Courtesy of International Code Council, Inc. and the International Plumbing Code 2000)*

SLOPE OF HORIZONTAL DRAINAGE PIPE

SIZE (inches)	MINIMUM SLOPE (inch per foot)
$2^{1}/_2$ or less	$^{1}/_4$
3 to 6	$^{1}/_8$
8 or larger	$^{1}/_{16}$

For SI: 1 inch = 25.4 mm, 1 inch per foot = 0.0833 mm/m.

FIGURE 1.18   Slope of horizontal drainage pipe. *(Courtesy of International Code Council, Inc. and the International Plumbing Code 2000)*

DRAINAGE FIXTURE UNITS FOR FIXTURE DRAINS OR TRAPS

FIXTURE DRAIN OR TRAP SIZE (inches)	DRAINAGE FIXTURE UNIT VALUE
$1^{1}/_4$	1
$1^{1}/_2$	2
2	3
$2^{1}/_2$	4
3	5
4	6

For SI: 1 inch = 25.4 mm.

FIGURE 1.19   Drainage fixture units for fixture drains or traps. *(Courtesy of International Code Council, Inc. and the International Plumbing Code 2000)*

MINIMUM CAPACITY OF SEWAGE PUMP
OR SEWAGE EJECTOR

DIAMETER OF THE DISCHARGE PIPE (inches)	CAPACITY OF PUMP OR EJECTOR (gpm)
2	21
$2^1/_2$	30
3	46

For SI: 1 inch = 25.4 mm, 1 gpm = 3.785 L/m.

FIGURE 1.20   Minimum capacity of sewage pump or sewage ejector. *(Courtesy of International Code Council, Inc. and the International Plumbing Code 2000)*

DRAINAGE FIXTURE UNITS FOR FIXTURES AND GROUPS

FIXTURE TYPE	DRAINAGE FIXTURE UNIT VALUE AS LOAD FACTORS	MINIMUM SIZE OF TRAP (inches)
Automatic clothes washers, commercial[a]	3	2
Automatic clothes washers, residential	2	2
Bathroom group as defined in Section 202 (1.6 gpf water closet)[f]	5	—
Bathroom group as defined in Section 202 (water closet flushing greater than 1.6 gpf)[f]	6	—
Bathtub[b] (with or without overhead shower or whirlpool attachments)	2	$1^1/_2$
Bidet	1	$1^1/_4$
Combination sink and tray	2	$1^1/_2$
Dental lavatory	1	$1^1/_4$
Dental unit or cuspidor	1	$1^1/_4$
Dishwashing machine,[c] domestic	2	$1^1/_2$
Drinking fountain	$1/_2$	$1^1/_4$
Emergency floor drain	0	2
Floor drains	2	2
Kitchen sink, domestic	2	$1^1/_2$
Kitchen sink, domestic with food waste grinder and/or dishwasher	2	$1^1/_2$
Laundry tray (1 or 2 compartments)	2	$1^1/_2$
Lavatory	1	$1^1/_4$
Shower	2	$1^1/_2$
Sink	2	$1^1/_2$
Urinal	4	Footnote d
Urinal, 1 gallon per flush or less	2[e]	Footnote d
Wash sink (circular or multiple) each set of faucets	2	$1^1/_2$
Water closet, flushometer tank, public or private	4[e]	Footnote d
Water closet, private (1.6 gpf)	3[e]	Footnote d
Water closet, private (flushing greater than 1.6 gpf)	4[e]	Footnote d
Water closet, public (1.6 gpf)	4[e]	Footnote d
Water closet, public (flushing greater than 1.6 gpf)	6[e]	Footnote d

For SI: 1 inch = 25.4 mm, 1 gallon = 3.785 L.

a. For traps larger than 3 inches, use Table 709.2.
b. A showerhead over a bathtub or whirlpool bathtub attachments does not increase the drainage fixture unit value.
c. See Sections 709.2 through 709.4 for methods of computing unit value of fixtures not listed in Table 709.1 or for rating of devices with intermittent flows.
d. Trap size shall be consistent with the fixture outlet size.
e. For the purpose of computing loads on building drains and sewers, water closets or urinals shall not be rated at a lower drainage fixture unit unless the lower values are confirmed by testing.
f. For fixtures added to a dwelling unit bathroom group, add the DFU value of those additional fixtures to the bathroom group fixture count.

FIGURE 1.21   Drainage fixture units. *(Courtesy of International Code Council, Inc. and the International Plumbing Code 2000)*

MAXIMUM DISTANCE OF FIXTURE TRAP FROM VENT

SIZE OF TRAP (inches)	SIZE OF FIXTURE DRAIN (inches)	SLOPE (inch per foot)	DISTANCE FROM TRAP (feet)
$1^1/_4$	$1^1/_4$	$^1/_4$	$3^1/_2$
$1^1/_4$	$1^1/_2$	$^1/_4$	5
$1^1/_2$	$1^1/_2$	$^1/_4$	5
$1^1/_2$	2	$^1/_4$	6
2	2	$^1/_4$	6
3	3	$^1/_8$	10
4	4	$^1/_8$	12

For SI: 1 inch = 25.4 mm, 1 foot = 304.8 mm, 1 inch per foot = 0.0833 mm/m.

FIGURE 1.22 Maximum distance of fixture trap from vent. *(Courtesy of International Code Council, Inc. and the International Plumbing Code 2000)*

COMMON VENT SIZES

PIPE SIZE (inches)	MAXIMUM DISCHARGE FROM UPPER FIXTURE DRAIN (dfu)
$1^1/_2$	1
2	4
$2^1/_2$ to 3	6

For SI: 1 inch = 25.4 mm.

FIGURE 1.23 Common vent sizes. *(Courtesy of International Code Council, Inc. and the International Plumbing Code 2000)*

SIZE AND LENGTH OF SUMP VENTS

DISCHARGE CAPACITY OF PUMP (gpm)	MAXIMUM DEVELOPED LENGTH OF VENT (feet)[a]					
	Diameter of vent (inches)					
	1¼[b]	1½	2	2½	3	4
10	No limit[b]	No limit	No limit	No limit	No limit	No limit
20	270	No limit	No limit	No limit	No limit	No limit
40	72	160	No limit	No limit	No limit	No limit
60	31	75	270	No limit	No limit	No limit
80	16	41	150	380	No limit	No limit
100	10[c]	25	97	250	No limit	No limit
150	Not permitted	10[c]	44	110	370	No limit
200	Not permitted	Not permitted	20	60	210	No limit
250	Not permitted	Not permitted	10	36	132	No limit
300	Not permitted	Not permitted	10[c]	22	88	380
400	Not permitted	Not permitted	Not permitted	10[c]	44	210
500	Not permitted	Not permitted	Not permitted	Not permitted	24	130

For SI: 1 inch = 25.4 mm, 1 foot = 304.8 mm, 1 gallons per minute = 3.785 L/m.

a. Developed length plus an appropriate allowance for entrance losses and friction due to fittings, changes in direction and diameter. Suggested allowances shall be obtained from NBS Monograph 31 or other approved sources. An allowance of 50 percent of the developed length shall be assumed if a more precise value is not available.

b. Actual values greater than 500 feet.

c. Less than 10 feet.

FIGURE 1.24 Size and length of sump vents. (*Courtesy of International Code Council, Inc. and the International Plumbing Code 2000*)

MINIMUM DIAMETER AND MAXIMUM LENGTH OF INDIVIDUAL BRANCH FIXTURE VENTS AND INDIVIDUAL FIXTURE HEADER VENTS FOR SMOOTH PIPES

DIAMETER OF VENT PIPE (inches)	INDIVIDUAL VENT AIRFLOW RATE (cubic feet per minute)																			
	Maximum developed length of vent (feet)																			
	1	2	3	4	5	6	7	8	9	10	11	12	13	14	15	16	17	18	19	20
$1/2$	95	25	13	8	5	4	3	2	1	1	1	1	1	1	1	1	1	1	1	1
$3/4$	100	88	47	30	20	15	10	9	7	6	5	4	3	3	3	2	2	2	2	1
1	—	—	100	94	65	48	37	29	24	20	17	14	12	11	9	8	7	7	6	6
$1^1/_4$	—	—	—	—	—	—	—	100	87	73	62	53	46	40	36	32	29	26	23	21
$1^1/_2$	—	—	—	—	—	—	—	—	—	—	—	100	96	84	75	67	60	54	49	45
2	—	—	—	—	—	—	—	—	—	—	—	—	—	—	—	—	—	—	—	100

For SI: 1 inch = 25.4 mm, 1 foot = 304.8 mm, 1 cfm = 0.4719 L/s.

FIGURE 1.25 Minimum diameter and maximum length of individual branch fixture vents and individual fixture header vents for smooth pipes. (*Courtesy of International Code Council, Inc. and the International Plumbing Code 2000*)

20

SIZE AND DEVELOPED LENGTH OF STACK VENTS AND VENT STACKS

DIAMETER OF SOIL OR WASTE STACK (inches)	TOTAL FIXTURE UNITS BEING VENTED (dfu)	MAXIMUM DEVELOPED LENGTH OF VENT (feet)[a] DIAMETER OF VENT (inches)										
		1¼	1½	2	2½	3	4	5	6	8	10	12
1¼	2	30										
1½	8	50	150									
1½	10	30	100									
2	12	30	75	200								
2	20	26	50	150								
2½	42		30	100	300							
3	10		42	150	360	1,040						
3	21		32	110	270	810						
3	53		27	94	230	680						
3	102		25	86	210	620						
4	43			35	85	250	980					
4	140			27	65	200	750					
4	320			23	55	170	640					
4	540			21	50	150	580					
5	190				28	82	320	990				
5	490				21	63	250	760				
5	940				18	53	210	670				
5	1,400				16	49	190	590				
6	500					33	130	400	1,000			
6	1,100					26	100	310	780			
6	2,000					22	84	260	660			
6	2,900					20	77	240	600	940		
8	1,800						31	95	240	720		
8	3,400						24	73	190	720		
8	5,600						20	62	160	610		
8	7,600						18	56	140	560		
10	4,000							31	78	310	960	
10	7,200							24	60	240	740	
10	11,000							20	51	200	630	
10	15,000							18	46	180	570	
12	7,300								31	120	380	940
12	13,000								24	94	300	720
12	20,000								20	79	250	610
12	26,000								18	72	230	500
15	15,000									40	130	310
15	25,000									31	96	240
15	38,000									26	81	200
15	50,000									24	74	180

FIGURE 1.26 Size and developed length of stack vents and vent stacks. *(Courtesy of International Code Council, Inc. and the International Plumbing Code 2000)*

SIZE OF DRAIN PIPES FOR WATER TANKS

TANK CAPACITY (gallons)	DRAIN PIPE (inches)
Up to 750	1
751 to 1,500	1½
1,501 to 3,000	2
3,001 to 5,000	2½
5,001 to 7,500	3
Over 7,500	4

For SI: 1 inch = 25.4 mm, 1 gallon = 3.785 L.

FIGURE 1.27 Size of drain pipes for water tanks. *(Courtesy of International Code Council, Inc. and the International Plumbing Code 2000)*

HANGER SPACING

PIPING MATERIAL	MAXIMUM HORIZONTAL SPACING (feet)	MAXIMUM VERTICAL SPACING (feet)
ABS pipe	4	10[b]
Aluminum tubing	10	15
Brass pipe	10	10
Cast-iron pipe[a]	5	15
Copper or copper-alloy pipe	12	10
Copper or copper-alloy tubing, $1^1/_4$-inch diameter and smaller	6	10
Copper or copper-alloy tubing, $1^1/_2$-inch diameter and larger	10	10
Cross-linked polyethylene (PEX) pipe	2.67 (32 inches)	10[b]
Cross-linked polyethylene/ aluminum/crosslinked polyethylene (PEX-AL-PEX) pipe	$2^2/_3$ (32 inches)	4
CPVC pipe or tubing, 1 inch or smaller	3	10[b]
CPVC pipe or tubing, $1^1/_4$ inches or larger	4	10[b]
Steel pipe	12	15
Lead pipe	Continuous	4
PB pipe or tubing	2.67 (32 inches)	4
Polyethylene/aluminum/polyethylene (PE-AL-PE) pipe	2.67 (32 inches)	4
PVC pipe	4	10[b]
Stainless steel drainage systems	10	10[b]

For SI: 1 inch = 25.4 mm, 1 foot = 304.8 mm.

a. The maximum horizontal spacing of cast-iron pipe hangers shall be increased to 10 feet where 10-foot lengths of pipe are installed.

b. Midstory guide for sizes 2 inches and smaller.

FIGURE 1.28   Hanger spacing. *(Courtesy of International Code Council, Inc. and the International Plumbing Code 2000)*

SIZES FOR OVERFLOW PIPES FOR WATER SUPPLY TANKS

MAXIMUM CAPACITY OF WATER SUPPLY LINE TO TANK (gpm)	DIAMETER OF OVERFLOW PIPE (inches)
0 - 50	2
50 - 150	$2^1/_2$
150 - 200	3
200 - 400	4
400 - 700	5
700 - 1,000	6
Over 1,000	8

For SI: 1 inch = 25.4 mm, 1 gallon per minute = 3.785 L/m.

FIGURE 1.29   Sizes for overflow pipes for water supply tanks. *(Courtesy of International Code Council, Inc. and the International Plumbing Code 2000)*

Materials	Type of Joints	Horizontal	Vertical
Cast Iron Hub and Spigot	Lead and Oakum	5 feet (1524 mm), except may be 10 feet (3048 mm) where 10 foot (3048 mm) lengths are installed [1, 2, 3]	Base and each floor not to exceed 15 feet (4572 mm)
	Compression Gasket	Every other joint, unless over 4 feet (1219 mm), then support each joint [1,2,3]	Base and each floor not to exceed 15 feet (4572 mm)
Cast Iron Hubless	Shielded Coupling	Every other joint, unless over 4 feet (1249 mm), then support each joint [1,2,3,4]	Base and each floor not to exceed 15 feet (4572 mm)
Copper Tube and Pipe	Soldered, Brazed or Welded	1-1/2 inch (40 mm) and smaller, 6 feet (1829 mm), 2 inch (50 mm) and larger, 10 feet (3048 mm)	Each floor, not to exceed 10 feet (3048 mm)
Steel and Brass Pipe for Water or DWV	Threaded or Welded	3/4 inch (20 mm) and smaller, 10 feet (3048 mm), 1 inch (25 mm) and larger, 12 feet (3658 mm)	Every other floor, not to exceed 25 feet (7620 mm) [5]
Steel, Brass and Tinned Copper Pipe for Gas	Threaded or Welded	1/2 inch (15 mm), 6 feet (1829 mm) 3/4 (20 mm) and 1 inch (25.4 mm), 8 feet (2436 mm), 1-1/4 inch (32 mm) and larger, 10 feet (3048 mm)	1/2 inch (12.7 mm), 6 feet (1829 mm), 3/4 (19 mm) and 1 inch (25.4 mm), 8 feet (2436 mm), 1-1/4 inch (32 mm) and larger, every floor level
Schedule 40 PVC and ABS DWV	Solvent Cemented	All sizes, 4 feet (1219 mm). Allow for expansion every 30 feet (9144 mm) [3, 6]	Base and each floor. Provide mid-story guides. Provide for expansion every 30 feet (9144 mm) [6]
CPVC	Solvent Cemented	1 inch (25 mm) and smaller, 3 feet (914 mm), 1-1/4 inch (32 mm) and larger, 4 feet (1219 mm)	Base and each floor. Provide mid-story guides [6]
Lead	Wiped or Burned	Continuous support	Not to exceed 4 feet (1219 mm)
Copper	Mechanical	In accordance with standards acceptable to the Administrative Authority	
Steel & Brass	Mechanical	In accordance with standards acceptable to the Administrative Authority	
PEX	Metal Insert and Metal Compression	32 inches (800 mm)	Base and each floor. Provide midstory guides

[1] Support adjacent to joint, not to exceed eighteen (18) inches (457 mm).
[2] Brace at not more than forty (40) foot (12192 mm) intervals to prevent horizontal movement.
[3] Support at each horizontal branch connection.
[4] Hangers shall not be placed on the coupling.
[5] Vertical water lines may be supported in accordance with recognized engineering principles with regard to expansion and contraction, when first approved by the Administrative Authority.
[6] See the appropriate IAPMO Installation Standard for expansion and other special requirements.

FIGURE 1.30   Horizontal and vertical use of materials and joints. *(Courtesy of International Code Council, Inc. and the International Plumbing Code 2000)*

MINIMUM REQUIRED AIR GAPS

FIXTURE	MINIMUM AIR GAP	
	Away from a wall[a] (inches)	Close to a wall (inches)
Lavatories and other fixtures with effective opening not greater than $1/2$ inch in diameter	1	$1^1/_2$
Sink, laundry trays, gooseneck back faucets and other fixtures with effective openings not greater than $3/4$ inch in diameter	1.5	2.5
Over-rim bath fillers and other fixtures with effective openings not greater than 1 inch in diameter	2	3
Drinking water fountains, single orifice not greater than $7/_{16}$ inch in diameter or multiple orifices with a total area of 0.150 square inch (area of circle $7/_{16}$ inch in diameter)	1	$1^1/_2$
Effective openings greater than 1 inch	Two times the diameter of the effective opening	Three times the diameter of the effective opening

For SI: 1 inch = 25.4 mm.

a. Applicable where walls or obstructions are spaced from the nearest inside edge of the spout opening a distance greater than three times the diameter of the effective opening for a single wall, or a distance greater than four times the diameter of the effective opening for two intersecting walls.

FIGURE 1.31 Minimum required air gaps. (Courtesy of International Code Council, Inc. and the International Plumbing Code 2000)

24

Minimum Airgaps for Water Distribution[4]		
Fixtures	When not affected by side walls[1] Inches (mm)	When affected by side walls[2] Inches (mm)
Effective openings[3] not greater than one-half (1/2) inch (12.7 mm) in diameter	1 (25.4)	1-1/2 (38)
Effective openings[3] not greater than three-quarters (3/4) inch (20 mm) in diameter	1-1/2 (38)	2-1/4 (57)
Effective openings[3] not greater than one (1) inch (25 mm) in diameter	2 (51)	3 (76)
Effective openings[3] greater than one (1) inch (25 mm) in diameter	Two (2) times diameter of effective opening	Three (3) times diameter of effective opening

[1] Side walls, ribs or similar obstructions do not affect airgaps when spaced from the inside edge of the spout opening a distance greater than three times the diameter of the effective opening for a single wall, or a distance greater than four times the effective opening for two intersecting walls.

[2] Vertical walls, ribs or similar obstructions extending from the water surface to or above the horizontal plane of the spout opening other than specified in Note 1 above. The effect of three or more such vertical walls or ribs has not been determined. In such cases, the airgap shall be measured from the top of the wall.

[3] The effective opening shall be the minimum cross-sectional area at the seat of the control valve or the supply pipe or tubing which feeds the device or outlet. If two or more lines supply one outlet, the effective opening shall be the sum of the cross-sectional areas of the individual supply lines or the area of the single outlet, whichever is smaller.

[4] Airgaps less than one (1) inch (25.4 mm) shall only be approved as a permanent part of a listed assembly that has been tested under actual backflow conditions with vacuums of 0 to 25 inches (635 mm) of mercury.

FIGURE 1.32 Minimum airgaps for water distribution. *(Courtesy of International Code Council, Inc. and the International Plumbing Code 2000)*

Minimum Required Air Chamber Dimensions

Nominal Pipe Diameter	Length of Pipe (ft.)	Flow Pressure P.S.I.G.	Velocity in Ft. Per. Sec.	Required Vol. in Cubic Inch	Air Chamber Phys. Size in Inches
1/2" (15 mm)	25	30	10	8	3/4" x 15"
1/2" (15 mm)	100	60	10	60	1" x 69.5"
3/4" (20 mm)	50	60	5	13	1" x 5"
3/4" (20 mm)	200	30	10	108	1.25" x 72.5"
1" (25 mm)	100	60	5	19	1.25" x 12.7 "
1" (25 mm)	50	30	10	40	1.25" x 27"
1-1/4" (32 mm)	50	60	10	110	1.25" x 54"
1-1/2" (40 mm)	200	30	5	90	2" x 27"
1-1/2" (40 mm)	50	60	10	170	2" x 50.5"
2" (50 mm)	100	30	10	329	3" x 44.5"
2" (50 mm)	25	60	10	150	2.5" x 31"
2" (50 mm)	200	60	5	300	3" x 40.5"

FIGURE 1.33   Minimum required air chamber dimensions. *(Courtesy of International Code Council, Inc. and the International Plumbing Code 2000)*

STACK SIZES FOR BEDPAN STEAMERS AND BOILING-TYPE STERILIZERS
(Number of Connections of Various Sizes Permitted to Various-sized Sterilizer Vent Stacks)

STACK SIZE (inches)	CONNECTION SIZE		
	1 1/2"		2"
1 1/2" [a]	1	or	0
2 [a]	2	or	1
2 [b]	1	and	1
3 [a]	4	or	2
3 [b]	2	and	2
4 [a]	8	or	4
4 [b]	4	and	4

For SI: 1 inch = 25.4 mm.

a. Total of each size.

b. Combination of sizes.

FIGURE 1.34   Stack sizes for bedpan steamers and boiling-type sterilizers. *(Courtesy of International Code Council, Inc. and the International Plumbing Code 2000)*

STACK SIZES FOR PRESSURE STERILIZERS
(Number of Connections of Various Sizes Permitted
to Various-sized Vent Stacks)

STACK SIZE (Inches)	CONNECTION SIZE			
	3/4"	1"	1 1/4"	1 1/2"
1 1/2 [a]	3 or	2 or	1	
1 1/2 [b]	2 and	1		
2 [a]	6 or	3 or	2 or	1
2 [b]	3 and	2		
2 [b]	2 and	1 and	1	
2 [b]	1 and	1 and		1
3 [a]	15 or	7 or	5 or	3
3 [b]		1 and	2 and	2
	1 and	5 and		1

For SI: 1 inch = 25.4 mm.

a. Total of each size.

b. Combination of sizes.

FIGURE 1.35 Stack sizes for pressure sterilizers. *(Courtesy of International Code Council, Inc. and the International Plumbing Code 2000)*

	Minimum Flow Rates
Oxygen	.71 CFM per outlet[1] (20 LPM)
Nitrous Oxide	.71 CFM per outlet[1] (20 LPM)
Medical Compressed Air	.71 CFM per outlet[1] (20 LPM)
Nitrogen	15 CFM (0.42 m³/min.) free air per outlet
Vacuum	1 SCFM (0.03 Sm³/min.) per inlet[2]
Carbon Dioxide	.71 CFM per outlet[1] (20 LPM)
Helium	.71 CFM per outlet (20 LPM)

[1] Any room designed for a permanently located respiratory ventilator or anesthesia machine shall have an outlet capable of a flow rate of 180 LPM (6.36 CFM) at the station outlet.

[2] For testing and certification purposes, individual station inlets shall be capable of a flow rate of 3 SCFM, while maintaining a system pressure of not less than 12 inches (305 mm) at the nearest adjacent vacuum inlet.

FIGURE 1.36 Minimum flow rates. *(Courtesy of International Code Council, Inc. and the International Plumbing Code 2000)*

Location of Gray-Water System

Minimum Horizontal Distance In Clear Required From:	Holding Tank		Irrigation/ Disposal Field	
	Feet	(mm)	Feet	(mm)
Building Structures[1]	5[2]	(1524 mm)	2[3]	(610 mm)
Property line adjoining private property	5	(1524 mm)	5	(1524 mm)
Water supply wells[4]	50	(15240 mm)	100	(30480 mm)
Streams and lakes[4]	50	(15240 mm)	50[5]	(15240 mm)
Sewage pits or cesspools	5	(1524 mm)	5	(1524 mm)
Disposal field and 100% expansion area	5	(1524 mm)	4[6]	(1219 mm)
Septic tank	0	(0)	5	(1524 mm)
On-site domestic water service line	5	(1524 mm)	5	(1524 mm)
Pressurized public water main	10	(3048 mm)	10[7]	(3048 mm)

Notes: When irrigation/disposal fields are installed in sloping ground, the minimum horizontal distance between any part of the distribution system and the ground surface shall be fifteen (15) feet (4572 mm).

[1] Including porches and steps, whether covered or uncovered, breezeways, roofed porte-cocheres, roofed patios, carports, covered walks, covered driveways and similar structures or appurtenances.

[2] The distance may be reduced to zero feet for above ground tanks when first approved by the Administrative Authority.

[3] Assumes a 45 degree (0.79 rad) angle from foundation.

[4] Where special hazards are involved, the distance required shall be increased as may be directed by the Administrative Authority.

[5] These minimum clear horizontal distances shall also apply between the irrigation/disposal field and the ocean mean higher high tide line.

[6] Plus two (2) feet (610 mm) for each additional foot of depth in excess of one (1) foot (305 mm) below the bottom of the drain line.

[7] For parallel construction/for crossings, approval by the Administrative Authority shall be required.

FIGURE 1.37   Location of gray water system. *(Courtesy of International Code Council, Inc. and the International Plumbing Code 2000)*

Design Criteria of Six Typical Soils

Type of Soil	Minimum square feet of irrigation/leaching area per 100 gallons of estimated graywater discharge per day	Maximum absorption capacity in gallons per square foot of irrigation/leaching area for a 24-hour period
Coarse sand or gravel	20	5.0
Fine sand	25	4.0
Sandy loam	40	2.5
Sandy clay	60	1.7
Clay with considerable sand or gravel	90	1.1
Clay with small amounts of sand or gravel	120	0.8

FIGURE 1.38   Design criteria of six typical soils. *(Courtesy of International Code Council, Inc. and the International Plumbing Code 2000)*

	Design Criteria of Six Typical Soils	
Type of Soil	Minimum square meters of irrigation/leaching area per liter of estimated graywater discharge per day	Maximum absorption capacity in liters per square meter of irrigation/leaching area for a 24-hour period
Coarse sand or gravel	0.005	203.7
Fine sand	0.006	162.9
Sandy loam	0.010	101.8
Sandy clay	0.015	69.2
Clay with considerable sand or gravel	0.022	44.8
Clay with small amounts of sand or gravel	0.030	32.6

FIGURE 1.39   Design criteria of six typical soils. *(Courtesy of International Code Council, Inc. and the International Plumbing Code 2000)*

RATES OF RAINFALL FOR VARIOUS CITIES

Rainfall rates, in inches per hour, are based on a storm of one-hour duration and a 100-year return period. The rainfall rates shown in the appendix are derived from Figure 1106.1.

Alabama:
Birmingham 3.8
Huntsville 3.6
Mobile 4.6
Montgomery 4.2

Alaska:
Fairbanks 1.0
Juneau 0.6

Arizona:
Flagstaff 2.4
Nogales 3.1
Phoenix 2.5
Yuma 1.6

Arkansas:
Fort Smith 3.6
Little Rock 3.7
Texarkana 3.8

California:
Barstow 1.4
Crescent City 1.5
Fresno 1.1
Los Angeles 2.1
Needles 1.6
Placerville 1.5
San Fernando 2.3
San Francisco 1.5
Yreka 1.4

Colorado:
Craig 1.5
Denver 2.4
Durango 1.8
Grand Junction 1.7
Lamar 3.0
Pueblo 2.5

Connecticut:
Hartford 2.7
New Haven 2.8
Putnam 2.6

Delaware:
Georgetown 3.0
Wilmington 3.1

District of Columbia:
Washington 3.2

Florida:
Jacksonville 4.3
Key West 4.3
Miami 4.7
Pensacola 4.6
Tampa 4.5

Georgia:
Atlanta 3.7
Dalton 3.4
Macon 3.9
Savannah 4.3
Thomasville 4.3

Hawaii:
Hilo 6.2
Honolulu 3.0
Wailuku 3.0

Idaho:
Boise 0.9
Lewiston 1.1
Pocatello 1.2

Illinois:
Cairo 3.3
Chicago 3.0
Peoria 3.3
Rockford 3.2
Springfield 3.3

Indiana:
Evansville 3.2
Fort Wayne 2.9
Indianapolis 3.1

Iowa:
Davenport 3.3
Des Moines 3.4
Dubuque 3.3
Sioux City 3.6

Kansas:
Atwood 3.3
Dodge City 3.3
Topeka 3.7
Wichita 3.7

Kentucky:
Ashland 3.0
Lexington 3.1
Louisville 3.2
Middlesboro 3.2
Paducah 3.3

Louisiana:
Alexandria 4.2
Lake Providence ... 4.0
New Orleans 4.8
Shreveport 3.9

Maine:
Bangor 2.2
Houlton 2.1
Portland 2.4

Maryland:
Baltimore 3.2
Hagerstown 2.8
Oakland 2.7
Salisbury 3.1

Massachusetts:
Boston 2.5
Pittsfield 2.8
Worcester 2.7

Michigan:
Alpena 2.5
Detroit 2.7
Grand Rapids 2.6
Lansing 2.8
Marquette 2.4
Sault Ste. Marie ... 2.2

Minnesota:
Duluth 2.8
Grand Marais 2.3
Minneapolis 3.1
Moorhead 3.2
Worthington 3.5

Mississippi:
Biloxi 4.7
Columbus 3.9
Corinth 3.6
Natchez 4.4
Vicksburg 4.1

Missouri:
Columbia 3.2
Kansas City 3.6
Springfield 3.4
St. Louis 3.2

Montana:
Ekalaka 2.5
Havre 1.6
Helena 1.5
Kalispell 1.2
Missoula 1.3

Nebraska:
North Platte 3.3
Omaha 3.8
Scottsbluff 3.1
Valentine 3.2

Nevada:
Elko 1.0
Ely 1.1
Las Vegas 1.4
Reno 1.1

New Hampshire:
Berlin 2.5
Concord 2.5
Keene 2.4

New Jersey:
Atlantic City 2.9
Newark 3.1
Trenton 3.1

New Mexico:
Albuquerque 2.0
Hobbs 3.0
Raton 2.5
Roswell 2.6
Silver City 1.9

New York:
Albany 2.5
Binghamton 2.3
Buffalo 2.3
Kingston 2.7
New York 3.0
Rochester 2.2

North Carolina:
Asheville 4.1
Charlotte 3.7
Greensboro 3.4
Wilmington 4.2

North Dakota:
Bismarck 2.8
Devils Lake 2.9
Fargo 3.1
Williston 2.6

Ohio:
Cincinnati 2.9
Cleveland 2.6
Columbus 2.8
Toledo 2.8

Oklahoma:
Altus 3.7
Boise City 3.3
Durant 3.8
Oklahoma City ... 3.8

Oregon:
Baker 0.9
Coos Bay 1.5
Eugene 1.3
Portland 1.2

Pennsylvania:
Erie 2.6
Harrisburg 2.8
Philadelphia 3.1
Pittsburgh 2.6
Scranton 2.7

FIGURE 1.40 Rates of rainfall. *(Courtesy of International Code Council, Inc. and the International Plumbing Code 2000)*

Rhode Island:
  Block Island 2.75
  Providence 2.6

South Carolina:
  Charleston 4.3
  Columbia 4.0
  Greenville 4.1

South Dakota:
  Buffalo 2.8
  Huron 3.3
  Pierre 3.1
  Rapid City 2.9
  Yankton 3.6

Tennessee:
  Chattanooga 3.5
  Knoxville 3.2
  Memphis 3.7
  Nashville 3.3

Texas:
  Abilene 3.6
  Amarillo 3.5
  Brownsville 4.5
  Dallas 4.0
  Del Rio 4.0
  El Paso 2.3
  Houston 4.6
  Lubbock 3.3
  Odessa 3.2
  Pecos 3.0
  San Antonio 4.2

Utah:
  Brigham City 1.2
  Roosevelt 1.3
  Salt Lake City 1.3
  St. George 1.7

Vermont:
  Barre 2.3
  Bratteboro 2.7
  Burlington 2.1
  Rutland 2.5

Virginia:
  Bristol 2.7
  Charlottesville 2.8
  Lynchburg 3.2
  Norfolk 3.4
  Richmond 3.3

Washington:
  Omak 1.1
  Port Angeles 1.1
  Seattle 1.4
  Spokane 1.0
  Yakima 1.1

West Virginia:
  Charleston 2.8
  Morgantown 2.7

Wisconsin:
  Ashland 2.5
  Eau Claire 2.9
  Green Bay 2.6
  La Crosse 3.1
  Madison 3.0
  Milwaukee 3.0

Wyoming:
  Cheyenne 2.2
  Fort Bridger 1.3
  Lander 1.5
  New Castle 2.5
  Sheridan 1.7
  Yellowstone Park .. 1.4

For SI: 1 inch =25.4 mm.

Source: National Weather Service, National Oceanic and Atmospheric Administration, Washington, D.C.

FIGURE 1.40 *(continued)* Rates of rainfall. *(Courtesy of International Code Council, Inc. and the International Plumbing Code 2000)*

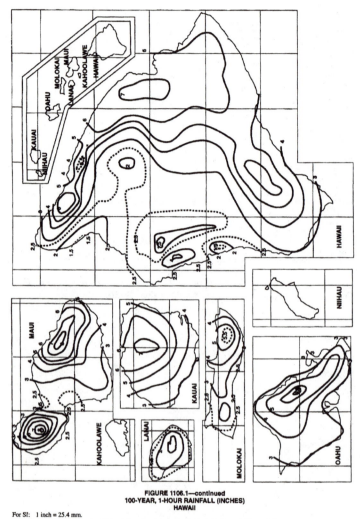

FIGURE 1106.1—continued
100-YEAR, 1-HOUR RAINFALL (INCHES)
HAWAII

For SI: 1 inch = 25.4 mm.

Source: National Weather Service, National Oceanic and Atmospheric Administration, Washington, DC.

FIGURE 1.41   Hawaii figures show a 100-year, one-hour rainfall rate. *(Courtesy of International Code Council, Inc. and the International Plumbing Code 2000)*

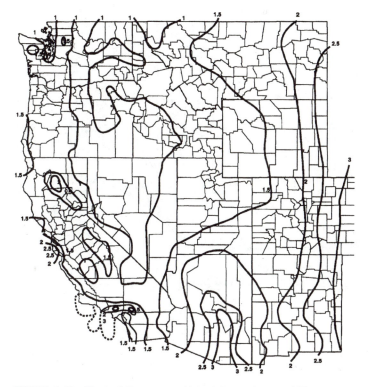

FIGURE 1.42 Chart of the western United States shows a 100-year, one-hour rainfall rate. *(Courtesy of International Code Council, Inc. and the International Plumbing Code 2000)*

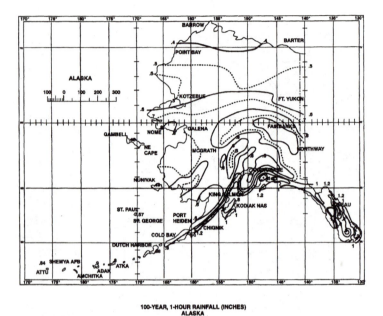

100-YEAR, 1-HOUR RAINFALL (INCHES)
ALASKA

For SI: 1 inch = 25.4 mm.

Source: National Weather Service, National Oceanic and Atmospheric Administration, Washington, DC.

FIGURE 1.43 Alaska's 100-year, one-hour rainfall rate. *(Courtesy of International Code Council, Inc. and the International Plumbing Code 2000)*

FIGURE 1.44 100-year, one-hour rainfall rate for the eastern United States. *(Courtesy of International Code Council, Inc. and the International Plumbing Code 2000)*

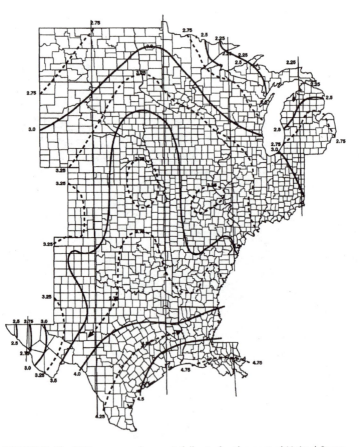

FIGURE 1.45 100-year, one-hour rainfall rate for the central United States. *(Courtesy of International Code Council, Inc. and the International Plumbing Code 2000)*

SIZE OF VERTICAL CONDUCTORS AND LEADERS

DIAMETER OF LEADER (inches)[a]	HORIZONTALLY PROJECTED ROOF AREA (square feet)											
	Rainfall rate (inches per hour)											
	1	2	3	4	5	6	7	8	9	10	11	12
2	2,880	1,440	960	720	575	480	410	360	320	290	260	240
3	8,800	4,400	2,930	2,200	1,760	1,470	1,260	1,100	980	880	800	730
4	18,400	9,200	6,130	4,600	3,680	3,070	2,630	2,300	2,045	1,840	1,675	1,530
5	34,600	17,300	11,530	8,650	6,920	5,765	4,945	4,325	3,845	3,460	3,145	2,880
6	54,000	27,000	17,995	13,500	10,800	9,000	7,715	6,750	6,000	5,400	4,910	4,500
8	116,000	58,000	38,660	29,000	23,200	19,315	16,570	14,500	12,890	11,600	10,545	9,660

For SI: 1 inch = 25.4 mm, 1 square foot = 0.0929 m^2.

a. Sizes indicated are the diameter of circular piping. This table is applicable to piping of other shapes provided the cross-sectional shape fully encloses a circle of the diameter indicated in this table.

FIGURE 1.46 Size of vertical conductors and leaders. (Courtesy of International Code Council, Inc. and the International Plumbing Code 2000)

SIZE OF HORIZONTAL STORM DRAINAGE PIPING

SIZE OF HORIZONTAL PIPING (inches)	HORIZONTALLY PROJECTED ROOF AREA (square feet)					
	Rainfall rate (inches per hour)					
	1	2	3	4	5	6
	1/8 unit vertical in 12 units horizontal (1-percent slope)					
3	3,288	1,644	1,096	822	657	548
4	7,520	3,760	2,506	1,800	1,504	1,253
5	13,360	6,680	4,453	3,340	2,672	2,227
6	21,400	10,700	7,133	5,350	4,280	3,566
8	46,000	23,000	15,330	11,500	9,200	7,600
10	82,800	41,400	27,600	20,700	16,580	13,800
12	133,200	66,600	44,400	33,300	26,650	22,200
15	218,000	109,000	72,800	59,500	47,600	39,650
	1/4 unit vertical in 12 units horizontal (2-percent slope)					
3	4,640	2,320	1,546	1,160	928	773
4	10,600	5,300	3,533	2,650	2,120	1,766
5	18,880	9,440	6,293	4,720	3,776	3,146
6	30,200	15,100	10,066	7,550	6,040	5,033
8	65,200	32,600	21,733	16,300	13,040	10,866
10	116,800	58,400	38,950	29,200	23,350	19,450
12	188,000	94,000	62,600	47,000	37,600	31,350
15	336,000	168,000	112,000	84,000	67,250	56,000
	1/2 unit vertical in 12 units horizontal (4-percent slope)					
3	6,576	3,288	2,295	1,644	1,310	1,096
4	15,040	7,520	5,010	3,760	3,010	2,500
5	26,720	13,360	8,900	6,680	5,320	4,450
6	42,800	21,400	13,700	10,700	8,580	7,140
8	92,000	46,000	30,650	23,000	18,400	15,320
10	171,600	85,800	55,200	41,400	33,150	27,600
12	266,400	133,200	88,800	66,600	53,200	44,400
15	476,000	238,000	158,800	119,000	95,300	79,250

For SI: 1 inch = 25.4 mm, 1 square foot = 0.0929 m².

FIGURE 1.47 Size of horizontal storm drainage piping. (*Courtesy of International Code Council, Inc. and the International Plumbing Code 2000*)

Sizing of Horizontal Rainwater Piping

Size of Pipe, Inches	Flow at 1/8"/ft. Slope, gpm	Maximum Allowable Horizontal Projected Roof Areas Square Feet at Various Rainfall Rates					
		1"/hr	2"/hr	3"/hr	4"/hr	5"/hr	6"/hr
3	34	3288	1644	1096	822	657	548
4	78	7520	3760	2506	1880	1504	1253
5	139	13,360	6680	4453	3340	2672	2227
6	222	21,400	10,700	7133	5350	4280	3566
8	478	46,000	23,000	15,330	11,500	9200	7670
10	860	82,800	41,400	27,600	20,700	16,580	13,800
12	1384	133,200	66,600	44,400	33,300	26,650	22,200
15	2473	238,000	119,000	79,333	59,500	47,600	39,650

Size of Pipe, Inches	Flow at 1/4"/ft. Slope, gpm	Maximum Allowable Horizontal Projected Roof Areas Square Feet at Various Rainfall Rates					
		1"/hr	2"/hr	3"/hr	4"/hr	5"/hr	6"/hr
3	48	4640	2320	1546	1160	928	773
4	110	10,600	5300	3533	2650	2120	1766
5	196	18,880	9440	6293	4720	3778	3146
6	314	30,200	15,100	10,066	7550	6040	5033
8	677	65,200	32,600	21,733	16,300	13,040	10,866
10	1214	116,800	58,400	38,950	29,200	23,350	19,450
12	1953	188,000	94,000	62,600	47,000	37,600	31,350
15	3491	336,000	168,000	112,000	84,000	67,250	56,000

Size of Pipe, Inches	Flow at 1/2"/ft. Slope, gpm	Maximum Allowable Horizontal Projected Roof Areas Square Feet at Various Rainfall Rates					
		1"/hr	2"/hr	3"/hr	4"/hr	5"/hr	6"/hr
3	68	6576	3288	2192	1644	1310	1096
4	156	15,040	7520	5010	3760	3010	2500
5	278	26,720	13,360	8900	6680	5320	4450
6	445	42,800	21,400	14,267	10,700	8580	7140
8	956	92,000	46,000	30,650	23,000	18,400	15,320
10	1721	165,600	82,800	55,200	41,400	33,150	27,600
12	2768	266,400	133,200	88,800	66,600	53,200	44,400
15	4946	476,000	238,000	158,700	119,000	95,200	79,300

Notes:
1. The sizing data for horizontal piping is based on the pipes flowing full.
2. For rainfall rates other than those listed, determine the allowable roof area by dividing the area given in the 1 inch/hour (25 mm/hour) column by the desired rainfall rate.

FIGURE 1.48 Sizing of horizontal rainwater piping. *(Courtesy of International Code Council, Inc. and the International Plumbing Code 2000)*

Size of Gutters

Diameter of Gutter in Inches

Maximum Rainfall in Inches per Hour

1/16"/ft. Slope	2	3	4	5	6
3	340	226	170	136	113
4	720	480	360	288	240
5	1250	834	625	500	416
6	1920	1280	960	768	640
7	2760	1840	1380	1100	918
8	3980	2655	1990	1590	1325
10	7200	4800	3600	2880	2400

Diameter of Gutter in Inches

Maximum Rainfall in Inches per Hour

1/8"/ft. Slope	2	3	4	5	6
3	480	320	240	192	160
4	1020	681	510	408	340
5	1760	1172	880	704	587
6	2720	1815	1360	1085	905
7	3900	2600	1950	1560	1300
8	5600	3740	2800	2240	1870
10	10,200	6800	5100	4080	3400

Diameter of Gutter in Inches

Maximum Rainfall in Inches per Hour

1/4"/ft. Slope	2	3	4	5	6
3	680	454	340	272	226
4	1440	960	720	576	480
5	2500	1668	1250	1000	834
6	3840	2560	1920	1536	1280
7	5520	3680	2760	2205	1840
8	7960	5310	3980	3180	2655
10	14,400	9600	7200	5750	4800

Diameter of Gutter in Inches

Maximum Rainfall in Inches per Hour

1/2"/ft. Slope	2	3	4	5	6
3	960	640	480	384	320
4	2040	1360	1020	816	680
5	3540	2360	1770	1415	1180
6	5540	3695	2770	2220	1850
7	7800	5200	3900	3120	2600
8	11,200	7460	5600	4480	3730
10	20,000	13,330	10,000	8000	6660

FIGURE 1.49 Size of gutters. *(Courtesy of International Code Council, Inc. and the International Plumbing Code 2000)*

SIZE OF SEMICIRCULAR ROOF GUTTERS

DIAMETER OF GUTTERS (inches)	HORIZONTALLY PROJECTED ROOF AREA (square feet)					
	RAINFALL RATE (inches per hour)					
	1	2	3	4	5	6
1/16 unit vertical in 12 units horizontal (0.5-percent slope)						
3	680	340	226	170	136	113
4	1,440	720	480	360	288	240
5	2,500	1,250	834	625	500	416
6	3,840	1,920	1,280	960	768	640
7	5,520	2,760	1,840	1,380	1,100	918
8	7,960	3,980	2,655	1,990	1,590	1,325
10	14,400	7,200	4,800	3,600	2,880	2,400
1/8 unit vertical in 12 units horizontal (1-percent slope)						
3	960	480	320	240	192	160
4	2,040	1,020	681	510	408	340
5	3,520	1,760	1,172	880	704	587
6	5,440	2,720	1,815	1,360	1,085	905
7	7,800	3,900	2,600	1,950	1,560	1,300
8	11,200	5,600	3,740	2,800	2,240	1,870
10	20,400	10,200	6,800	5,100	4,080	3,400
1/4 unit vertical in 12 units horizontal (2-percent slope)						
3	1,360	680	454	340	272	226
4	2,880	1,440	960	720	576	480
5	5,000	2,500	1,668	1,250	1,000	834
6	7,680	3,840	2,560	1,920	1,536	1,280
7	11,040	5,520	3,860	2,760	2,205	1,840
8	15,920	7,960	5,310	3,980	3,180	2,655
10	28,800	14,400	9,600	7,200	5,750	4,800
1/2 unit vertical in 12 units horizontal (4-percent slope)						
3	1,920	960	640	480	384	320
4	4,080	2,040	1,360	1,020	816	680
5	7,080	3,540	2,360	1,770	1,415	1,180
6	11,080	5,540	3,695	2,770	2,220	1,850
7	15,600	7,800	5,200	3,900	3,120	2,600
8	22,400	11,200	7,460	5,600	4,480	3,730
10	40,000	20,000	13,330	10,000	8,000	6,660

For SI: 1 inch = 25.4 mm, 1 square foot = 0.0929 m².

FIGURE 1.50 Size of semicircular roof gutters. (*Courtesy of International Code Council, Inc. and the International Plumbing Code 2000*)

Controlled Flow Maximum Roof Water Depth

Roof Rise,*		Max Water Depth at Drain,	
Inches	(mm)	Inches	(mm)
Flat	(Flat)	3	(76)
2	(51)	4	(102)
4	(102)	5	(127s)
6	(152)	6	(152)

*Vertical measurement from the roof surface at the drain to the highest point of the roof surface served by the drain, ignoring any local depression immediately adjacent to the drain.

FIGURE 1.51 Controlled flow maximum roof water depth. *(Courtesy of International Code Council, Inc. and the International Plumbing Code 2000)*

The visual graphics here should serve you well in your career. Knowing and understanding your local code is very important, so spend time with your code book to gain a complete understanding of your local codes. Keep in mind that the information in this chapter is based in the International Code. If you work with the Uniform code you may discover some differences between local requirements and those shown here.

chapter

2

MATHEMATICS FOR THE TRADE

A or a	Area, acre
AWG	American Wire Gauge
B or b	Breadth
bbl	Barrels
bhp	Brake horsepower
BM	Board measure
Btu	British thermal units
BWG	Birmingham Wire Gauge
B & S	Brown and Sharpe Wire Gauge (American Wire Gauge)
C of g	Center of gravity
cond	Condensing
cu	Cubic
cyl	Cylinder
D or d	Depth, diameter
dr	Dram
evap	Evaporation
F	Coefficient of friction; Fahrenheit
F or f	Force, factor of safety
ft (or ')	Foot
ft lb	Foot pound
fur	Furlong
gal	Gallon
gi	Gill
ha	Hectare
H or h	Height, head of water
HP	horsepower
IHP	Indicated horsepower
in (or ")	Inch
L or l	Length
lb	Pound
lb/sq in.	Pounds per square inch
mi	Mile
o.d.	Outside diameter (pipes)
oz	Ounces
pt	Pint
P or p	Pressure, load
psi	Pounds per square inch
R or r	Radius
rpm	Revolutions per minute
sq ft	Square foot
sq in.	Square inch
sq yd	Square yard
T or t	Thickness, temperature
temp	Temperature
V or v	Velocity
vol	Volume
W or w	Weight
W. I.	Wrought iron

FIGURE 2.1   Abbreviations.

Circumference of a circle = π × diameter or 3.1416 × diameter

Diameter of a circle = circumference × 0.31831

Area of a square = length × width

Area of a rectangle = length × width

Area of a parallelogram = base × perpendicular height

Area of a triangle = ½ base × perpendicular height

Area of a circle = π radius squared or diameter squared × 0.7854

Area of an ellipse = length × width × 0.7854

Volume of a cube or rectangular prism = length × width × height

Volume of a triangular prism = area of triangle × length

Volume of a sphere = diameter cubed × 0.5236 (diameter × diameter × diameter × 0.5236)

Volume of a cone = π × radius squared × ⅓ height

Volume of a cylinder = π × radius squared × height

Length of one side of a square × 1.128 = the diameter of an equal circle

Doubling the diameter of a pipe or cylinder increases its capacity 4 times

The pressure (in lb/sq in.) of a column of water = the height of the column (in feet) × 0.434

The capacity of a pipe or tank (in U.S. gallons) = the diameter squared (in inches) × the length (in inches) × 0.0034

1 gal water = 8½ lb = 231 cu in.

1 cu ft water = 62½ lb = 7½ gal

FIGURE 2.2   Useful formulas.

Sine	sin =	side opposite / hypotenuse
Cosine	cos =	side adjacent / hypotenuse
Tangent	tan =	side opposite / side adjacent
Cosecant	csc =	hypotenuse / side opposite
Secant	sec =	hypotenuse / side adjacent
Cotangent	cot =	side adjacent / side opposite

FIGURE 2.3   Trigonometry.

Pentagon	5 sides
Hexagon	6 sides
Heptagon	7 sides
Octagon	8 sides
Nonagon	9 sides
Decagon	10 sides

FIGURE 2.4   Polygons.

The capacity of pipes is as the square of their diameters. Thus, doubling the diameter of a pipe increases its capacity four times. The area of a pipe wall may be determined by the following formula:

Area of pipe wall = 0.7854 × [(o.d. × o.d.) − (i.d. × i.d.)]

FIGURE 2.5   Piping.

The approximate weight of a piece of pipe may be determined by the following formulas:

Cast-iron pipe: weight = $(A^2 − B^2)$ × length × 0.2042
Steel pipe: weight = $(A^2 − B^2)$ × length × 0.2199
Copper pipe: weight = $(A^2 − B^2)$ × length × 0.2537
A = outside diameter of the pipe in inches
B = inside diameter of the pipe in inches

FIGURE 2.6   Determining pipe weight.

The formula for calculating expansion or contraction in plastic piping is:

$$L = Y \times \frac{T - F}{10} \times \frac{L}{100}$$

L = expansion in inches
Y = constant factor expressing inches of expansion per 100°F temperature change
    per 100 ft of pipe
T = maximum temperature (°F)
F = minimum temperature (°F)
L = length of pipe run in feet

FIGURE 2.7   Expansion in plastic piping.

Parallelogram	Area = base × distance between the two parallel sides
Pyramid	Area = ½ perimeter of base × slant height + area of base
	Volume = area of base × ⅓ of the altitude
Rectangle	Area = length × width
Rectangular prism	Volume = width × height × length
Sphere	Area of surface = diameter × diameter × 3.1416
	Side of inscribed cube = radius × 1.547
	Volume = diameter × diameter × diameter × 0.5236
Square	Area = length × width
Triangle	Area = one-half of height times base
Trapezoid	Area = one-half of the sum of the parallel sides × the height
Cone	Area of surface = one-half of circumference of base × slant height + area of base
	Volume = diameter × diameter × 0.7854 × one-third of the altitude
Cube	Volume = width × height × length
Ellipse	Area = short diameter × long diameter × 0.7854
Cylinder	Area of surface = diameter × 3.1416 × length + area of the two bases
	Area of base = diameter × diameter × 0.7854
	Area of base = volume ÷ length
	Length = volume ÷ area of base
	Volume = length × area of base
	Capacity in gallons = volume in inches ÷ 231
	Capacity of gallons = diameter × diameter × length × 0.0034
	Capacity in gallons = volume in feet × 7.48
Circle	Circumference = diameter × 3.1416
	Circumference = radius × 6.2832
	Diameter = radius × 2
	Diameter = square root of = (area ÷ 0.7854)
	Diameter = square root of area × 1.1233

FIGURE 2.8   Area and other formulas.

The formulas for pipe radiation of heat are as follows:

$$L = \frac{144}{OD \times 3.1416} \times R \div 12$$

D = outside diameter (OD) of pipe
L = length of pipe needed in feet
R = square feet of radiation needed

FIGURE 2.9 Formulas for pipe radiation of heat.

Temperature may be expressed according to the Fahrenheit (F) scale or the Celsius (C) scale. To convert °C to °F or °F to °C, use the following formulas:

°F = 1.8 × °C + 32
°C = 0.55555555 × °F − 32
°C = °F − 32 ÷ 1.8
°F = °C. × 1.8 + 32

FIGURE 2.10 Temperature conversion.

To figure the final temperature when two different temperatures of water are mixed together, use the following formula:

$$\frac{(A \times C) + (B \times D)}{A + B}$$

A = weight of lower temperature water
B = weight of higher temperature water
C = lower temperature
D = higher temperature

FIGURE 2.11 Computing water temperature.

Radiation

---

3 ft of 1-in. pipe equal 1 ft² R.
2⅓ lineal ft of 1¼-in. pipe equal 1 ft² R.
Hot water radiation gives off 150 Btu/ft² R/hr.
Steam radiation gives off 240 Btu/ft² R/hr.
On greenhouse heating, figure ⅔ ft² R/ft² glass.
1 ft² of direct radiation condenses 0.25 lb water/hr.

FIGURE 2.12 Radiant heat facts.

−100°–30°		
°C	*Base temperature*	°F
−73	−100	−148
−68	−90	−130
−62	−80	−112
−57	−70	−94
−51	−60	−76
−46	−50	−58
−40	−40	−40
−34.4	−30	−22
−28.9	−20	−4
−23.3	−10	14
−17.8	0	32
−17.2	1	33.8
−16.7	2	35.6
−16.1	3	37.4
−15.6	4	39.2
−15.0	5	41.0
−14.4	6	42.8
−13.9	7	44.6
−13.3	8	46.4
−12.8	9	48.2
−12.2	10	50.0
−11.7	11	51.8
−11.1	12	53.6
−10.6	13	55.4
−10.0	14	57.2
31°–71°		
°C	*Base temperature*	°F
−0.6	31	87.8
0	32	89.6
0.6	33	91.4
1.1	34	93.2
1.7	35	95.0
2.2	36	96.8
2.8	37	98.6
3.3	38	100.4
3.9	39	102.2
4.4	40	104.0
5.0	41	105.8
5.6	42	107.6

FIGURE 2.13 Temperature conversion.

Vacuum in inches of mercury	Boiling point
29	76.62
28	99.93
27	114.22
26	124.77
25	133.22
24	140.31
23	146.45
22	151.87
21	156.75
20	161.19
19	165.24
18	169.00
17	172.51
16	175.80
15	178.91
14	181.82
13	184.61
12	187.21
11	189.75
10	192.19
9	194.50
8	196.73
7	198.87
6	200.96
5	202.25
4	204.85
3	206.70
2	208.50
1	210.25

FIGURE 2.14   Boiling points of water based on pressure.

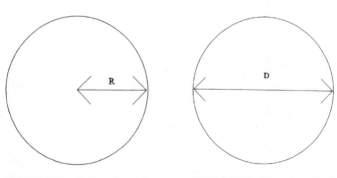

FIGURE 2.15 Radius of a circle.   **FIGURE 2.16** Diameter of a circle.

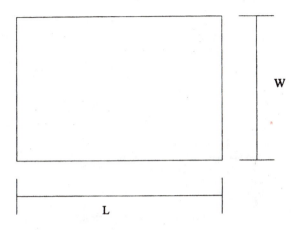

$$A = L \times W$$

FIGURE 2.17 Area of a rectangle.

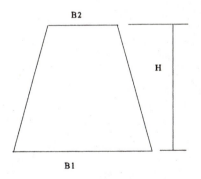

A= Area, B1= First Base, B2 = Second Base, H = Height

To find the area of a trapezoid, use the following formula:

$$A = \frac{1}{2} (B1 + B2) H$$

Note: Perform math function in parenthesis first.

FIGURE 2.18 Area of a trapezoid.

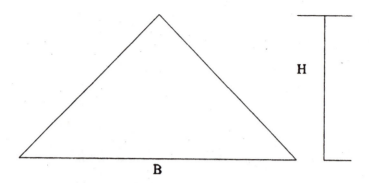

$$A = Area, \ B = Base, \ H = Height$$

$$A = \frac{1}{2} BH$$

FIGURE 2.19   Area of a triangle.

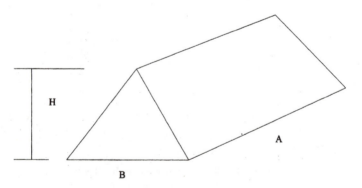

$$V = Volume, \ A = Length \ of \ Prism, \ B = Base \ of \ Triangle, \ H = Height$$

$$V = \frac{ABH}{2}$$

FIGURE 2.20   Area of a triangular prism.

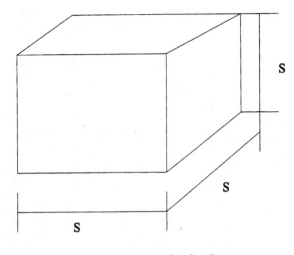

Volume is S cubed or V = S x S x S

FIGURE 2.21 Volume of a cube.

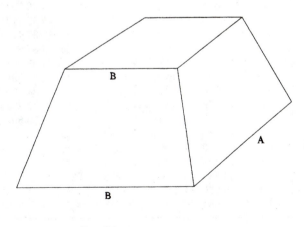

$$V = \frac{(B + B)}{2} \, AH$$

FIGURE 2.22 Volume of a trapezoidal prism.

Set	Travel	Set	Travel	Set	Travel
2	2.828	¼	15.907	½	28.987
¼	3.181	½	16.261	¾	29.340
½	3.531	¾	16.614	21	29.694
¾	3.888	12	16.968	¼	30.047
3	4.242	¼	17.321	½	30.401
¼	4.575	½	17.675	¾	30.754
½	4.949	¾	18.028	22	31.108
¾	5.302	13	18.382	¼	31.461
4	5.656	¼	18.735	½	31.815
¼	6.009	½	19.089	¾	32.168
½	6.363	¾	19.442	23	32.522
¾	6.716	14	19.796	¼	32.875
5	7.070	¼	20.149	½	33.229
¼	7.423	½	20.503	¾	33.582
½	7.777	¾	20.856	24	33.936
¾	8.130	15	21.210	¼	34.289
6	8.484	¼	21.563	½	34.643
¼	8.837	½	21.917	¾	34.996
½	9.191	¾	22.270	25	35.350
¾	9.544	16	22.624	¼	35.703
7	9.898	¼	22.977	½	36.057
¼	10.251	½	23.331	¾	36.410
½	10.605	¾	23.684	26	36.764
¾	10.958	17	24.038	¼	37.117
8	11.312	¼	24.391	½	37.471
¼	11.665	½	24.745	¾	37.824
½	12.019	¾	25.098	27	38.178
¾	12.372	18	25.452	¼	38.531
9	12.726	¼	25.805	½	38.885
¼	13.079	½	26.159	¾	39.238
½	13.433	¾	26.512	28	39.592
¾	13.786	19	26.866	¼	39.945
10	14.140	¼	27.219	½	40.299
¼	14.493	½	27.573	¾	40.652
½	14.847	¾	27.926	29	41.006
¾	15.200	20	28.280	¼	41.359
11	15.554	¼	28.635	½	41.713

FIGURE 2.23   Set and travel relationships in inches for 45° offsets.

Inches	Decimal of an inch	Inches	Decimal of an inch
$\frac{1}{64}$	.015625	$\frac{33}{64}$	.515625
$\frac{1}{32}$	.03125	$\frac{17}{32}$	.53125
$\frac{3}{64}$	.046875	$\frac{35}{64}$	.546875
$\frac{1}{16}$	.0625	$\frac{9}{16}$	.5625
$\frac{5}{64}$	.078125	$\frac{37}{64}$	.578125
$\frac{3}{32}$	.09375	$\frac{19}{32}$	.59375
$\frac{7}{64}$	.109375	$\frac{39}{64}$	.609375
$\frac{1}{8}$	.125	$\frac{5}{8}$	.625
$\frac{9}{64}$	.140625	$\frac{41}{64}$	.640625
$\frac{5}{32}$	.15625	$\frac{21}{32}$	.65625
$\frac{11}{64}$	.171875	$\frac{43}{64}$	.671875
$\frac{3}{16}$	.1875	$\frac{11}{16}$	.6875
$\frac{13}{64}$	.203125	$\frac{45}{64}$	.703125
$\frac{7}{32}$	.21875	$\frac{23}{32}$	.71875
$\frac{15}{64}$	.234375	$\frac{47}{64}$	.734375
$\frac{1}{4}$	.25	$\frac{3}{4}$	.75
$\frac{17}{64}$	.265625	$\frac{49}{64}$	.765625
$\frac{9}{32}$	.28125	$\frac{25}{32}$	.78125
$\frac{19}{64}$	.296875	$\frac{51}{64}$	.796875
$\frac{5}{16}$	.3125	$\frac{13}{16}$	.8125
$\frac{21}{64}$	.328125	$\frac{53}{64}$	.828125
$\frac{11}{32}$	.34375	$\frac{27}{32}$	.84375
$\frac{23}{64}$	.359375	$\frac{55}{64}$	.859375
$\frac{3}{8}$	.375	$\frac{7}{8}$	.875
$\frac{35}{64}$	.390625	$\frac{57}{64}$	.890625
$\frac{13}{32}$	.40625	$\frac{22}{32}$	.90625
$\frac{27}{64}$	.421875	$\frac{59}{64}$	.921875
$\frac{7}{16}$	.4375	$\frac{15}{16}$	.9375
$\frac{29}{64}$	.453125	$\frac{61}{64}$	.953125
$\frac{15}{32}$	.46875	$\frac{31}{32}$	.96875
$\frac{31}{64}$	.484375	$\frac{63}{64}$	.984375
$\frac{1}{2}$	.5	1	1

FIGURE 2.24 Simple offsets.

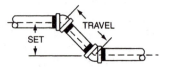

FIGURE 2.25 Calculated 45° offsets.

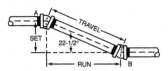

FIGURE 2.26 Decimal equivalents of fractions of an inch.

To find side*	When known side is	Multiply Side	For 60° ells by	For 45° ells by	For 30° ells by	For 22½° ells by	For 11¼° ells by	For 5⅝° ells by
T	S	S	1.155	1.414	2.000	2.613	5.125	10.187
S	T	T	.866	.707	.500	.383	.195	.098
R	S	S	.577	1.000	1.732	2.414	5.027	10.158
S	R	R	1.732	1.000	.577	.414	.198	.098
T	R	R	2.000	1.414	1.155	1.082	1.019	1.004
R	T	T	.500	.707	.866	.924	.980	.995

*S = set, R = run, T = travel.

FIGURE 2.27 Multipliers for calculating simple offsets.

Metric	U.S.
144 in.²	1 ft²
9 ft²	1 yd²
1 yd²	1296 in.²
4840 yd²	1 a
640 a	1 mi²

FIGURE 2.28 Square measure.

1 cm²	0.1550 in.²
1 dm²	0.1076 ft²
1 ms2	1.196 yd²
1 A (are)	3.954 rd²
1 ha	2.47 a (acres)
1 km²	0.386 mi²
1 in.²	6.452 cm²
1 ft²	9.2903 dm²
1 yd²	0.8361 m²
1 rd²	0.2529 A (are)
1 a (acre)	0.4047 ha
1 mi²	2.59 km²

FIGURE 2.29 Square measures of length and area.

144 in.²	1 ft²	272.25 ft²
9 ft²	1 yd²	
30¼ yd²	1 rd²	272.25 ft²
160 rd²	1 a	4840 yd²
		43,560 ft²)
640 a	1 m²	3,097,600 yd²
36 mi²	1 township	

100 mm²	1 cm²
100 cm²	1 dm²
100 dm²	1 m²

FIGURE 2.30   Metric square measure.

FIGURE 2.31a   Square measures.

Square feet	Square meters
1	0.925
2	0.1850
3	0.2775
4	0.3700
5	0.4650
6	0.5550
7	0.6475
8	0.7400
9	0.8325
10	0.9250
25	2.315
50	4.65
100	9.25

FIGURE 2.31b   Square feet to square meters.

Fraction	Square root
1/8	.3535
1/4	.5000
3/8	.6124
1/2	.7071
5/8	.7906
3/4	.8660
7/8	.9354

FIGURE 2.32   Square roots of fractions.

Fraction	Cube root
1/8	.5000
1/4	.6300
3/8	.7211
1/2	.7937
5/8	.8550
3/4	.9086
7/8	.9565

FIGURE 2.33   Cube roots of fractions.

Number	Cube	Number	Cube	Number	Cube
1	1	36	46,656	71	357,911
2	8	37	50.653	72	373,248
3	27	38	54,872	73	389,017
4	64	39	59,319	74	405,224
5	125	40	64,000	75	421,875
6	216	41	68,921	76	438,976
7	343	42	74,088	77	456,533
8	512	43	79,507	78	474,552
9	729	44	85,184	79	493,039
10	1,000	45	91,125	80	512,000
11	1,331	46	97,336	81	531,441
12	1,728	47	103,823	82	551,368
13	2,197	48	110,592	83	571,787
14	2,477	49	117,649	84	592,704
15	3,375	50	125,000	85	614,125
16	4,096	51	132,651	86	636,056
17	4,913	52	140,608	87	658,503
18	5,832	53	148,877	88	681,472
19	6,859	54	157,464	89	704,969
20	8,000	55	166,375	90	729,000
21	9,621	56	175,616	91	753,571
22	10,648	57	185,193	92	778,688
23	12,167	58	195,112	93	804,357
24	13,824	59	205,379	94	830,584
25	15,625	60	216,000	95	857,375
26	17,576	61	226,981	96	884,736
27	19,683	62	238,328	97	912,673
28	21,952	63	250,047	98	941,192
29	24,389	64	262,144	99	970,299
30	27,000	65	274,625	100	1,000,000
31	29,791	66	287,496		
32	32,768	67	300,763		
33	35,937	68	314,432		
34	39,304	69	328,500		
35	42,875	70	343,000		

FIGURE 2.34 Cubes of numbers.

Number	Square root	Number	Square root	Number	Square root
1	1.00000	36	6.00000	71	8.42614
2	1.41421	37	6.08276	72	8.48528
3	1.73205	38	6.16441	73	8.54400
4	2.00000	39	6.24499	74	8.60232
5	2.23606	40	6.32455	75	8.66025
6	2.44948	41	6.40312	76	8.71779
7	2.64575	42	6.48074	77	8.77496
8	2.82842	43	6.55743	78	8.83176
9	3.00000	44	6.63324	79	8.88819
10	3.16227	45	6.70820	80	8.94427
11	3.31662	46	6.78233	81	9.00000
12	3.46410	47	6.85565	82	9.05538
13	3.60555	48	6.92820	83	9.11043
14	3.74165	49	7.00000	84	9.16515
15	3.87298	50	7.07106	85	9.21954
16	4.00000	51	7.14142	86	9.27361
17	4.12310	52	7.21110	87	9.32737
18	4.24264	53	7.28010	88	9.38083
19	4.35889	54	7.34846	89	9.43398
20	4.47213	55	7.41619	90	9.48683
21	4.58257	56	7.48331	91	9.53939
22	4.69041	57	7.54983	92	9.59166
23	4.79583	58	7.61577	93	9.64365
24	4.89897	59	7.68114	94	9.69535
25	5.00000	60	7.74596	95	9.74679
26	5.09901	61	7.81024	96	9.79795
27	5.19615	62	7.87400	97	9.84885
28	5.29150	63	7.93725	98	9.89949
29	5.38516	64	8.00000	99	9.94987
30	5.47722	65	8.06225	100	10.00000
31	5.56776	66	8.12403		
32	5.65685	67	8.18535		
33	5.74456	68	8.24621		
34	5.83095	69	8.30662		
35	5.91607	70	8.36660		

FIGURE 2.35 Square roots of numbers.

Number	Square	Number	Square	Number	Square
1	1	36	1296	71	5041
2	4	37	1369	72	5184
3	9	38	1444	73	5329
4	16	39	1521	74	5476
5	25	40	1600	75	5625
6	36	41	1681	76	5776
7	49	42	1764	77	5929
8	64	43	1849	78	6084
9	81	44	1936	79	6241
10	100	45	2025	80	6400
11	121	46	2116	81	6561
12	144	47	2209	82	6724
13	169	48	2304	83	6889
14	196	49	2401	84	7056
15	225	50	2500	85	7225
16	256	51	2601	86	7396
17	289	52	2704	87	7569
18	324	53	2809	88	7744
19	361	54	2916	89	7921
20	400	55	3025	90	8100
21	441	56	3136	91	8281
22	484	57	3249	92	8464
23	529	58	3364	93	8649
24	576	59	3481	94	8836
25	625	60	3600	95	9025
26	676	61	3721	96	8216
27	729	62	3844	97	9409
28	784	63	3969	98	9604
29	841	64	4096	99	9801
30	900	65	4225	100	10000
31	961	66	4356		
32	1024	67	4489		
33	1089	68	4624		
34	1156	69	4761		
35	1225	70	4900		

FIGURE 2.36 Squares of numbers.

Diameter	Circumference	Diameter	Circumference
1/8	0.3927	10	31.41
1/4	0.7854	10½	32.98
3/8	1.178	11	34.55
1/2	1.570	11½	36.12
5/8	1.963	12	37.69
3/4	2.356	12½	39.27
7/8	2.748	13	40.84
1	3.141	13½	42.41
1⅛	3.534	14	43.98
1¼	3.927	14½	45.55
1⅜	4.319	15	47.12
1½	4.712	15½	48.69
1⅝	5.105	16	50.26
1¾	5.497	16½	51.83
1⅞	5.890	17	53.40
2	6.283	17½	54.97
2¼	7.068	18	56.54
2½	7.854	18½	58.11
2¾	8.639	19	56.69
3	9.424	19½	61.26
3¼	10.21	20	62.83
3½	10.99	20½	64.40
3¾	11.78	21	65.97
4	12.56	21½	67.54
4½	14.13	22	69.11
5	15.70	22½	70.68
5½	17.27	23	72.25
6	18.84	23½	73.82
6½	20.42	24	75.39
7	21.99	24½	76.96
7½	23.56	25	78.54
8	25.13	26	81.68
8½	26.70	27	84.82
9	28.27	28	87.96
9½	29.84	29	91.10
		30	94.24

FIGURE 2.37   Circumference of circle.

Diameter	Area	Diameter	Area
⅛	0.0123	10	78.54
¼	0.0491	10½	86.59
⅜	0.1104	11	95.03
½	0.1963	11½	103.86
⅝	0.3068	12	113.09
¾	0.4418	12½	122.71
⅞	0.6013	13	132.73
1	0.7854	13½	143.13
1⅛	0.9940	14	153.93
1¼	1.227	14½	165.13
1⅜	1.484	15	176.71
1½	1.767	15½	188.69
1⅝	2.073	16	201.06
1¾	2.405	16½	213.82
1⅞	2.761	17	226.98
2	3.141	17½	240.52
2¼	3.976	18	254.46
2½	4.908	18½	268.80
2¾	5.939	19	283.52
3	7.068	19½	298.60
3¼	8.295	20	314.16
3½	9.621	20½	330.06
3¾	11.044	21	346.36
4	12.566	21½	363.05
4½	15.904	22	380.13
5	19.635	22½	397.60
5½	23.758	23	415.47
6	28.274	23½	433.73
6½	33.183	24	452.39
7	38.484	24½	471.43
7½	44.178	25	490.87
8	50.265	26	530.93
8½	56.745	27	572.55
9	63.617	28	615.75
9½	70.882	29	660.52
		30	706.86

FIGURE 2.38 Area of circle.

COLUMN	1		2	3	4	5	6	7	8	9	10
Line	Description		Lb. per square inch (psi)	Gal. per min. through section	Length of section (feet)	Trial pipe size (inches)	Equivalent length of fittings and valves (feet)	Total equivalent length Col. 4 and Col. 6 (100 feet)	Friction loss per 100 feet of trial size pipe (psi)	Friction loss in equivalent length Col. 8 x Col. 7 (psi)	Excess pressure over friction losses (psi)
a	Service and cold water distribution piping[a]	Minimum pressure available at main	55.00								
b		Highest pressure required at a fixture (Section 604.3)	15.00								
c		Meter loss 2" meter	11.00								
d		Tap in main loss 2" tap (Table E103A) 1.61									
e		Static head loss 21 × 0.43 psi 9.03									
f		Special fixture loss backflow preventer	9.00								
g		Special fixture loss—Filter	0.00								
h		Special fixture loss—Other	0.00								
i		Total overall losses and requirements (sum of Lines b through h) 45.64									
j		Pressure available to overcome pipe friction (Line a minus Lines b to h)	9.36								
	Designation		FU								
	Pipe section (from diagram)	AB	294	108.0	54	2½	12	0.66	3.3	2.18	
	Cold water distribution piping	BC	264	108.0	8	2½	2.5	0.105	3.2	0.34	
		CD	132	77.0	13	2½	8	0.21	1.9	0.40	
		CF	132	77.0	150	2½	12	1.62	1.9	3.08	
		DE	132	77.0	150	2½	14.5	1.645	1.9	3.12	
k	Total pipe friction losses (cold)						9.36	6.24	6.24		
l	Difference (Line j minus Line k)										3.12
	Pipe section (from diagram)	A'B'	294	108.0	54	2½	9.6	0.64	3.3	2.1	
	Hot water distribution piping	B'C'	24	38.0	8	2	9.0	0.18	1.4	0.24	
		C'D'	12	28.6	13	1½	5	0.18	3.2	0.58	
		C'F'[b]	12	28.6	150	1½	14	1.64	3.2	5.25	
		D'E'[b]	12	28.6	150	1½	7	1.57	3.2	5.02	
k	Total pipe friction losses (hot)						9.36	7.94	7.94		
l	Difference (Line j minus Line k)										1.42

For SI: 1 inch = 25.4 mm, 1 foot = 304.8 mm, 1 psi = 6.895 kPa, 1 gpm = 3.785 L/m.

a. To be considered as pressure gain for fixtures below main (to consider separately, omit from "i" and add to "j").

b. To consider separately, in k use C-F only if greater loss than above.

FIGURE 2.39 This table can help resolve any problems that might arise in relation to pipe sizing. (Courtesy of International Code Council, Inc. and International Plumbing Code 2000)

LOAD VALUES ASSIGNED TO FIXTURES[a]

FIXTURE	OCCUPANCY	TYPE OF SUPPLY CONTROL	LOAD VALUES, IN WATER SUPPLY FIXTURE UNITS (wsfu)		
			Cold	Hot	Total
Bathroom group	Private	Flush tank	2.7	1.5	3.6
Bathroom group	Private	Flush valve	6.0	3.0	8.0
Bathtub	Private	Faucet	1.0	1.0	1.4
Bathtub	Public	Faucet	3.0	3.0	4.0
Bidet	Private	Faucet	1.5	1.5	2.0
Combination fixture	Private	Faucet	2.25	2.25	3.0
Dishwashing machine	Private	Automatic		1.4	1.4
Drinking fountain	Offices, etc.	3/8" valve	0.25		0.25
Kitchen sink	Private	Faucet	1.0	1.0	1.4
Kitchen sink	Hotel, restaurant	Faucet	3.0	3.0	4.0
Laundry trays (1 to 3)	Private	Faucet	1.0	1.0	1.4
Lavatory	Private	Faucet	0.5	0.5	0.7
Lavatory	Public	Faucet	1.5	1.5	2.0
Service sink	Offices, etc.	Faucet	2.25	2.25	3.0
Shower head	Public	Mixing valve	3.0	3.0	4.0
Shower head	Private	Mixing valve	1.0	1.0	1.4
Urinal	Public	1" flush valve	10.0		10.0
Urinal	Public	3/4" flush valve	5.0		5.0
Urinal	Public	Flush tank	3.0		3.0
Washing machine (8 lbs.)	Private	Automatic	1.0	1.0	1.4
Washing machine (8 lbs.)	Public	Automatic	2.25	2.25	3.0
Washing machine (15 lbs.)	Public	Automatic	3.0	3.0	4.0
Water closet	Private	Flush valve	6.0		6.0
Water closet	Private	Flush tank	2.2		2.2
Water closet	Public	Flush valve	10.0		10.0
Water closet	Public	Flush tank	5.0		5.0
Water closet	Public or private	Flushometer tank	2.0		2.0

For SI: 1 inch = 25.4 mm, 1 pound = 0.454 kg.

a. For fixtures not listed, loads should be assumed by comparing the fixture to one listed using water in similar quantities and at similar rates. The assigned loads for fixtures with both hot and cold water supplies are given for separate hot and cold water loads and for total load, the separate hot and cold water loads being three-fourths of the total load for the fixture in each case.

FIGURE 2.40 Every fixture involved in plumbing has a load value. They are determined here. (Courtesy of International Code Council, Inc. and International Plumbing Code 2000)

Table for Estimating Demand

SUPPLY SYSTEMS PREDOMINANTLY FOR FLUSH TANKS			SUPPLY SYSTEMS PREDOMINANTLY FOR FLUSH VALVES		
Load	Demand		Load	Demand	
(Water supply fixture units)	(Gallons per minute)	(Cubic feet per minute)	(Water supply fixture units)	(Gallons per minute)	(Cubic feet per minute)
1	3.0	0.04104			
2	5.0	0.0684			
3	6.5	0.86892			
4	8.0	1.06944			
5	9.4	1.256592	5	15.0	2.0052
6	10.7	1.430376	6	17.4	2.326032
7	11.8	1.577424	7	19.8	2.646364
8	12.8	1.711104	8	22.2	2.967696
9	13.7	1.831416	9	24.6	3.288528
10	14.6	1.951728	10	27.0	3.60936
11	15.4	2.058672	11	27.8	3.716304
12	16.0	2.13888	12	28.6	3.823248
13	16.5	2.20572	13	29.4	3.930192
14	17.0	2.27256	14	30.2	4.037136
15	17.5	2.3394	15	31.0	4.14408
16	18.0	2.90624	16	31.8	4.241024
17	18.4	2.459712	17	32.6	4.357968
18	18.8	2.513184	18	33.4	4.464912
19	19.2	2.566656	19	34.2	4.571856
20	19.6	2.620128	20	35.0	4.6788
25	21.5	2.87412	25	38.0	5.07984
30	23.3	3.114744	30	42.0	5.61356
35	24.9	3.328632	35	44.0	5.88192
40	26.3	3.515784	40	46.0	6.14928
45	27.7	3.702936	45	48.0	6.41664
50	29.1	3.890088	50	50.0	6.684

FIGURE 2.41 This table will let a user estimate demand. (Courtesy of International Code Council, Inc. and International Plumbing Code 2000)

Table for Estimating Demand—cont'd

SUPPLY SYSTEMS PREDOMINANTLY FOR FLUSH TANKS			SUPPLY SYSTEMS PREDOMINANTLY FOR FLUSH VALVES		
Load	Demand		Load	Demand	
(Water supply fixture units)	(Gallons per minute)	(Cubic feet per minute)	(Water supply fixture units)	(Gallons per minute)	(Cubic feet per minute)
60	32.0	4.27776	60	54.0	7.21872
70	35.0	4.6788	70	58.0	7.75344
80	38.0	5.07984	80	61.2	8.181216
90	41.0	5.48088	90	64.3	8.595624
100	43.5	5.81508	100	67.5	9.0234
120	48.0	6.41664	120	73.0	9.75864
140	52.5	7.0182	140	77.0	10.29336
160	57.0	7.61976	160	81.0	10.82808
180	61.0	8.15448	180	85.5	11.42964
200	65.0	8.6892	200	90.0	12.0312
225	70.0	9.3576	225	95.5	12.76644
250	75.0	10.0260	250	101.0	13.50168
275	80.0	10.6944	275	104.5	13.96956
300	85.0	11.3628	300	108.0	14.43744
400	105.0	14.0364	400	127.0	16.97736
500	124.0	16.57632	500	143.0	19.11624
750	170.0	22.7256	750	177.0	23.66136
1,000	208.0	27.80544	1,000	208.0	27.80544
1,250	239.0	31.94952	1,250	239.0	31.94952
1,500	269.0	35.95992	1,500	269.0	35.95992
1,750	297.0	39.70296	1,750	297.0	39.70296
2,000	325.0	43.446	2,000	325.0	43.446
2,500	380.0	50.7984	2,500	380.0	50.7984
3,000	433.0	57.88344	3,000	433.0	57.88344
4,000	535.0	70.182	4,000	525.0	70.182
5,000	593.0	79.27224	5,000	593.0	79.27224

For SI: 1 gpm = 3.785 L/m, 1 cfm = 0.4719 L/s.

FIGURE 2.42 The table for estimating demand for flush tanks and valves continues. (Courtesy of International Code Council, Inc. and International Plumbing Code 2000)

LOSS OF PRESSURE THROUGH TAPS AND TEES IN POUNDS PER SQUARE INCH (psi)

GALLONS PER MINUTE	SIZE OF TAP OR TEE (inches)						
	5/8	3/4	1	1¼	1½	2	3
10	1.35	0.64	0.18	0.08			
20	5.38	2.54	0.77	0.31	0.14		
30	12.1	5.72	1.62	0.69	0.33		
40		10.2	3.07	1.23	0.58	0.10	
50		15.9	4.49	1.92	0.91	0.18	
60			6.46	2.76	1.31	0.28	
70			8.79	3.76	1.78	0.40	0.10
80			11.5	4.90	2.32	0.55	0.13
90			14.5	6.21	2.94	0.72	0.16
100			17.94	7.67	3.63	0.91	0.21
120			25.8	11.0	5.23	1.12	0.30
140			35.2	15.0	7.12	1.61	0.41
150				17.2	8.16	2.20	0.47
160				19.6	9.30	2.52	0.54
180				24.8	11.8	2.92	0.68
200				30.7	14.5	3.62	0.84
225				38.8	18.4	4.48	1.06
250				47.9	22.7	5.6	1.31
275					27.4	7.00	1.59
300					32.6	10.1	1.88

For SI: 1 inch = 25.4 mm, 1 psi = 6.895 kPa, 1 gpm = 3.785 L/m.

FIGURE 2.43 Pressure can be lost in taps and tees. This examines the numbers. (*Courtesy of International Code Council, Inc. and International Plumbing Code 2000*)

ALLOWANCE IN EQUIVALENT LENGTH OF PIPE FOR FRICTION LOSS IN VALVES AND THREADED FITTINGS (feet)

FITTING OR VALVE	PIPE SIZES (inches)							
	1/2	3/4	1	1 1/4	1 1/2	2	2 1/2	3
45-degree elbow	1.2	1.5	1.8	2.4	3.0	4.0	5.0	6.0
90-degree elbow	2.0	2.5	3.0	4.0	5.0	7.0	8.0	10.0
Tee, run	0.6	0.8	0.9	1.2	1.5	2.0	2.5	3.0
Tee, branch	3.0	4.0	5.0	6.0	7.0	10.0	12.0	15.0
Gate valve	0.4	0.5	0.6	0.8	1.0	1.3	1.6	2.0
Balancing valve	0.8	1.1	1.5	1.9	2.2	3.0	3.7	4.5
Plug-type cock	0.8	1.1	1.5	1.9	2.2	3.0	3.7	4.5
Check valve, swing	5.6	8.4	11.2	14.0	16.8	22.4	28.0	33.6
Globe valve	15.0	20.0	25.0	35.0	45.0	55.0	65.0	80.0
Angle valve	8.0	12.0	15.0	18.0	22.0	28.0	34.0	40.0

For SI: 1 inch = 25.4 mm, 1 foot = 304.8 mm, 1 degree = 0.0175 rad.

FIGURE 2.44 This chart examines the allowances involved in friction loss in valves and threaded fittings. *(Courtesy of International Code Council, Inc. and International Plumbing Code 2000)*

PRESSURE LOSS IN FITTINGS AND VALVES EXPRESSED AS EQUIVALENT LENGTH OF TUBE[a] (feet)

NOMINAL OR STANDARD SIZE (Inches)	FITTINGS						VALVES		
	Standard Ell		90-Degree Tee		Coupling	Ball	Gate	Butterfly	Check
	90 Degree	45 Degree	Side Branch	Straight Run					
$3/8$	0.5	—	1.5	—	—	—	—	—	1.5
$1/2$	1	0.5	2	—	—	—	—	—	2
$5/8$	1.5	0.5	2	—	—	—	—	—	2.5
$3/4$	2	0.5	3	—	—	—	—	—	3
1	2.5	1	4.5	—	—	0.5	—	—	4.5
$1 1/4$	3	1	5.5	0.5	0.5	0.5	—	—	5.5
$1 1/2$	4	1.5	7	0.5	0.5	0.5	—	—	6.5
2	5.5	2	9	0.5	0.5	0.5	0.5	7.5	9
$2 1/2$	7	2.5	12	0.5	0.5	—	1	10	11.5
3	9	3.5	15	1	1	—	1.5	15.5	14.5
$3 1/2$	9	3.5	14	1	1	—	2	—	12.5
4	12.5	5	21	1	1	—	2	16	18.5
5	16	6	27	1.5	1.5	—	3	11.5	23.5
6	19	7	34	2	2	—	3.5	13.5	26.5
8	29	11	50	3	3	—	5	12.5	39

For SI: 1 inch = 25.4 mm, 1 foot = 304.8 mm, 1 degree = 0.0175 rad.

a. Allowances are for streamlined soldered fittings and recessed threaded fittings. For threaded fittings, double the allowances shown in the table. The equivalent lengths presented above are based on a C factor of 150 in the Hazen-Williams friction loss formula. The lengths shown are rounded to the nearest half-foot.

FIGURE 2.45 You can determine pressure losses as equivalent lengths from this table. (*Courtesy of International Code Council, Inc. and International Plumbing Code 2000*)

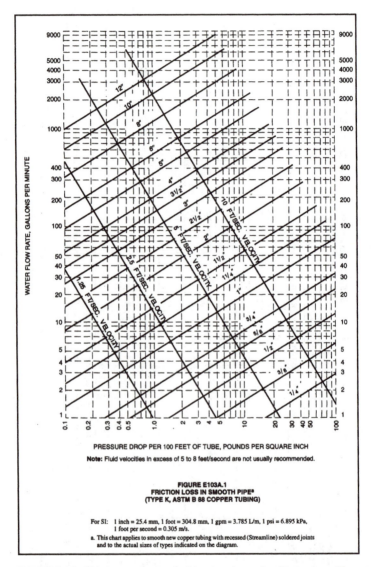

PRESSURE DROP PER 100 FEET OF TUBE, POUNDS PER SQUARE INCH

Note: Fluid velocities in excess of 5 to 8 feet/second are not usually recommended.

FIGURE E103A.1
FRICTION LOSS IN SMOOTH PIPE^a
(TYPE K, ASTM B 88 COPPER TUBING)

For SI: 1 inch = 25.4 mm, 1 foot = 304.8 mm, 1 gpm = 3.785 L/m, 1 psi = 6.895 kPa,
1 foot per second = 0.305 m/s.

a. This chart applies to smooth new copper tubing with recessed (Streamline) soldered joints
and to the actual sizes of types indicated on the diagram.

FIGURE 2.46   This is one of several tables that determine friction loss. *(Courtesy of International Code Council, Inc. and International Plumbing Code 2000)*

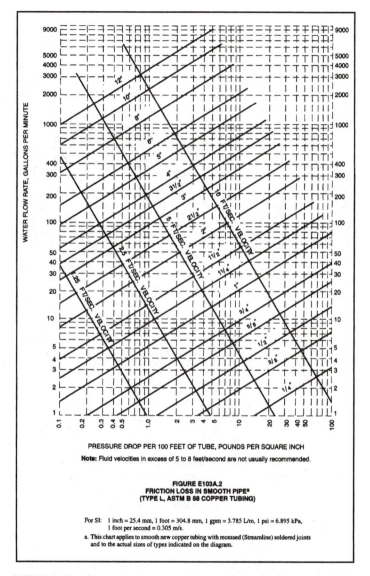

PRESSURE DROP PER 100 FEET OF TUBE, POUNDS PER SQUARE INCH

Note: Fluid velocities in excess of 5 to 8 feet/second are not usually recommended.

FIGURE E103A.2
FRICTION LOSS IN SMOOTH PIPE[a]
(TYPE L, ASTM B 88 COPPER TUBING)

For SI: 1 inch = 25.4 mm, 1 foot = 304.8 mm, 1 gpm = 3.785 L/m, 1 psi = 6.895 kPa,
1 foot per second = 0.305 m/s.

a. This chart applies to smooth new copper tubing with recessed (Streamline) soldered joints
   and to the actual sizes of types indicated on the diagram.

FIGURE 2.47   This is one of several tables that determine friction loss. *(Courtesy of International Code Council, Inc. and International Plumbing Code 2000)*

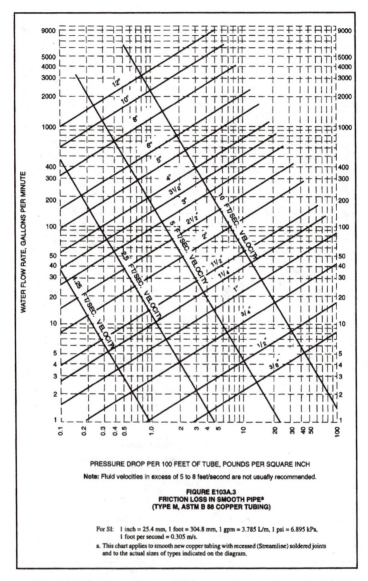

PRESSURE DROP PER 100 FEET OF TUBE, POUNDS PER SQUARE INCH

Note: Fluid velocities in excess of 5 to 8 feet/second are not usually recommended.

**FIGURE E103A.3
FRICTION LOSS IN SMOOTH PIPE[a]
(TYPE M, ASTM B 88 COPPER TUBING)**

For SI: 1 inch = 25.4 mm, 1 foot = 304.8 mm, 1 gpm = 3.785 L/m, 1 psi = 6.895 kPa,
1 foot per second = 0.305 m/s.

a. This chart applies to smooth new copper tubing with recessed (Streamline) soldered joints
and to the actual sizes of types indicated on the diagram.

FIGURE 2.48   This is one of several tables that determine friction loss. *(Courtesy of International Code Council, Inc. and International Plumbing Code 2000)*

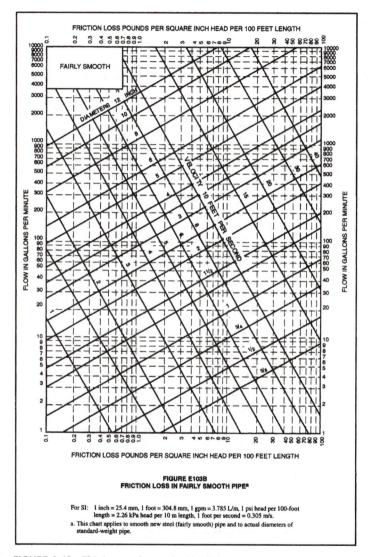

FRICTION LOSS POUNDS PER SQUARE INCH HEAD PER 100 FEET LENGTH

FIGURE E103B
FRICTION LOSS IN FAIRLY SMOOTH PIPEª

For SI: 1 inch = 25.4 mm, 1 foot = 304.8 mm, 1 gpm = 3.785 L/m, 1 psi head per 100-foot
length = 2.26 kPa head per 10 m length, 1 foot per second = 0.305 m/s.

a. This chart applies to smooth new steel (fairly smooth) pipe and to actual diameters of
standard-weight pipe.

FIGURE 2.49 This is one of several table that determine friction loss. *(Courtesy of International Code Council, Inc. and International Plumbing Code 2000)*

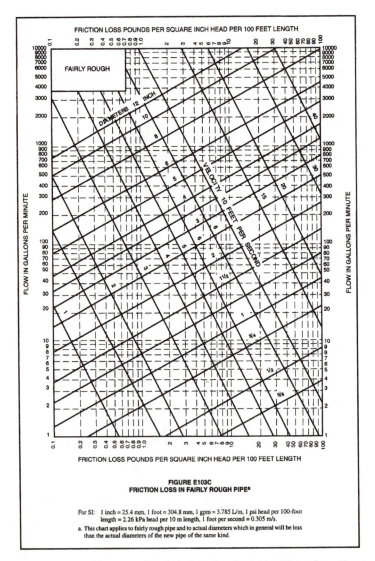

FRICTION LOSS POUNDS PER SQUARE INCH HEAD PER 100 FEET LENGTH

FRICTION LOSS POUNDS PER SQUARE INCH HEAD PER 100 FEET LENGTH

FIGURE E103C
FRICTION LOSS IN FAIRLY ROUGH PIPE[a]

For SI: 1 inch = 25.4 mm, 1 foot = 304.8 mm, 1 gpm = 3.785 L/m, 1 psi head per 100-foot
length = 2.26 kPa head per 10 m length, 1 foot per second = 0.305 m/s.

a. This chart applies to fairly rough pipe and to actual diameters which in general will be less
than the actual diameters of the new pipe of the same kind.

FIGURE 2.50 This is one of several tables that determine friction loss. *(Courtesy of International Code Council, Inc. and International Plumbing Code 2000)*

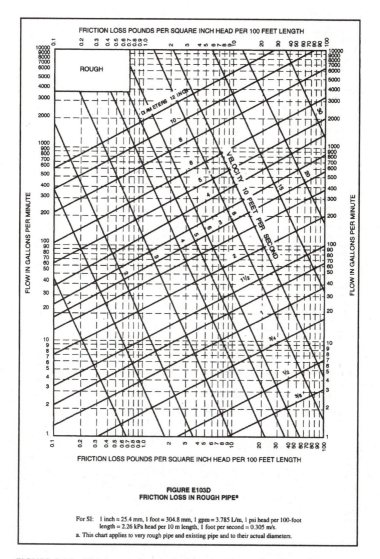

**FIGURE E103D
FRICTION LOSS IN ROUGH PIPE[a]**

For SI: 1 inch = 25.4 mm, 1 foot = 304.8 mm, 1 gpm = 3.785 L/m, 1 psi head per 100-foot
length = 2.26 kPa head per 10 m length, 1 foot per second = 0.305 m/s.

a. This chart applies to very rough pipe and existing pipe and to their actual diameters.

FIGURE 2.51   This is one of several tables that determine friction loss. *(Courtesy of International Code Council, Inc. and International Plumbing Code 2000)*

Inch	mm
1/2	15
3/4	20
1	25

Water Supply Fixture Units (WSFU) and Minimum Fixture Branch Pipe Sizes[3]

Appliances, Appurtenances or Fixtures[2]	Minimum Fixture Branch Pipe Size[1,4]	Private	Public	Assembly[5]
Bathtub or Combination Bath/Shower (fill)	1/2"	4.0	4.0	
3/4" Bathtub Fill Valve	3/4"	10.0	10.0	
Bidet	1/2"	1.0		
Clotheswasher	1/2"	4.0	4.0	
Dental Unit, cuspidor	1/2"		1.0	
Dishwasher, domestic	1/2"	1.5	1.5	
Drinking Fountain or Watercooler	1/2"	0.5	0.5	0.75
Hose Bibb	1/2"	2.5	2.5	
Hose Bibb, each additional[7]	1/2"	1.0	1.0	
Lavatory	1/2"	1.0	1.0	1.0
Lawn Sprinkler, each head[5]		1.0	1.0	
Mobile Home, each (minimum)		12.0		
Sinks				
Bar	1/2"	1.0	2.0	
Clinic Faucet	1/2"		3.0	
Clinic Flushometer Valve				
with or without faucet	1"		8.0	
Kitchen, domestic	1/2"	1.5	1.5	
Laundry	1/2"	1.5	1.5	
Service or Mop Basin	1/2"	1.5	3.0	
Washup, each set of faucets	1/2"		2.0	
Shower	1/2"	2.0	2.0	
Urinal, 1.0 GPF	3/4"	3.0	4.0	5.0
Urinal, greater than 1.0 GPF	3/4"	4.0	5.0	6.0
Urinal, flush tank	1/2"	2.0	2.0	3.0
Washfountain, circular spray	3/4"		4.0	
Water Closet, 1.6 GPF Gravity Tank	1/2"	2.5	2.5	3.5
Water Closet, 1.6 GPF Flushometer Tank	1/2"	2.5	2.5	3.5
Water Closet, 1.6 GPF Flushometer Valve	1"	5.0	5.0	6.0
Water Closet, greater than 1.6 GPF Gravity Tank	1/2"	3.0	5.5	7.0
Water Closet, greater than 1.6 GPF Flushometer Valve	1"	7.0	8.0	10.0

Notes:

1. Size of the cold branch outlet pipe, or both the hot and cold branch outlet pipes.

2. Appliances, Appurtenances or Fixtures not included in this Table may be sized by reference to fixtures having a similar flow rate and frequency of use.

3. The listed fixture unit values represent their total load on the cold water service. The separate cold water and hot water fixture unit value for fixtures having both cold and hot water connections shall each be taken as three-quarters (3/4) of the listed total value of the fixture.

4. The listed minimum supply branch pipe sizes for individual fixtures are the nominal (I.D.) pipe size.

5. For fixtures or supply connections likely to impose continuous flow demands, determine the required flow in gallons per minute (GPM) and add it separately to the demand (in GPM) for the distribution system or portions thereof.

6. Assembly [Public Use (See Table 4-1)].

7. Reduced fixture unit loading for additional hose bibbs as used is to be used only when sizing total building demand and for pipe sizing when more than one hose bibb is supplied by a segment of water distributing pipe. The fixture branch to each hose bibb shall be sized on the basis of 2.5 fixture units.

FIGURE 2.52 This shows the relationship between fixture units and fixture branch pipes for a water supply. *(Reprinted from the 2000 Uniform Plumbing Code (UPC) with the permission of the International Association of Plumbing and Mechanical Officials (IAPMO))*

Allowance in equivalent length of pipe for friction loss in valves and threaded fittings.*

Equivalent Length of Pipe for Various Fittings

Diameter of fitting Inches	90° Standard Elbow Feet	45° Standard Elbow Feet	Standard Tee 90° Feet	Coupling or Straight Run of Tee Feet	Gate Valve Feet	Globe Valve Feet	Angle Valve Feet
3/8	1.0	0.6	1.5	0.3	0.2	8	4
1/2	2.0	1.2	3.0	0.6	0.4	15	8
3/4	2.5	1.5	4.0	0.8	0.5	20	12
1	3.0	1.8	5.0	0.9	0.6	25	15
1-1/4	4.0	2.4	6.0	1.2	0.8	35	18
1-1/2	5.0	3.0	7.0	1.5	1.0	45	22
2	7.0	4.0	10.0	2.0	1.3	55	28
2-1/2	8.0	5.0	12.0	2.5	1.6	65	34
3	10.0	6.0	15.0	3.0	2.0	80	40
4	14.0	8.0	21.0	4.0	2.7	125	55
5	17.0	10.0	25.0	5.0	3.3	140	70
6	20.0	12.0	30.0	6.0	4.0	165	80

FIGURE 2.53 Pipe length in various fittings is described. *(Reprinted from the 2000 Uniform Plumbing Code (UPC) with the permission of the International Association of Plumbing and Mechanical Officials (IAPMO))*

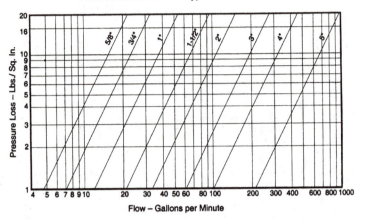

FIGURE 2.54 English and metric unit information about friction loss. *(Reprinted from the 2000 Uniform Plumbing Code (UPC) with the permission of the International Association of Plumbing and Mechanical Officials (IAPMO))*

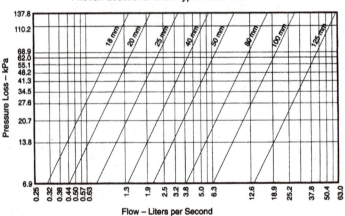

FIGURE 2.55 English and metric unit information about friction loss. *(Reprinted from the 2000 Uniform Plumbing Code (UPC) with the permission of the International Association of Plumbing and Mechanical Officials (IAPMO))*

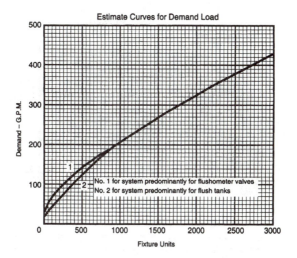

FIGURE 2.56 English and metric lengths. *(Reprinted from the 2000 Uniform Plumbing Code (UPC) with the permission of the International Association of Plumbing and Mechanical Officials (IAPMO))*

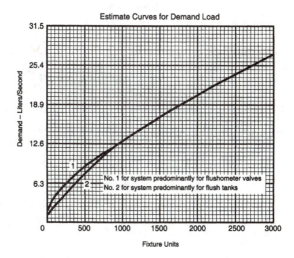

FIGURE 2.57 English and metric lengths. *(Reprinted from the 2000 Uniform Plumbing Code (UPC) with the permission of the International Association of Plumbing and Mechanical Officials (IAPMO))*

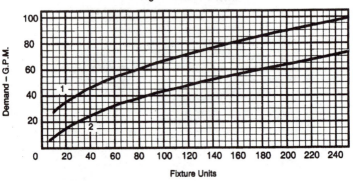

FIGURE 2.58 Enlarged scale of demand loads. *(Reprinted from the 2000 Uniform Plumbing Code (UPC) with the permission of the International Association of Plumbing and Mechanical Officials (IAPMO))*

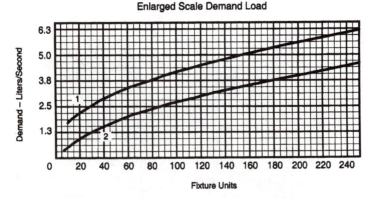

FIGURE 2.59 Enlarged scale of demand loads. *(Reprinted from the 2000 Uniform Plumbing Code (UPC) with the permission of the International Association of Plumbing and Mechanical Officials (IAPMO))*

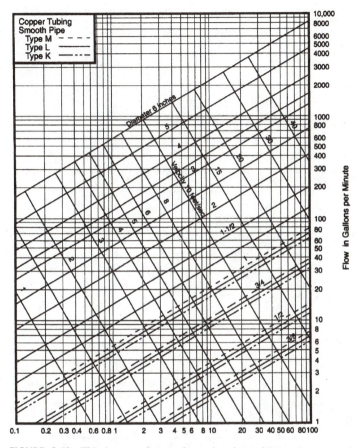

FIGURE 2.60 This is one of several graphs about friction loss. *(Reprinted from the 2000 Uniform Plumbing Code (UPC) with the permission of the International Association of Plumbing and Mechanical Officials (IAPMO))*

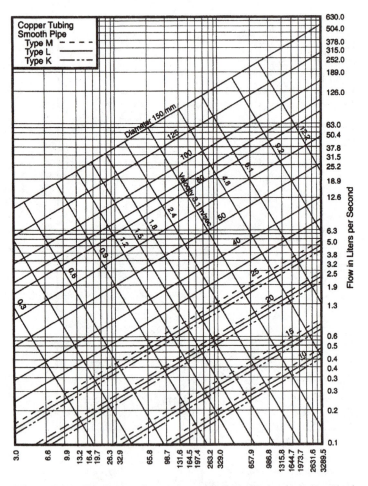

FIGURE 2.61 This is one of several graphs about friction loss. *(Reprinted from the 2000 Uniform Plumbing Code (UPC) with the permission of the International Association of Plumbing and Mechanical Officials (IAPMO))*

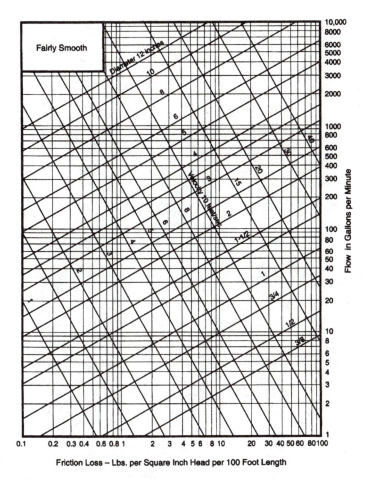

Friction Loss – Lbs. per Square Inch Head per 100 Foot Length

FIGURE 2.62 This is one of several graphs about friction loss. *(Reprinted from the 2000 Uniform Plumbing Code (UPC) with the permission of the International Association of Plumbing and Mechanical Officials (IAPMO))*

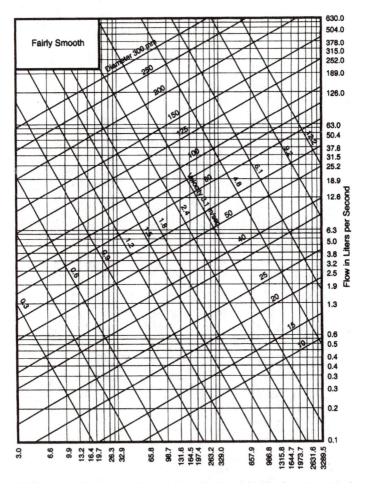

FIGURE 2.63 This is one of several graphs about friction loss. *(Reprinted from the 2000 Uniform Plumbing Code (UPC) with the permission of the International Association of Plumbing and Mechanical Officials (IAPMO))*

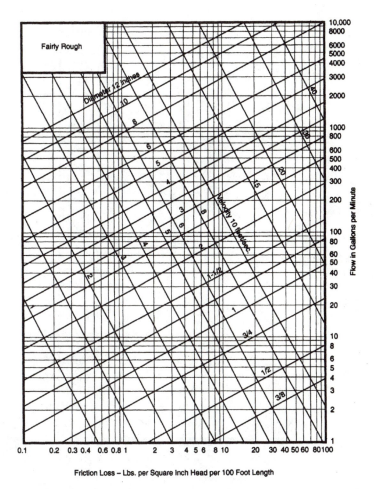

Friction Loss – Lbs. per Square Inch Head per 100 Foot Length

FIGURE 2.64 This is one of several graphs about friction loss. *(Reprinted from the 2000 Uniform Plumbing Code (UPC) with the permission of the International Association of Plumbing and Mechanical Officials (IAPMO))*

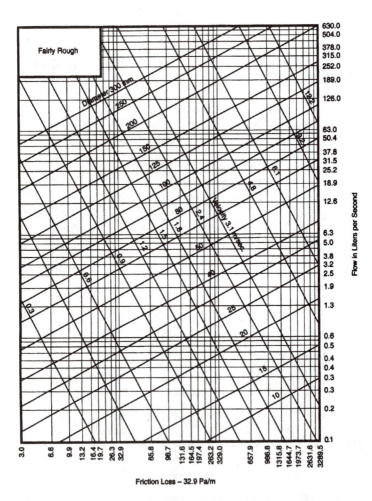

FIGURE 2.65 This is one of several graphs about friction loss. *(Reprinted from the 2000 Uniform Plumbing Code (UPC) with the permission of the International Association of Plumbing and Mechanical Officials (IAPMO))*

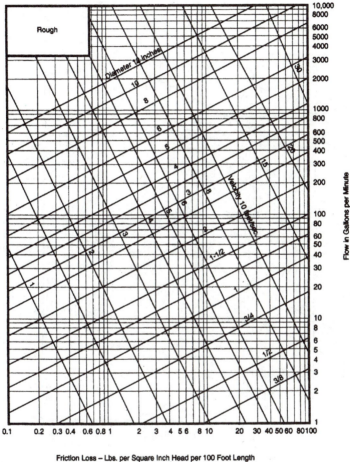

Friction Loss – Lbs. per Square Inch Head per 100 Foot Length

FIGURE 2.66 This is one of several graphs about friction loss. *(Reprinted from the 2000 Uniform Plumbing Code (UPC) with the permission of the International Association of Plumbing and Mechanical Officials (IAPMO))*

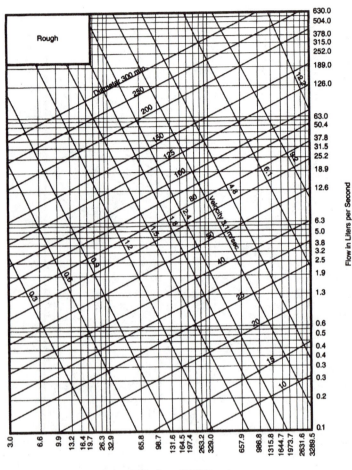

Friction Loss – 32.9 Pa/m

FIGURE 2.67   This is one of several graphs about friction loss. *(Reprinted from the 2000 Uniform Plumbing Code (UPC) with the permission of the International Association of Plumbing and Mechanical Officials (IAPMO))*

Sizing of Grease Interceptors

Number of meals per peak hour[1]	x	Waste flow rate[2]	x	retention time[3]	x	storage factor[4]	=	Interceptor size (liquid capacity)

1. **Meals Served at Peak Hour**

2. **Waste Flow Rate**
   a. With dishwashing machine ....6 gallon (22.7 L) flow
   b. Without dishwashing machine ....5 gallon (18.9 L) flow
   c. Single service kitchen ....2 gallon (7.6 L) flow
   d. Food waste disposer ....1 gallon (3.8 L) flow

3. **Retention Times**
   Commercial kitchen waste
      Dishwasher...2.5 hours
   Single service kitchen
      Single serving...1.5 hours

4. **Storage Factors**
   Fully equipped commercial kitchen ....8 hour operation: 1
   ....16 hour operation: 2
   ....24 hour operation: 3
   Single Service Kitchen 1.5

FIGURE 2.68 This chart shows information about grease interceptors and how to size them. *(Reprinted from the 2000 Uniform Plumbing Code (UPC) with the permission of the International Association of Plumbing and Mechanical Officials (IAPMO))*

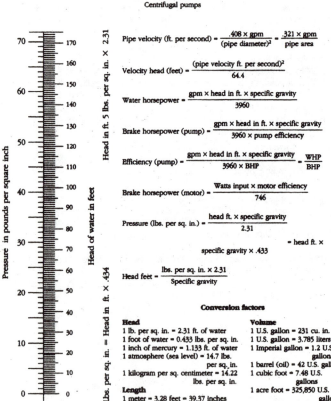

Engineering data
Formulas and conversion factors
Centrifugal pumps

Pipe velocity (ft. per second) = $\dfrac{.408 \times \text{gpm}}{(\text{pipe diameter})^2}$ = $\dfrac{.321 \times \text{gpm}}{\text{pipe area}}$

Velocity head (feet) = $\dfrac{(\text{pipe velocity ft. per second})^2}{64.4}$

Water horsepower = $\dfrac{\text{gpm} \times \text{head in ft.} \times \text{specific gravity}}{3960}$

Brake horsepower (pump) = $\dfrac{\text{gpm} \times \text{head in ft.} \times \text{specific gravity}}{3960 \times \text{pump efficiency}}$

Efficiency (pump) = $\dfrac{\text{gpm} \times \text{head in ft.} \times \text{specific gravity}}{3960 \times \text{BHP}}$ = $\dfrac{\text{WHP}}{\text{BHP}}$

Brake horsepower (motor) = $\dfrac{\text{Watts input} \times \text{motor efficiency}}{746}$

Pressure (lbs. per sq. in.) = $\dfrac{\text{head ft.} \times \text{specific gravity}}{2.31}$

= head ft. × specific gravity × .433

Head feet = $\dfrac{\text{lbs. per sq. in.} \times 2.31}{\text{Specific gravity}}$

Conversion factors

Head
1 lb. per sq. in. = 2.31 ft. of water
1 foot of water = 0.433 lbs. per sq. in.
1 inch of mercury = 1.133 ft. of water
1 atmosphere (sea level) = 14.7 lbs. per sq. in.
1 kilogram per sq. centimeter = 14.22 lbs. per sq. in.

Length
1 meter = 3.28 feet = 39.37 inches

Power
1 horsepower = 745.7 watts
1 kilowatt = 1000 watts
1 kilowatt = 1.341 HP
100 boiler HP requires 7 gpm feed water approximately.

Volume
1 U.S. gallon = 231 cu. in.
1 U.S. gallon = 3.785 liters
1 Imperial gallon = 1.2 U.S. gallons
1 barrel (oil) = 42 U.S. gallons
1 cubic foot = 7.48 U.S. gallons
1 acre foot = 325,850 U.S. gallons
1 cubic meter = 264.2 U.S. gallons

Weight
1 U.S. gallon water weighs 8.35 lbs.
1 cubic foot water weighs

FIGURE 6.69 Formulas and conversion factors for centrifugal pumps.

MULTIPLY	BY	TO OBTAIN
Gallons/minute	8.0208	Cubic feet/hour
Gallons water/minute	6.0086	Tons of water/24 hours
Inches	2.540	Centimeters
Inches of mercury	0.03342	Atmospheres
Inches of mercury	1.133	Feet of water
Inches of mercury	0.4912	Pounds/square inch
Inches of water	0.002458	Atmospheres
Inches of water	0.07355	Inches of mercury
Inches of water	5.202	Pounds/square feet
Inches of water	0.03613	Pounds/square inch
Liters	1000	Cubic centimeters
Liters	61.02	Cubic inches
Liters	0.2642	Gallons
Miles	5280	Feet
Miles/hour	88	Feet/minute
Miles/hour	1.467	Feet/second
Millimeters	0.1	Centimeters
Millimeters	0.03937	Inches
Million gallon/day	1.54723	Cubic feet/second
Pounds of water	0.01602	Cubic feet
Pounds of water	27.68	Cubic inches
Pounds of water	0.1198	Gallons
Pounds/cubic inch	1728	Pounds/cubic feet
Pounds/square foot	0.01602	Feet of water
Pounds/square inch	0.06804	Atmospheres
Pounds/square inch	2.307	Feet of water
Pounds/square inch	2.036	Inches of mercury
Quarts (dry)	67.20	Cubic inches
Quarts (liquid)	57.75	Cubic inches
Square feet	144	Square inches
Square miles	640	Acres
Square yards	9	Square feet
Temperature (°C) + 273	1	Abs. temperature (°C)
Temperature (°C) + 17.28	1.8	Temperature (°F)
Temperature (°F) + 460	1	Abs. temperature (°F)
Temperature (°F) - 32	5/9	Temperature (°C)
Tons (short)	2000	Pounds
Tons of water/24 hours	83.333	Pounds water/hour
Tons of water/24 hours	0.16643	Gallons/minute
Tons of water/24 hours	1.3349	Cubic feet/hour

FIGURE 2.70 A useful set of tables to keep on hand. *(Reprinted from the 2000 Uniform Plumbing Code (UPC) with the permission of the International Association of Plumbing and Mechanical Officials (IAPMO))*

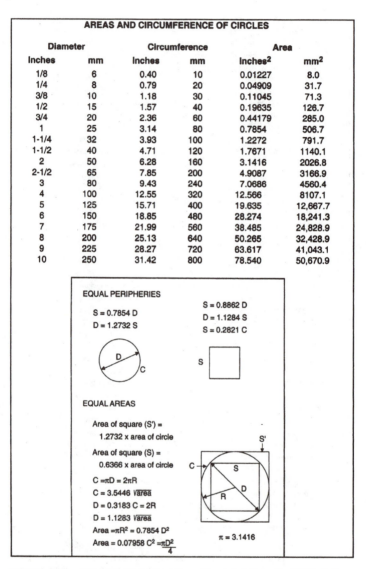

AREAS AND CIRCUMFERENCE OF CIRCLES

Diameter		Circumference		Area	
Inches	mm	Inches	mm	Inches^2	mm^2
1/8	6	0.40	10	0.01227	8.0
1/4	8	0.79	20	0.04909	31.7
3/8	10	1.18	30	0.11045	71.3
1/2	15	1.57	40	0.19635	126.7
3/4	20	2.36	60	0.44179	285.0
1	25	3.14	80	0.7854	506.7
1-1/4	32	3.93	100	1.2272	791.7
1-1/2	40	4.71	120	1.7671	1140.1
2	50	6.28	160	3.1416	2026.8
2-1/2	65	7.85	200	4.9087	3166.9
3	80	9.43	240	7.0686	4560.4
4	100	12.55	320	12.566	8107.1
5	125	15.71	400	19.635	12,667.7
6	150	18.85	480	28.274	18,241.3
7	175	21.99	560	38.485	24,828.9
8	200	25.13	640	50.265	32,428.9
9	225	28.27	720	63.617	41,043.1
10	250	31.42	800	78.540	50,670.9

EQUAL PERIPHERIES

$S = 0.7854\ D$

$D = 1.2732\ S$

$S = 0.8862\ D$

$D = 1.1284\ S$

$S = 0.2821\ C$

EQUAL AREAS

Area of square (S') =
  1.2732 x area of circle

Area of square (S) =
  0.6366 x area of circle

$C = \pi D = 2\pi R$

$C = 3.5446\ \sqrt{area}$

$D = 0.3183\ C = 2R$

$D = 1.1283\ \sqrt{area}$

Area $= \pi R^2 = 0.7854\ D^2$

Area $= 0.07958\ C^2 = \dfrac{\pi D^2}{4}$

$\pi = 3.1416$

FIGURE 2.71 More useful information. *(Reprinted from the 2000 Uniform Plumbing Code (UPC) with the permission of the International Association of Plumbing and Mechanical Officials (IAPMO))*

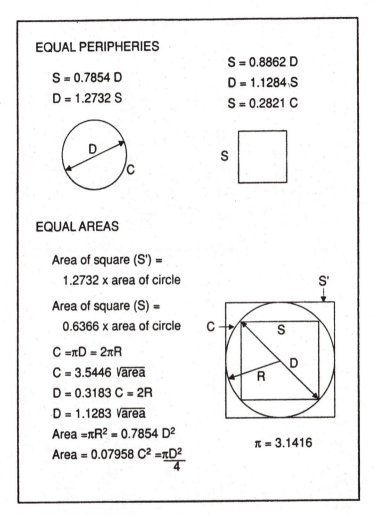

EQUAL PERIPHERIES

S = 0.7854 D
D = 1.2732 S

S = 0.8862 D
D = 1.1284 S
S = 0.2821 C

EQUAL AREAS

Area of square (S') =
    1.2732 x area of circle

Area of square (S) =
    0.6366 x area of circle

C = πD = 2πR
C = 3.5446 $\sqrt{area}$
D = 0.3183 C = 2R
D = 1.1283 $\sqrt{area}$
Area = πR^2 = 0.7854 D^2
Area = 0.07958 C^2 = $\dfrac{πD^2}{4}$

π = 3.1416

FIGURE 2.72 Mathematical formulas.

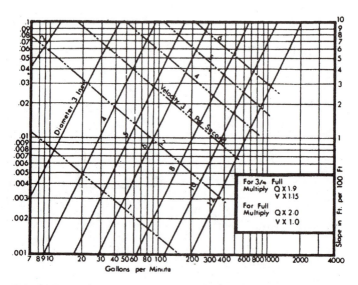

FIGURE 2.73 Flow in partially filled pipes.

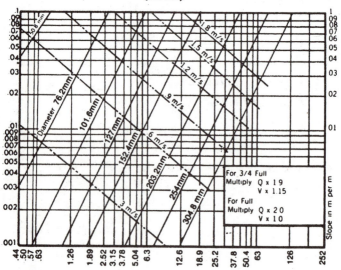

FIGURE 2.74 Flow in partially filled pipes.

Circumference = Diameter × 3.1416
Circumference = Radius × 6.2832
Diameter = Radius × 2
Diameter = Square root of (area ÷ .7854)
Diameter = Square root of area × 1.1283
Diameter = Circumference × .31831
Radius = Diameter ÷ 2
Radius = Circumference × .15915
Radius = Square root of area × .56419
Area = Diameter × Diameter × .7854
Area = Half of the circumference × half of the diameter
Area = Square of the circumference × .0796
Arc length = Degrees × radius × .01745
Degrees of arc = Length ÷ (radius × .01745)
Radius of arc = Length ÷ (degrees × .01745)
Side of equal square = Diameter × .8862
Side of inscribed square = Diameter × .7071
Area of sector = Area of circle × degrees of arc ÷ 360

FIGURE 2.75   Formulas for a circle.

1. Circumference of a circle = π × diameter or 3.1416 × diameter
2. Diameter of a circle = Circumference × .31831
3. Area of a square = Length × width
4. Area of a rectangle = Length × width
5. Area of a parallelogram = Base × perpendicular height
6. Area of a triangle = ½ base × perpendicular height
7. Area of a circle = π × radius squared or diameter squared × .7854
8. Area of an ellipse = Length × width × .7854
9. Volume of a cube or rectangular prism = Length × width × height
10. Volue of a triangular prism = Area of triangle × length
11. Volume of a sphere = Diameter cubed × .5236 or (dia. × dia. × dia. × .5236)
12. Volume of a cone = π × radius square × ⅓ height
13. Volume of a cylinder = π × radius squared × height
14. Length of one side of a square × 1.128 = Diameter of an equal circle
15. Doubling the diameter of a pipe or cylinder increases its capacity 4 times
16. The pressure (in lbs. per sq. inch) of a column of water = Height of the column (in feet) × .434
17. The capacity of a pipe or tank (in U.S. gallons) = Diameter squared (in inches) × the length (in inches) × .0034
18. A gallon of water = 8⅓ lb. = 231 cu. inches
19. A cubic foot of water = 62½ lb. = 7½ gallons

FIGURE 2.76   Useful formulas.

Area = Base × distance between the two parallel sides

FIGURE 2.77 Parallelograms.

Area = Length × width

FIGURE 2.78 Rectangles

Area of surface = Diameter × diameter × 3.1416
Side of inscribed cube = Radius × 1.547
Volume = Diameter × diameter × diameter × .5236

FIGURE 2.79 Spheres.

Area = One-half of height times base

FIGURE 2.80 Triangles.

Area = One-half of the sum of the parallel sides × the height

FIGURE 2.81 Trapezoids.

Volume = Width × height × length

FIGURE 2.82 Cubes.

Area of surface = One half of circumference of base × slant height + area
of base.
Volume = Diameter × diameter × .7854 × one-third of the altitude.

FIGURE 2.83 Cone calculation.

Volume = Width × height × length

FIGURE 2.84 Volume of a rectangular prism.

Area = Length × width

FIGURE 2.85 Finding the area of a square.

Area = ½ perimeter of base × slant height + area of base
Volume = Area of base × ⅓ of the altitude

FIGURE 2.86 Finding the area and volume of a pyramid.

These comprise the numerous figures having more than four sides, names according to the number of sides, thus:

Figure	Sides
Pentagon	5
Hexagon	6
Heptagon	7
Octagon	8
Nonagon	9
Decagon	10

To find the area of a polygon: Multiply the sum of the sides (perimeter of the polygon) by the perpendicular dropped from its center to one of its sides, and half the product will be the area. This rule applies to all regular polygons.

FIGURE 2.87 Polygons.

Multiply Length × Width × Thickness
Example: 50 ft. × 10 ft. × 8 in.
50' × 10' × ⁸⁄₁₂' = 333.33 cu. feet
To convert to cubic yards, divide by 27 cu. ft. per cu. yd.
Example: 333.33 ÷ 27 = 12.35 cu. yd.

FIGURE 2.88 Estimating volume.

Area = Width of side × 2.598 × width of side

FIGURE 2.89 Hexagons.

Area of surface = Diameter × 3.1416 × length + area of the two bases
Area of base = Diameter × diameter × .7854
Area of base = Volume ÷ length
Length = Volume ÷ area of base
Volume = Length × area of base
Capacity in gallons = Volume in inches ÷ 231
Capacity of gallons = Diameter × diameter × length × .0034
Capacity in gallons = Volume in feet × 7.48

FIGURE 2.90 Cylinder formulas.

Area = Short diameter × long diameter × .7854

FIGURE 2.91 Ellipse calculation.

Deg. F. = Deg. C. × 1.8 + 32

FIGURE 2.92 Temperature conversion.

To change	To	Multiply by
Feet of water	Pounds per square inch	0.434
Feet of water	Pounds per square foot	62.5
Feet of water	Inches of mercury	0.8824
Atmospheres	Pounds per square inch	14.696
Atmospheres	Inches of mercury	29.92
Atmospheres	Feet of water	34
Long tons	Pounds	2240
Short tons	Pounds	2000
Short tons	Long tons	0.89295

FIGURE 2.93 Useful multipliers.

To figure the final temperature when two different temperatures of water are mixed together, use the following formula:

$$\frac{(A \times C) + (B \times D)}{A + B}$$

A = Weight of lower temperature water
B = Weight of higher temperature water
C = Lower temperature
D = Higher temperature

FIGURE 2.94 Temperature calculation.

Temperature can be expressed according to the Fahrenheit scale or the Celsius scale. To convert C to F or F to C, use the following formulas:

$$°F = 1.8 \times °C + 32$$

$$°C = 0.55555555 \times °F - 32$$

FIGURE 2.95 Temperature conversion.

Deg. C. = Deg. F. − 32 ÷ 1.8

FIGURE 2.96 Temperature conversion.

Temperature can be expressed according to the Fahrenheit scale or the Celsius scale. To convert C to F or F to C, use the following formulas:

$$°F = 1.8 \times °C + 32$$

$$°C = 0.55555555 \times °F - 32$$

FIGURE 2.97 Finding the area of a pipe.

The capacity of pipes is as the square of their diameters. Thus, doubling the diameter of a pipe increases its capacity four times.

FIGURE 2.98 A piping fact.

The formula for calculating expansion or contraction in plastic piping is:

$$L = Y \times \frac{T - F}{10} \times \frac{L}{100}$$

L = Expansion in inches
Y = Constant factor expressing inches of expansion per 100°F temperature change per 100 ft. of pipe
T = Maximum temperature (0°F)
F = Minimum temperature (0°F)
L = Length of pipe run in feet

FIGURE 2.99 Expansion in plastic piping.

Inch scale	Metric scale
¼"	1:50
⅛"	1:100

FIGURE 2.100 Scales used for building plans.

Inch scale	Metric scale
⅟₁₆"	1:200

FIGURE 2.101 Scale used for site plans.

(Surface area ÷ R value) × (temperature inside − temperature outside)

Surface area of a material (in square feet) divided by its "R" value and multiplied by the difference in Fahrenheit degrees between inside and outside temperature equals heat loss in BTU's per hour.

FIGURE 2.102 Calculating heat loss per hour with R-value.

$$L = \frac{144}{D \times 3.1414} \times R \div 12$$

D = O.D. of pipe
L = length of pipe needed in ft.
R = sq. ft. of radiation needed

FIGURE 2.103 Formulas for pipe radiation.

- 3 feet of 1-in. pipe equal 1 square foot of radiation.
- 2⅓ linear feet of 1¼ in. pipe equal 1 square foot of radiation.
- Hot water radiation gives off 150 BTU per square foot of radiation per hour.
- Steam radiation gives off 240 BTU per square foot of radiation per hour.
- On greenhouse heating, figure ⅔ square foot of radiation per square foot of glass.
- One square foot of direct radiation condenses .25 pound of water per hour.

FIGURE 2.104   Radiant heat facts.

The approximate weight of a piece of pipe can be determined by the following formulas:

Cast Iron Pipe: weight = $(A^2 - B^2) \times C \times .2042$

Steel Pipe: weight = $(A^2 - B^2) \times C \times .2199$

Copper Pipe: weight = $(A^2 - B^2) \times C \times .2537$

A = outside diameter of the pipe in inches

B = inside diameter of the pipe in inches

C = length of the pipe in inches

FIGURE 2.105   Finding the weight of piping.

3

NEW INSTALLATIONS

nstalling water-distribution systems is usually easier than installing a drain-waste-and-vent (DWV) system. For one thing, the
size of the pipes being installed with a water system is smaller.
Also, the requirements for pipe grading are less stringent with water
systems. Code requirements for water systems are easier for most
plumbers to understand. When all elements are considered, a residential water system is pretty simple. Commercial jobs are more
complex, but piping diagrams are supplied for commercial jobs that
make the installations easy if you are good at following directions.

Sizing water systems can get tricky. Most residential jobs are fairly
simple, but sometimes a plumber can run into some complicated
sizing situations. Since most residential water systems are designed
by plumbers, rather than engineers and architects, it is essential that
residential plumbers know how to interpret the plumbing code for
sizing requirements.

Sizing is not the only part of installing a water system that can get
tricky. Deciding on what pipe will work best could be confusing. Trying to route pipes in a cost-effective layout is important to the profit
picture of a plumber. There's a lot to learn about water systems, so
let's started.

POTABLE WATER

What purposes require potable water? The important needs for potable water include drinking, cooking, bathing, and the preparation of food and medicine. Potable water is also used for other activities, but the ones listed are the uses dealt with in the plumbing code. Fixtures used for any of the above purposes must be plumbed in a way that only potable water is accessible to them.

What are the requirements for providing hot water? Hot water must be provided in buildings where people work and for all permanent residences. How about cold water? Cold water must be supplied to every building that contains plumbing fixtures and is used for human occupancy.

How much water pressure is required? The water pressure for a water distribution system must be enough to provide proper flow rates at each of the fixtures. Flow rates for various fixtures are determined by referring to the tables or text in your local code.

Generally, 40 psi is considered adequate water pressure. If the incoming pressure is 80 psi or more, the pressure must be controlled with a pressure-reducing valve. This valve will be installed between the water service and the main water distribution pipe. The device will allow the water pressure to be regulated to a lower pressure rating. There are, of course, exceptions to most rules. Check your local code for the requirements in your jurisdiction.

Water Conservation

Water conservation is an issue that most codes deal with. The codes restrict the flow rates of certain fixtures, like showers, sinks, and lavatories. An unmodified shower head can normally produce a flow of five gallons per minute. This flow rate is often reduced to three gallons per minute with the insertion of a water-saver device. The device is a small disk with holes in it.

Other water-saver regulations apply to buildings with public restrooms. It is frequently required that all public-use lavatories be equipped with self-closing faucets. These faucets are restricted in their flow rates, and they will cut themselves off after use.

Urinals are supposed to have flow rates of not more than 1.5 gallons per flush. Toilets generally are not allowed to use more than four gallons of water during their flush cycle. Many new toilets use far less water. It is common for newer toilets to use less than two gallons of water per flush.

fastfacts

➤ *Potable water is required for cooking.*

➤ *Potable water is required for drinking.*

➤ *Potable water is required for the preparation of food.*

➤ *Potable water is required for the preparation of medicine.*

➤ *Potable water is required for bathing.*

Anti-Scald Devices

Anti-scald devices are required on some plumbing fixtures. Different codes require these valves on different types of fixtures, and they sometimes set different limits for the maximum water temperature. Anti-scald devices are valves or faucets that are specially equipped to avoid burns from hot water. These devices come in different configurations, but they all have the same goal: to avoid scalding.

Whenever a gang shower is installed, such as in a school gym, anti-scald shower valves should be installed. Some codes require anti-scald valves on residential showers. The maximum temperature allowed for hot water in some codes is 110 degrees F.

CHOOSING THE TYPE OF PIPE AND FITTINGS TO BE USED

The three primary types of pipe used in above-grade water systems are cross-linked polyethylene (PEX), copper, and CPVC. All of these pipes will provide adequate service as water distribution carriers. The pipe you choose will be largely a matter of personal preference. Many homeowners opt for CPVC. They make this choice because they do not have to solder the joints. Soldering seems difficult to inexperienced people and they assume it will be easier to glue plastic pipe together than it will to solder copper joints. Most people outside of the trade never think of PEX; they simply don't know it exists. Copper pipe and tubing has long been the standard for professional plumbers. Currently, many plumbers are using PEX more and more often.

CPVC PIPE

CPVC pipe has been used as potable water pipe for many years. The fittings for the pipe are installed with a solvent weld. CPVC does not require you to have any soldering skills or equipment. The pipe has some flexibility and is easy to snake through floor joists. It is suitable for hot and cold water applications and can be cut with a roller-cutter or hacksaw. Almost anyone can install it.

Professional plumbers I know, and have known, do not show great affection for CPVC. To make a waterproof joint, the pipe and fittings must be properly prepared. This preparation is very much like the procedures used with PVC drainage pipe. There is a cleaning solution that should be applied to the pipe and fittings. Then, there is a primer that is applied to both the pipe and the fittings. Joints are made with a solvent or glue.

Going through all the steps of making the joint is a slow process. Since professionals seek to complete their jobs as quickly as possible, CPVC is not an ideal choice. Not only do you lose time with all the prep work, the joints cannot be disturbed for some time. If they are jarred before they have cured, the joints may suffer from voids that will leak. The amount of time you must wait for the joint to cure will vary depending upon the temperature.

The time lost waiting for the joints to set up is a drawback for professionals. When working with PEX or copper, the time spent waiting to work with new joints is greatly reduced. This reason alone is enough to cause professionals to use a different pipe, but it is not the only reason professionals choose other pipes.

Even after a CPVC joint is made, it is not extremely strong. Any stress on the joint can cause it to break loose and leak. CPVC becomes brittle in cold weather. If the pipe is dropped on a hard surface, it can develop small cracks. These cracks often go unnoticed until the pipe is installed and tested. When the pipe is tested and leaks, it must be replaced. This means more lost time for the professional. Copper and PEX are not subject to these same cracks when dropped.

If CPVC pipe or fittings have any water on them, the glue may not make a solid joint. With PEX, water on the pipe has no affect on the integrity of the joint. Water on the outside of copper pipe will turn to steam and normally does not cause a leak. The simple act of hanging the pipe is also a factor in making a choice. When you hang the pipe, you will probably use a hammer to drive the hanger into place. If the hammer slips and hits your pipe, CPVC may shatter. Copper may dent, but it will not shatter. PEX will bounce right back into shape from a hammer blow. These may be small differences, but they add up to stack the deck against CPVC for professionals.

There is one advantage that CPVC has over copper for professional plumbing applications. It is not adversely affected by acidic water. When potable water is being provided by a well, acid in the water can cause copper pipe to leak. The acid causes pinhole leaks in copper after some period of time. Since CPVC is plastic, it is not affected by the acid.

PEX PIPE

PEX pipe is, in my personal opinion, a fine choice for water systems. If you are going to install PEX pipe, learn how to do it right. This type of pipe is not like polyethylene. Special fittings and clamps should be used to ensure good joints. PEX slides over a ridged insert fitting and is held in place with a special clamp. These clamps are not adjustable, stainless-steel clamps, like those used on polyethylene pipe. The clamps used on PEX are solid metal and are installed with a special crimping tool. There are many advantages to working with PEX.

PEX comes coiled in a roll. The pipe is extremely flexible and can be run very much like electrical wire. After holes are drilled, the pipe can be pulled through the holes in long lengths. The pipe is approved for hot and cold water and can be cut with a hacksaw, although special cutters do a neater job. In cold climates, PEX offers another advantage; it can expand a great deal before splitting during freezing conditions. Acidic water will not eat holes in PEX pipe. From my perspective, it is difficult to think of a reason why you shouldn't use PEX.

When houses are plumbed with PEX, the piping design is often different than when the house is plumbed with a rigid pipe. CPVC and copper installations incorporate the use of a main water pipe and several branches to feed individual fixtures. With PEX, most plumbers run individual lengths of pipe from each fixture to a common manifold. By installing the pipe in this way, there are no joints concealed in the walls or ceilings. This reduces the likelihood of a leak that will be difficult to access when the house is completed.

Basically, the water service or a main water pipe is run to the manifold location. The manifold receives its water from the water service or main and distributes it through the individual pipes. This type of installation requires more pipe than traditional installations, but PEX pipe is cheap when compared with the cost of copper. It is the fittings that are expensive, but in a manifold installation, there are very few fittings used.

Manifold systems are convenient for service plumbers and home-owners. The fact that PEX pipe can be installed with very few concealed joints is an attribute I find attractive. PEX pipe may not always be the best pipe for a job, but it often is.

COPPER PIPE

Copper has long been the leader in water distribution systems. While copper is expensive and has some drawbacks, it remains the most common type of pipe in potable water systems. Many home-owners specify copper in their building plans, and old-school plumbers swear by it. As a seasoned plumber, I like copper and use it frequently. However, I seem to be using more and more PEX pipe as people become more accepting of it. Copper is a proven performer with a solid track record.

When soldering joints on potable water pipes, you must use a low-lead solder. In the old days, 50/50 solder was used, but not today. Buy lead-free solder for your joints. The old 50/50 solder is still available for making connections on non-potable pipes, but don't use it on potable water systems.

Copper is easy to cut when you use roller-cutters. Cutting the pipe simply entails placing the pipe between the cutting wheel and the rollers, tightening the handle and turning the cutter. As you rotate the cutter around the pipe, tighten the handle to maintain steady pressure. After a few turns, the pipe will be cut. Be careful of jagged pieces of copper after cutting the pipe. Be especially careful when you sand the pipe to clean it. If the pipe does not cut evenly, there may be sharp, jagged edges protruding from the edge of the pipe. The shrouds of copper can cut you or become imbedded in your skin.

Copper pipe is durable and is approved for hot and cold water. Once you know how to solder, copper is easy to work with. The pipe installs quickly and produces a neat looking job. Since the pipe is rigid, it can be installed to allow the draining of the water system. This is a factor in seasonal cottages and other circumstances where the water is turned off for the winter.

THE PIPE DECISION

The decision on which pipe to use is up to you. As a professional, it would be a toss up between PEX and copper. My decision between the two would be based on the working conditions and type of plumbing being installed. Personally, I would not use CPVC. Consider all your options and make your choice based on factors you feel strongly about.

MATERIALS APPROVED FOR WATER DISTRIBUTION

There are a number of materials approved for water distribution, but only a few are used frequently in new plumbing systems. While I don't have any formal statistics in front of me, I would guess that copper is still the number one choice as a water distribution material. I would think PEX pipe comes in second, and that CPVC runs a distant third.

Materials approved for use in water distribution systems may vary slightly from code to code. The information I am about to give you is valid in many jurisdictions, but remember to always check your local code before installing plumbing.

Galvanized Steel Pipe

Galvanized steel pipe is an approved water distribution pipe. It has been used for a very long time, and you will still find it in many older plumbing systems. It is not, however, used much in modern plumbing. This pipe is heavy and requires threaded joints. The pipe tends to rust and deteriorate more quickly than other approved water distribution pipes. There are few, if any, occasions when galvanized steel pipe is used for modern plumbing installations.

Brass Pipe

Brass pipe is approved for water distribution, but like galvanized pipe, it is rarely used.

Copper

We talked about the use of copper as a water distribution pipe and is, of course, approved for that application. There are, however, different types of copper. By types, I mean ratings of thickness. The two types most commonly used for water distribution are type-L and type-M. Type-K copper is also fine for water distribution, but it is more expensive and usually the extra-thick pipe is not needed. Type-M copper, the thinnest of the three, is not approved in all areas for water distribution. For years type-M was the standard, but now type-L is required by some administrative authorities.

PEX

PEX is approved for water distribution, and we have already discussed its use.

CPVC

Chlorinated polyvinyl chloride, or CPVC as it is commonly called, is another approved material that we have discussed. You may find this pipe ideal, but I don't favor it. Try it and see what you think, but experience has shown me that I do better with other types of piping.

SIZING

The task of sizing water pipes strikes fear into many plumbers. They take one look at the friction-loss charts in their code books, see some math formulas that read like a foreign language, and run the other way. Sizing water pipes can be intimidating.

There are two approaches to sizing water pipe: the engineer's approach and the plumber's approach. Large jobs typically are required to be designed by an engineer or other suitable professional.

Your local code book will provide information on how you can use formulas, friction charts, and other complicated aids to size water pipes, but they will also give you some easier ways around the problem. Tables are provided that show you the minimum pipe size for the fixture supplies to various types of fixtures. All you have to do is look at your code book, and you can tell quickly what the minimum size requirements are for your fixture water supplies. Up to the first thirty inches of a section of pipe connecting to a fixture is considered a fixture supply. Once the supply is more than thirty inches away from the fixture, it is considered a water distribution pipe. If you want to know what the required flow rates and pressures for the various fixtures are, you can simply refer to your code book. All of this information is available to you in an easy-to-understand format.

The next step is not quite as easy, but it still is not difficult. There are several methods for sizing water pipe. Most code books disclaim any responsibility for the examples they give on sizing. The examples are just that—examples, not carved in stone procedures. One of the easiest ways to size water pipe is with the use of the fixture-unit method, and that is the one we are going to concentrate on. Let me show you how it works.

The Fixture-Unit Method

The fixture-unit method of sizing water distribution pipes is pretty easy to understand. You will need to know the total number of fixture units that will be placed on your pipes. This information can be obtained from listings in your code book.

The total developed length of your water pipes will have to be known. If you are on the job, you can measure this distance with a tape measure. When you are working from blueprints, a scale rule will give you the numbers you need. The measurement will begin at the location of the water meter.

You will also need to know what the water pressure on the system will be. If you are not controlling the water pressure, as you might with a pressure-reducing valve or a well system, you can call the municipal water department to obtain the pressure rating on the water main.

To pinpoint the accuracy of your sizing, you should learn the difference in elevation between the water meter and the highest plumbing fixture. However, by being generous with your pipe sizes, you can get by without this information most of the time.

The fixture units assigned to fixtures for the purpose of sizing water pipe will normally include both the hot and cold water demand of the fixture. The ratings for fixtures used by the public will be different than those installed for private use. All of the information you need for this method of sizing is easily obtained from you local code book.

Now, lets assume we are going to size the water pipe for a small house. The house has one full bathroom and a kitchen. It also has a laundry hook-up and one hose bibb. We need to know what fixture-unit (FU) ratings to assign to these fixtures. A quick look at a code book might give us the following ratings:

- *Bathtub: 2 FU*
- *Lavatory: 1 FU*
- *Toilet: 3 FU*
- *Sink: 2 FU*
- *Hose bibb: 3 FU*
- *Laundry hook-up: 2 FU*

Okay, now we know our total number of fixture units. The working pressure on this water system will be 50 psi. Our total developed length of piping, from the water meter to the most remote fixture is 95 feet. We now have all we really need to size our water pipe, with a little help from a table in our code book. When we refer to our sizing table in the code book, we will see different tables for different pressure ratings. We choose the table that matches our rating of 50 psi.

There will be columns of numbers to identify fixture units and developed lengths of pipe. We will see sizes for building water supplies and branches. Even the sizes for water meters will be available. All we have to do is put our plan into action.

Our water meter for this example is a ¾ inch meter. First we have to size the water service. How many fixture units will be on the water service? When you count all our fixture units, the total is thirteen. Our pipe has to run 95 feet, so what size does it have to be. By looking at the sizing table in your code book, you will see quickly what size it must be. In the case of this example, the required size is a ¾ inch water service. I was able to determine this just by running my finger down and across the sizing table.

The sizing table is based on the length of the pipe run, the water pressure, and the size of the water meter. Knowing those three variables, you can size pipe quickly and easily. The same cross-referencing that was done to size the water service also works when sizing water distribution pipes.

For example, assume that the most remote fixture in the house is forty feet from the end of the water service. When I look at my table, I see that a ¾ inch pipe is required to begin my water distribution system. As the main ¾ inch pipe reaches fixtures and lowers the number of fixture units remaining in the run, the pipe size can be reduced. As a rule-of-thumb, when there are only two fixtures remaining to be served on a residential water main, the pipe size can drop down to a ½ inch pipe.

Let's say my ¾ inch pipe is running down the center of the home. The kitchen sink is perpendicular to the main water line and lies twenty-two feet away. The sink as a fixture-unit demand of two. By checking my sizing table, I see that the pipe branching off the main to serve the sink can be a ½ inch pipe.

Once you sit down with your code book and work with the fixture-unit method of pipe sizing, you will see how easy it is. Now we are done with our sizing exercise, and we are going to move on to some techniques used to install water distribution pipes.

Information Needed to Size a Water System with a Fixture-Unit Method

✔ The total number of fixture units loaded on the system must be known.

✔ The total developed length of the piping in the system must be known.

✔ The water pressure present on the system must be known to size the system.

✔ The fixture units assigned to fixtures is critical is the sizing of a system.

INSTALLING THE POTABLE WATER SYSTEM

Aside from how the connections are made, CPVC and copper systems will be installed with the same principals. PEX systems could be installed along the same lines as copper, but a manifold installation makes more sense with PEX. I should mention that PEX pipe is available in semi-rigid lengths, in addition to coils.

Is a water service part of a water distribution system? It is part of the water system, but it is not normally considered to be a part of a water-distribution system. This can be confusing, so let me explain.

A water service does convey potable water to a building, but it is not considered a water-distribution pipe. Water services fall into a category all of their own. Water-distribution pipes are the pipes found inside a building. Water services run the majority of their length outside of a building's foundation and are usually buried in the ground. Water-distribution pipes are installed within the foundation of the building and don't normally run underground or outside of a foundation. However, they may run underground, such as in the case of a one story building where the water-distribution pipes are buried under a concrete floor. In any event, water services and water-distribution pipes are dealt with differently by the plumbing codes.

Why are water-distribution pipes considered different from water services? Water distribution pipes are dealt with differently because they serve different needs. Water-distribution pipes pick up where water services leave off. The pipes distributing water to plumbing fixtures do not have to meet the same challenges that underground water services do.

For one thing, the water pressure on a water-distribution pipe is often less than that of a water service. Most homes have water pressure ranges between 40 psi and 60 psi. A water service from a city main could easily have an internal pressure of 80 psi, or more.

Water services are also buried in the ground. Water-distribution pipes rarely are. The underground installation can affect the types of pipes that are approved for use. Another big difference between water service pipes and water-distribution pipes is the temperature of the water they contain. Water services deliver cold water to a building. Water-distribution pipes normally distribute both hot and cold water. Since most codes require the same type of pipe to be used for the cold water lines as is used for hot water pipes, the temperature ranges can disqualify some types of materials. This is a significant factor when choosing a water-distribution material.

VALVES

Many valves are required in a legal water-distribution system. The types of valves used and the places where they are required is mandated by the code. For example, a full-open valve is required near the origination of a water service. In the case of a city water hook-up, the valve is supposed to be installed near the curb of the street. Water services extending from a well are not required to have such a valve at their origination. For those of you that don't know, a full-open valve is a valve like a gate valve or ball valve. When these valves are open, water has the full diameter of the valve to flow through, rather than a restricted opening, like those in valves with washers and seats.

A full-open valve is required on the discharge side of every water meter. The main water distribution pipe must be fitted with a full-open valve where it meets the water service. In non-residential buildings, a full-open valve must be installed at the origination of every riser pipe. In these non-residential buildings a full-open valve is required at the top of all drop-feed pipes.

If a water supply pipe is feeding a water tank, like a well pressure tank, a full-open valve is required. All water supplies to water heaters must be equipped with full-open valves. If a building contains multiple dwellings, the water supplies for each dwelling must be equipped with full-open valves. I know this disclaimer is starting to get old, but be aware that not all codes have the same requirements. Check your local code for applicable restrictions. Since I assume by now you know you have to verify permissible plumbing standards with the use of your local code, I am going to stop reminding you.

The valves we are about to discuss are not required to be full-open valves. The valves used in the following locations are usually either stop-and-waste valves or supply-stop valves. Valves are required on all supply pipes feeding sillcocks and hose bibbs. If a water supply is serving an appliance, like an ice maker or dishwasher, a valve is required.

Some types of buildings are required to have individual cut-offs on the supplies to all plumbing fixtures. Other types of buildings are not. It is generally standard procedure to install cut-offs on all water supplies, with the possible exception of the pipes feeding tub and shower valves.

When a valve is required by the code, that valve must be accessible. It may be concealed by a door that opens or some other means of accessible concealment, but it must be accessible. Stop-and-waste valves are normally not approved for use below ground.

BACKFLOW PROTECTION

Backflow protection for water systems has become a major issue in the plumbing trade. The forms of protection required cover such issues as cross-connections, backflow, and vacuums. If backflow preventers are not installed, entire water mains and water supplies can be at risk of contamination.

There are many types of cross-connections possible. For example, hot water may pass into cold water pipes through a cross-connection. This shouldn't contaminate the system, but it can create a health risk. If someone turns on what they believe to be cold water and discover it's hot water, they could be burned. Normally, this type of cross-connection is more of a nuisance than a threat, but it does have the potential for producing injuries.

A much more serious type of cross-connection might occur in a commercial photography processing plant. Can you imagine what would happen if a pipe conveying photography chemicals were allowed to mix its contents with the potable water system? The results could be quite serious, indeed.

Cross-connection are normally created by accident, but sometimes they are designed. If a cross-connection of some sorts is to be installed, it must be protected with the proper devices to avoid unwanted mixing. Few, in any, codes will allow a cross-connection between a private water source, like a well, and a municipal water supply. Check your local code requirements for backflow-protection devices.

AN AIR GAP

An air gap is the open air space between the flood level rim of a fixture and the bottom of a potable water outlet. For example, there is an air gap between the outlet of a spout on a bathtub and the flood level rim of the tub, or at least there is supposed to be.

By having such an air gap, non-potable water, such as dirty bath water, cannot be sucked back into the plumbing system through the potable water outlet. If the tub spout was mounted in a way that it was submerged in the tub water, the dirty water might be pulled into the water pipes if a vacuum was formed in the plumbing system.

The amount of air gap required on various fixtures is determined by the plumbing code. For example, you might find in a table or in the text of your code that a tub spout must not be installed in a way to make its opening closer than two inches to the flood level rim. The same ruling for lavatories might require the spout of the faucet to remain at least one inch about the flood level rim.

There is another type of air gap used in plumbing. This type is a fitting that is used to accept the waste from a dishwasher for conveyance to the sanitary drainage system. The device allows the water draining out of the dishwasher to pass through open air space on its way to the drainage system. Since the water passes through open air, it is impossible for wastewater from the drainage system to be siphoned back into the dishwasher.

BACKFLOW PREVENTERS

Backflow preventers are devices that prohibit the contents of a pipe from flowing in a direction opposite of its intended flow. These protection devices come in many shapes and sizes and costs range from a few dollars to thousands of dollars. Backflow preventers must be installed in accessible locations.

When a potable water pipe is connected to a boiler to provide it with a supply of water, the potable water pipe must be equipped with a backflow preventer. If the boiler has chemicals in it, like antifreeze, the connection must be made through a reduced-pressure-principal backflow preventer of an air gap. Anytime a connection is subject to back pressure, the connection must be protected with a reduced-pressure-principal backflow preventer.

CHECK VALVES

Check valves have an integral section, something of a flap, that allows liquids to flow freely in one direction and not at all in the opposite direction. Check valves are required on potable supply pipes that are connected to automatic fire sprinkler systems and standpipes.

VACUUM BREAKERS

Vacuum breakers are installed to prevent a vacuum being formed in a plumbing system that could result in contamination. Some of these devices mount in-line on water distribution pipes and others screw onto the threads of silcocks and laundry-tray faucets.

When vacuum breakers are required, the critical level of the device is usually required to be set at least six inches above the flood level rim of the fixture it is serving. What is the critical level? The critical level is the point at which the vacuum breaker would be submerged, prior to backflow.

Normally, any type of fixture that is equipped with threads for a garden hose must be fitted with a vacuum breaker. Some silcocks are available with integral vacuum breakers. For the ones that are not, small vacuum breakers that screw onto the hose threads are available. After the devices are secured on the threads, a set screw is turned into the side of the device. The screw is tightened until its head breaks off. This insures that the safety device cannot be removed easily.

Since vacuum breakers have an opening that leads to the potable water system, they cannot be installed where toxic fumes or vapors may enter the plumbing system through the openings. One example of a place where they could not be installed is under a range hood.

The starting point for your potable water system will be where the water service enters the house. You will connect your main cold water pipe to the water service. In most places you will be required to install a backflow preventer on the water service pipe. If your water service is polyethylene pipe, it must be converted to a pipe rated for hot and cold water applications. Polyethylene cannot handle the high temperature of hot water. The code requires you to use the same type of pipe for both the hot and cold water. Since polyethylene cannot be used for hot water, it cannot be used as an interior water distribution pipe in a house that has hot water.

You can convert the polyethylene pipe to the pipe of your choice with an insert adapter. The insert portion will fit inside the polyethylene pipe and be held in place by stainless-steel clamps. The other end of the fitting will be capable of accepting the type of pipe you are using for water distribution. In most cases the adapter will be threaded on the conversion end. The threads allow you to mate any type of female adapter to the insert-by-male adapter.

Once the conversion is made, install a gate valve on the water distribution pipe. When required, install a backflow preventer after the gate valve. It is a good idea to install another gate valve after the backflow preventer. By isolating the backflow preventer with the two gate valves, you can cut the water off on both sides of the backflow preventer. This option will be appreciated if you must repair or replace the backflow preventer.

As you bring your pipe out of the second gate valve, consider installing a sillcock near the rising water main. If the location is suitable, you will use a minimum amount of pipe for the sillcock. The main water distribution pipe for most houses will be a ¾ inch pipe. If you have already designed your water distribution layout, install the pipe accordingly. If you haven't made a design, do it now.

The three-quarter pipe should go to the inlet of the water heater, undiminished in size. You may choose to branch off the main to supply fixtures with water as the main goes to the water heater. This is acceptable and economical.

Your water pipe will normally be installed in the floor or ceiling joists so they can be hidden from plain view. When possible, keep the pipe at least two inches from the top or the bottom of the joists. Keep in mind that your local building code will prohibit you from drilling, cutting, or notching joists close to their edges. Keeping pipes near the middle of studs and joists reduces the risk of the pipes being punctured by nails or screws. If the pipe passes through a stud or joist where it might be punctured, protect it with a nail plate.

Proper pipe support is always an issue that all plumbers must be aware of. Copper tubing with a diameter of 1¼ inches or less should be supported at intervals not to exceed six feet. Larger copper can run ten feet between supports. These recommendations are for pipes installed horizontally. If the pipes are run vertically, both sizes should be supported at intervals of no more than ten feet.

If your material of choice is CPVC, it should be supported every three feet, whether installed vertically or horizontally. PEX pipe that is installed vertically is required to be supported at intervals of four feet. When installed horizontally, PEX pipe should have a support every 32 inches.

Avoid placing water pipes in outside walls and areas that will not be heated, such as garages and attics. The main delivery pipe for your hot water will originate at the water heater. When piping the water heater, the cold water pipe should be equipped with a gate valve before it enters the water heater. Many localities require the installation of a vacuum breaker at the water heater. These devices prohibit the back-siphonage of the water heater's contents into the cold water pipes.

Do not put a cut-off valve on the hot water pipe leaving the water heater. If a valve was installed on the hot water side, it could become a safety hazard. Should the valve become closed, the water heater could build excessive pressure. The combination of a closed valve and a faulty relief valve could result in an exploding water heater.

During the rough-in you will not normally be installing valves on various fixtures. Most fixtures get their valves during the finish plumbing stage. If you install your sillcocks or hose bibbs during the rough-in, install stop-and-waste valves in the pipe supplying water to the sillcock or hose bibb. Stop-and-waste valves have an arrow on the side of the valve body. Install the valve so the arrow is pointing in the direction that the water is flowing to the fixture.

Rough-In Numbers

Refer to a rough-in book (available from plumbing suppliers for various brands of fixtures) to establish the proper location for your water pipes to serve the fixtures. The examples given here are common rough-in numbers, but your fixtures may require a different rough-in. The water supplies for sinks and lavatories are usually placed twenty-one inches above the sub-floor. Kitchen water pipes are normally set eight inches apart, with the center point being the center of the sink. Lavatory supplies are set four inches apart, with the center being the center of the lavatory bowl.

Most toilet supplies will rough-in at six inches above the sub-floor and six inches to the left of the closet flange, as you face the back wall. Shower heads are routinely placed 6 feet 6 inches above the sub-floor. They should be centered above the bathing unit. Tub spouts are centered over the tub's drain and roughed-in four inches above the tub's flood level rim. The faucet for a bathtub is commonly set twelve inches above the tub and centered on the tub's drain. Shower faucets are generally set four feet above the sub-floor and centered in the shower wall. If you go to your supplier and ask for a rough-in book, the supplier should be able to give you exact information for roughing-in your fixtures.

Secure all pipes near the rough-in for fixtures. When you rough-in for a shower head, use a wing ell. These ells have half-inch, female threads to accept the threads of a shower arm. The wing ell has an ear on each side that allows you to screw the ell to a piece of backing in the wall. Securing the ell will hold it in place for the later installation of the shower arm. Where necessary, install backing in the walls to secure all water pipes. It is very important to have all pipes firmly secured.

If you are soldering joints, remember to open all valves and faucets before soldering them. Failure to do this could damage the washers and internal parts once heat is applied. Consider installing air chambers to reduce the risk of noisy pipes later. The air chamber can be made from the same pipe you are piping the house with. Air chambers should be at least one pipe size larger than the pipes they serve. A standard air chamber will be about twelve inches tall.

Installing cut-off valves in the pipes feeding a bathtub or shower is optional. Most codes do not require the installation of these valves, but they come in handy when you have to work on the tub or shower faucets in later years. Stop-and-waste valves are typically used for this application. Be sure to install the valves where you will have access to them after the house is finished. Avoid extremely long, straight runs of water pipe. If you are running the pipe a long way, install offsets in it. The offsets reduce the risk of an annoying water hammer later. A little thought during the rough-in can save you a lot of trouble in the future.

A MANIFOLD SYSTEM

The rules for plumbing a water distribution system are basically the same for all approved materials. But, when running PEX, it often makes more sense to use a manifold than to run a main with many branches. One of the most desirable effects of this type of installation is the lack of concealed joints. All the concealed piping is run in solid lengths, without fittings.

To plumb a manifold system, run your water main to the manifold. The manifold can be purchased from a plumbing supplier. It will come with cut-off valves for the hot and cold water pipes. Your cold water comes into one end of the manifold and goes out the other to the water heater. The hot water comes into the manifold from the water heater. The manifold is divided into hot and cold sections. The pipe from each fixture connects to the manifold. When there is a demand for water at a fixture, it is satisfied with the water from the manifold. Once you have both the hot and cold supply to the manifold, all you have to do is run individual pipes to each fixture.

You can snake coiled PEX pipe through a house in much the same way as you would electrical wires. By installing your water distribution system in this way, you save many hours of labor. You also eliminate concealed joints, and can cut off any fixture with the valves on the manifold. This type of system is efficient, economical, and the way of the future.

TIPS FOR UNDERGROUND WATER PIPES

Let me give you a few tips for underground water pipes. When you are ready to run water pipe below grade, there are a few rules you must obey. Pipe used for the water service must have a minimum working pressure of 160 psi at 73.4 degrees F. If the water pressure in your area exceeds 160 psi, the pipe shall be rated to handle the highest pressure available to the pipe. If the water service is a plastic pipe, it must terminate within five feet of its point of entry into the home. If the water service pipe is run in the same trench as the sewer, special installation procedures are required (refer to your local plumbing code).

When running the water service and the sewer in a common trench, you must keep the bottom of the water service at least twelve inches above the top of the sewer. The water service must be placed on a solid shelf of firm material to one side of the trench. The sewer and the water service should be separated by undisturbed or compacted earth. The water service must be buried at a depth to protect it from freezing temperatures. The depth required will vary from state to state. Check with the code enforcement office for the proper depth in your area.

Copper pipe run under concrete should be type "L" or type "K" copper. Where copper pipe will come into direct contact with concrete it should be sleeved to protect the copper. Concrete can cause a damaging reaction when in direct contact with copper. The pipe may also vibrate during use and wear a hole into itself by rubbing against rough concrete. All water pipe should be installed to prevent abrasive surfaces from coming into contact with the pipe.

Rules for underground piping are different than those that apply to above-ground systems. Consult your local plumbing code to identify differences which affect plumbers in your area.

INSTALLING DRAIN-WASTE-AND-VENT SYSTEMS

Installing drain-waste-and-vent (DWV) systems is a bit more complex than installing water pipes. Code regulations for drains and vents are much more numerous than they are for water piping. A plumber has to know what fittings can be used to turn from a horizontal run to a vertical run. It's necessary to know the minimum pipe size for piping installed underground. How far above a roof does a vent pipe have to extend before it terminates? The answer depends on where you

fastfacts

➤ *Water service pipe must have a minimum pressure rating of 160 psi at 73.4 degrees F.*

➤ *The bottom of a water service pipe must be at least 12 inches above the top of a sewer.*

➤ *When a water service is installed in a trench with a sewer, the water service pipe must be installed on a solid shelf of firm material to one side of the trench.*

➤ *A water service pipe must be protected from freezing temperatures.*

➤ *Water service pipes should be protected from abrasive surfaces.*

live, and you must be in tune with your local plumbing code to avoid getting rejection slips when your work is inspected.

Code regulations are not the only thing that complicates the installation of a DWV system. Since drains and vents are much larger in diameter than most water pipes, it is harder to find open routes to install the plumbing in. There is also a matter of pipe grade that must be maintained with a DWV system. This doesn't apply to water pipes.

Some plumbers prefer installing DWV systems over water systems. Others find the job of installing drains and vents to be a bother. I don't dislike either type of piping. In my opinion, water pipe is much easier to work with and install, but I enjoy running DWV pipes.

There is no question that DWV systems require planning. It's pretty easier to change directions with a water pipe, but this is not always the case with a drain or vent. Thinking ahead is crucial to the success of a job. This is true of any type of plumbing, and the rule certainly applies to DWV systems. Since many drainage systems start in the ground, that is where we will begin.

UNDERGROUND PLUMBING

Underground plumbing is frequently overlooked in books dealing with plumbing. While underground plumbing receives little recognition, it is instrumental in the successful operation of some plumbing systems. Underground plumbing is installed before a house is built. When the plumbing is not installed accurately, you will have to break up a new concrete floor to relocate it your pipes. This is a fast way to lose your job or make an enemy out of your customer.

Since the placement of underground plumbing is critical, you must be acutely aware of how it is installed. Miscalculating a measurement in the above-grade plumbing may mean cutting out the plumbing, but the same mistake with underground plumbing will be much more troublesome to correct.

An underground plumbing system is often called groundwork among professional plumbers. The groundwork is routinely installed after the footings for a home are poured and before the concrete slab is installed. When installing underground plumbing you will have to take note of code differences between groundworks and above-grade plumbing. For example, while it is perfectly correct to run a 1½ inch drain for an above-grade bathtub, you must run a 2 inch drain if it is under concrete. Let's take a moment to look at some of the code variations you will have to adapt to with interior, underground plumbing.

CODE CONSIDERATIONS

Underground plumbing is quite different than above-grade plumbing. There are several rules pertaining to groundworks that do not apply to above-grade plumbing. This is true of both the DWV system and the water distribution system. The following information will cover highlights of common differences. This information may not apply in your jurisdiction and the information is not conclusive of all code requirements. The following code considerations are the ones I encounter most often with underground plumbing.

The differences for DWV installations are minimal, but they must be observed. Galvanized steel pipe is an approved material for a DWV system when the pipe is located above grade. But, galvanized drain pipes may not be installed below ground. This rule has little effect on you since most DWV systems today are plumbed with schedule 40 plastic pipe.

The minimum pipe size allowed under concrete is two inches. This rule could cause the unsuspecting plumber trouble. When you check your sizing charts and see that 1½ inch pipe is an approved size, you might choose to use it under concrete. If you do, you will be in violation of most codes. While 1½ inch pipe is fine above grade, it is illegal below grade. The type of fittings allowed are also different. When working above grade, a short-turn quarter bend is allowed when changing directions from horizontal to vertical. This is not true below concrete. Long-sweep quarter bends must be used for all ninety-degree turns under the concrete.

When running drains and vents above grade the pipes must be supported every four feet. This is done with some style of pipe hanger. When installing the pipes underground, you will not use pipe hangers. Instead, you will support the pipe on firm earth, or with sand or gravel. The pipe must be supported firmly and evenly.

When your groundworks leave the foundation of a house, the pipe will have to be protected by a sleeve. The sleeve can be a piece of schedule 40 plastic pipe, but it must be two pipe sizes larger than the pipe passing through the sleeve. This rule applies to any pipe passing through the foundation or under the footing.

It will be rare that any residential drains under concrete will run for 50 feet. If they do, you must install a clean-out in the pipe. Pipes with a diameter of 4 inches or less must be equipped with a clean-out every 50 feet in a horizontal run. The clean-outs must extend to the finished floor grade.

As your building drain becomes a sewer there are yet more rules to adhere to. The home's sewer must have a clean-out for every one-hundred feet it runs. There should also be a clean-out within five feet of the house where the pipe makes its exit. It can usually be either inside or outside of the foundation wall. These clean-outs must extend to the finished grade. If the sewer takes a turn of more than 45 degrees, a clean-out must be installed. Clean-outs should be the same size as the pipes they serve, and they must be installed so that they will open in the direction of the flow for the drainage system.

With pipes of a 3 inch, or larger, diameter, there must be a clear space of 18 inches in front of the clean-out. On smaller pipes, the clear distance must be at least 12 inches. The sewer must be supported on a firm base of earth, sand, or gravel as it travels the length of the trench.

PLACEMENT

The placement of groundworks is critical in the plumbing of a building. Before you install the underground plumbing you must do some careful planning. The first step is laying out the plumbing.

When the groundworks are installed, the sewer and water service may or may not already be installed. If they are already installed, your job is a little easier. The pipes will be stubbed into the foundation, ready to be connected. If these pipes are installed, you can start with them and run the underground plumbing. If the water service and sewer are not installed, you will have to locate the proper spot for them.

Refer to your blueprints for the location of the water and sewer entrance. If the pipe locations are not noted on the plans, use common sense. Determine where the water service and sewer will be coming to the house from. If you are working with municipal water and sewer, the public works department of your town or city can help you. The public works department will tell you where your connections will be made to the mains and how deep they will be. If you will be connecting to a septic system and well, find the locations for these systems.

You have two considerations on picking the location for the water service and sewer. The first is the most convenient location inside the home for plumbing purposes. The second is an exit point that will allow the successful connection to the main sewer and water service outside the home. Make your decision with a priority on connecting to the mains outside the home. You can adjust the interior plumbing to work to the incoming pipes, but you may not be able to adjust the existing outside conditions.

Once you have a location picked for the sewer and water service, you are ready to lay out the remainder of the groundworks. The locations for your underground pipes will be partially determined by the blueprints. The blueprints will show fixture locations. These fixture locations will determine where you must have pipes in place. In addition to any grade-level plumbing, you will have to look to the plumbing on upper floors. Where there are plumbing fixtures above the level of the concrete floor, you will have to rough-in pipes for their drains.

When you turn pipes up out of the concrete for drains and vents, location of the pipes can be crucial. Many of these pipes will be intended to be inside of walls to be built. If your pipe placement is off by even 1 inch, the pipe may miss the wall location. When this happens you have a pipe sticking up through the floor in the wrong place. It may be in a hall or a bedroom, but it will have to be moved. Moving the pipe will require the breaking and repairing of the new concrete floor. To avoid these problems, make careful measurements and check all measurements twice.

When you have decided on all the pipes you will need in the underground plumbing, you can lay out your ditches. You will normally have to dig ditches to lay the pipes in. The easiest way to mark your ditches is with lime or flour. By placing these white substances on the ground, in the path of your ditch, you can dig the ditches accurately. The ditches will have to be graded to allow the proper pitch on your pipes. The standard pitch for household plumbing is one quarter of an inch of fall for every foot the pipe travels.

INSTALLATION

With all the planning and layout done, you are ready to start the installation. The best place to start is with the sewer. The height of the exit point will be dictated by the depth of the connection at the main sewer. If the sewer is not already stubbed into the foundation, you will have to establish the proper depth for your sewer leaving the foundation. There are times when if you take the sewer out under the footing, the drain will be too low to make the final connection at the main. Before tunneling under the footing or cutting a hole in the foundation, determine the proper depth for your hole. This can be done by measuring the distance from the main sewer to the foundation. If the main connection is 60 feet away, the drain will drop 15 inches from the time it leaves the house until it reaches the final destination. This determination is made by dividing the distance by four, to allow for a ¼ inch of fall per foot.

The sewer pipe should be covered by at least 12 inches of dirt where it leaves the foundation. With this fact taken into consideration, the main connection must be at least 27 inches deep. This is arrived at by taking the twelve inches of depth needed for cover and adding it to the 15 inches of drop in the pipe's grade. It is best to allow a few extra inches to insure a good connection point.

I will start the instructions assuming the water service and sewer have not been run to the house when you start plumbing. With the depth of the hole known for the sewer, tunnel under the footing, or cut a hole in the foundation wall. Install a sleeve for the sewer that is at least two pipe sizes larger than the sewer pipe. Install a cap on one end of the pipe to be used for the sewer. Extend the capped end of the sewer pipe through the sleeve and about 5 feet beyond the foundation. You are now ready to install the groundworks for all the interior plumbing.

In almost every circumstance, the pipe used for underground DWV plumbing will be schedule 40 plastic pipe. Working with this pipe is easy. The pipe can be cut with saws or roller-type cutters. Most plumbers cut the pipe with a saw. The type of saw varies. I use a hacksaw on pipe up through a four-inch diameter. A hand saw, like carpenters use, is effective in cutting the pipe and is easier to use on pipe over 3 inches in diameter. There are saws made especially for cutting schedule 40 plastic pipe, and some plumbers use roller-cutters.

It is important to cut the pipe evenly. If the pipe is cut crooked, it will not seat completely in the fitting. This can cause the joint to leak. After making a square cut on the pipe, clear the pipe of any burrs and rough pieces of plastic. You can usually do this with your hand, but the burrs and plastic can be sharp. It is safer to use a pocket knife or similar tool to remove the rough spots. The pipe and the interior of the fitting's hub must be clean and dry. Wipe these areas with a cloth if necessary to clean and dry the surface.

Apply an approved cleaning solution to the end of the pipe and the interior of the fitting hub. Next, apply an approved primer, if required, to the pipe and fitting hub. Now apply the solvent or glue to the pipe and hub. Insert the pipe into the hub until it is seated completely in the hub. Turn the pipe a quarter of a turn to assure a good seal. This type of pipe and glue makes a quick joint. If you have made a mistake, you will have trouble removing the pipe from the fitting. If the glue sets up for long, you will have to cut the fitting off the pipe to correct mistakes.

Many plumbers dry-fit their joints to confirm their measurements. Dry-fitting is putting the pipe and fittings together without glue to check alignment and measurements. I have done plumbing for a long time and don't dry-fit. I never did it in the early years, but many plumbers prefer to dry-fit each joint. This procedure makes the job go much slower and can cause its own problems. If you are dry-fitting the pipe to get an accurate feel for your measurements, the pipe must seat all the way into the fitting. When this is accomplished, it can be difficult to get the pipe back out of the fitting.

While I never dry-fit, I do sometimes verify my measurements with an easier method. I place the fitting beside the pipe, with the hub positioned as it will be when glued. I can then take a measurement to the center of my fitting to confirm the length of the pipe. By holding the fitting beside the pipe, instead of inserting the pipe, I don't have to fight to get the pipe out of the fitting. Once a joint is made with schedule 40 plastic pipe, it should not be moved for a minute or so. It doesn't take long for the glue to set up and you can continue working without fear of breaking the joint.

The first fitting to install on the building drain is a clean-out. The clean-out will usually be made with a wye and an eighth bend. The pipe extending from the eighth bend must extend high enough for the clean-out to be accessible above the concrete floor. Many plumbers put this clean-out in a vertical stack that serves plumbing above the first floor. The clean-out can be stopped at the finished floor level, but it must be accessible above the concrete.

After the clean-out is installed, you can go on about your business. As you run your pipes to the appropriate locations, they must be supported on solid ground or an approved fill, like sand or gravel. Maintain an even grade on the pipes as they are installed. The minimum grade should be a quarter of an inch for every foot the pipe runs. Don't apply too much grade to the pipe. If the pipe falls too quickly, the drains may stop up when used. The fast grade will drain the pipes of water, but leave solids behind. These solids accumulate to form a stoppage in the pipe and fittings.

Support all the fittings installed in the system. If you are installing a 3 x 2 wye, the 2 inch portion will be higher above the ground than the three inch portion. Place dirt, sand, or gravel under these fittings to support them. Don't allow dirt, mud, or water to get in the fittings or on the ends of the pipe. A small piece of dirt is all it takes to cause a leak in the joint.

To know where to position your pipes for fixtures, vents, and stacks, it helps to put up string where the walls will be installed. You should also know what the finished floor level will be. The blue-prints will show wall locations, and the general contractor, concrete sub-contractor, or foreman will be able to give you the finished floor grade.

You must know the level of the finished floor so that your pipes will not get too high and wind up above the floor. Once you have the grade level for the finished floor, mark the level on the foundation. Drive stakes into the ground and stretch a string across the area where the plumbing is being installed. The string should be positioned to be at the same level as the finished floor. You can use this string as a guide to monitor the height of your pipes.

Using the blueprints, mark wall locations with stakes in the ground. You can use a single string and keep your pipes turning up through the concrete to one side of the string, or you can use two strings to simulate the actual wall. Using two strings will allow you to position your pipes in the center of the wall.

Plumbing measurements are generally given from the center of a drain or vent. If your water closet is supposed to be roughed in twelve inches off the back wall, the measurement is made from the edge of the wall to the center of the drain.

As you near completion on your underground plumbing, secure all your pipes that need to stay in the present position. The pipes you will have to be the most concerned with are pipes coming up through the concrete and pipes for bathtubs, showers, and floor drains. The best way to secure these pipes is with stakes. The stakes should not be made of wood. In many regions termites may come to the house for the wood under the floor. Use steel or copper stakes. The stakes should be on both sides of the pipe to ensure it is not moved by other tradesmen.

When the DWV pipe is installed, cap all the pipes. You can use temporary plastic caps or rubber caps, but cap the pipes. You do not want foreign objects getting into the drains and clogging them. Where you have a tub or a shower to be installed on the concrete floor, you will need a trap box. This is just a box to keep concrete away from your pipe and to allow the installation of a trap when the bathing unit is set. Put a spacer-cap over the pipe turned up for a toilet to be set on the concrete floor. This spacer will keep concrete away from the pipe so that you can install a closet flange when the time comes to set the toilet. Make sure all pipes are secured before you consider the job finished.

ABOVE-GROUND ROUGH-INS

When you are installing above-ground rough-ins, you don't have to dig any ditches. This is good. But, you do have to drill holes. Once a building is framed and under roof, you can start the installation of above-ground drains and vents. If the building has underground plumbing, you will be tying into the pipes you installed earlier. Not all houses have groundworks, so the framing stage might be your first visit to a job for installing pipes.

Houses without groundworks still require water services and sewers. Refer to the information given on these two components of a plumbing system in the earlier paragraphs. Make sure that you don't set yourself up with a building sewer that will be too high to connect to a main sewer or septic tank. When you are comfortable with your exit locations, you can begin to lay out and install your DWV rough-in.

Hole Sizes

Hole sizes are important. Some codes require you to keep the hole size to a minimum to reduce the effect of fire spreading through a home. If you drill oversized holes, they can act as a chimney for fires in a building. Choose a drill bit that is just slightly larger than the pipe you will be installing. In the case of 2 inch pipe, a standard drill bit size will be $2\frac{5}{16}$ of an inch.

When cutting a hole for a shower drain, the hole size will be determined by the shower drain. Keep the hole as small as possible, but large enough to allow the shower drain to fit in it. For a tub waste, the standard hole will be 15 inches long and 4 inches wide. This allows adequate space for the installation of the tub waste. In some jurisdictions, after the tub waste is installed, the hole must be covered with sheet metal to eliminate the risk of a draft during a house fire.

Running Pipe

When your design is made and your holes are open, you are ready to run pipe. This is where you must exercise your knowledge of the plumbing code. There are many rules pertaining to DWV systems. If you purchase a code book you will see pages upon pages of rules and regulations. The task seems overwhelming with all the rules to

follow. If you don't lose your composure, the job is not all that difficult. By following a few rules and basic plumbing principals, you can
install DWV systems with ease.

One of the most common mistakes made in drainage piping is a
failure to remember to allow for the pipe's grade. When you draw
the design on paper you are not likely to think of the size of the floor
joists and the grade you will need for the pipes. Since many houses
have their plumbing concealed by a ceiling, it is important to keep
the pipes above the ceiling level. Not many people want an ugly
drain pipe hanging below the ceiling in their formal dining room.
These small details or the planning will make a noticeable difference
in the outcome of the job.

Once you know the starting and ending points of your drains and
vents, you can project the space needed for adequate grade. Generally the grade is set at ¼ inch per foot for drains and vents. Drains
fall downward, toward the final destination. Vents pitch upward,
toward the roof of the house. With a twelve-foot piece of pipe, the
low end will be three inches lower than the high end. When you are
drilling through floor joists, most holes will be kept at least 1½ inches
above the bottom and below the top of the joist. If you follow this
rule-of-thumb, a 2 x 8 joist will have 4½ inches for you to work with.

A 2 x 8 has a planed width of 7½ inches. From the 7½ inches, you
deduct 3 inches for the top and bottom margin. This leaves 4½ inches
for you to work with. What does all this mean to you? It means that
with a 2 x 8 joist system, you can run a 2 inch pipe for about 10 feet
before you are in trouble. The pipe diameter uses 2 inches of the
remaining 4½ inches. This leaves 2½ inches to manipulate for grade.
At a ¼ inch per foot, you can run 10 feet with the 2 inch pipe. With
1½ inch pipe, you could go twelve feet. Three inch pipe will be
restricted to a distance of about 6 feet.

Under extreme conditions, you can run farther by drilling closer to
the top or the bottom of the floor joist. Before drilling any closer than
1 inch to either edge, consult the carpenters. They will probably have
to install headers or a small piece of steel to strengthen the joists.
Many experienced plumbers have trouble with running out of space
for their pipes. They don't look ahead and do the math before drilling
the holes. After drilling several joists, they realize they cannot get to
where they want to go. This results in a change in layout and a bunch
of joists with holes drilled in them that cannot be used. Plan your path
methodically and you will not have these embarrassing problems.

When you have your holes drilled, running the pipe will be easy.
Depending on the code you are using, every fixture must have a
vent. Except for jurisdictions using a combination-waste-and-vent

code, you must provide a vent for every fixture. This does not mean every fixture must have an individual back-vent. Most codes allow the use of wet vents. Using wet vents will save you time and money.

Wet Vents

Wet vents are pipes that serve two purposes. They are a drain for one fixture and a vent for another. Toilets are often wet vented with a lavatory. This involves placing a fitting within a prescribed distance from the toilet that serves as a drain for the lavatory. As the drain proceeds to the lavatory, it will turn into a dry vent after it extends above the trap arm. Exact distances and specifications are set forth in local plumbing codes.

Dry Vents

Many of your fixtures will be vented with dry vents. These are vents that do no receive the drainage discharge of a fixture. Since the pipes only carry air, they are called dry vents. There are many types of dry vents, they include: common vents, individual vents, circuit vents, vent stacks, and relief vents, to name a few. Don't let this myriad of vents intimidate you. When plumbing an average house, venting the drains does not have to be complicated.

If you don't understand wet vents, you can run all dry vents. This may cost a little more in material, but it can make the job easier for you to understand. Remember, every fixture needs a vent. How you vent the fixture is up to you, but as long as a legal vent is installed, you will be okay. You must install at least one three-inch vent that will penetrate the roof of the home. You can have more than one 3 inch vent, but you must have at least one.

After you have your mandatory 3 inch vent, most bathroom groups can be vented with a 2 inch vent. The majority of individual fixtures can be vented with 1½ inch pipe. Most secondary vents can be tied into the main 3 inch vent before the main vent leaves the attic. In a standard application, the average house will have one 3 inch vent going through the roof and a 1½ inch vent that serves the kitchen sink, going through the roof on its own.

If you wanted to, you could vent each fixture with an individual vent and take all of them through the roof. This would be a waste of time and money, but you would be in compliance with the plumbing code. It is more logical to tie most of the smaller vents back into the main vent, before it exits the roof. Before you can change the direction of a vent, it must be at least 6 inches above the flood level rim of the fixture it is serving.

Yoke vents are not common in residential plumbing, but branch vents, vent stacks, and stack vents are. Even island vents are used in some residential jobs. You should become familiar with the various types of vents so that you can install efficient systems.

When you start to install the drains, you must pay attention to pipe size, pitch, pipe support, and the fittings used. The sizing charts in your local plumbing code will help you to identify the proper pipe sizes. If you grade all of your pipes with the standard quarter of an inch per foot, you won't have a problem there. Support all the horizontal drains at 4 foot intervals or in compliance with your local code requirements. The last thing to learn about is the use of various fittings.

When your system has each fixture vented, use P-traps for tubs, showers, sinks, lavatories, and washing machine drains. When your pipes are rising vertically, use a sanitary tee to make the connection between the drain, vent, and trap arm. Use long-sweep quarter bends for horizontal changes in direction. You can use short-turn quarter bends above grade when the pipe is changing from horizontal to vertical, but if you want to keep it simple, use long-sweep quarter bends for all 90 degree turns.

Use wyes with eighth bends to create a stack or branch that changes from horizontal to vertical. Remember to install clean-outs near the base of each stack and at the end of horizontal runs when feasible. Keep turns in the piping at a minimum. The less turns the pipe makes now, the fewer problems you will have with drain stoppages later. Never tie a vent into a stack below a fixture if the stack receives that fixture's drainage. When your pipes might be hit by a nail or drywall screw, install a metal plate to protect the pipe. These nail plates are nailed or driven onto studs and floor joists to prevent damage to your pipes.

Keep your vents within the maximum distance allowed from the fixture. A 1½ inch drain cannot have a distance of more than five feet between the fixture's trap and the vent. For two-inch pipe, the distance is eight feet for a 1½ trap and six feet for a 2 inch trap. Three-inch pipe will allow you to develop a maximum distance of ten feet. With four-inch pipe, you can go twelve feet. If your fixture must be beyond these limits from the stack vent, install a relief vent.

A relief vent is a dry vent that comes off the horizontal drain as it goes to the fixture. The relief vents rises up at least 6 inches above the flood level rim and ties back into the stack vent. If you are faced with a sink in an island counter, you have some creative plumbing to do for the vent. Since the sink is in an island, there will be no walls to conceal a normal vent. Under these circumstances you must employ the use of·an island vent.

When your vents penetrate a roof, they must extend at least twelve inches above the roof. In some areas, the extension requirement is two feet. If the roof is used for any purpose other than weather protection, the vent must rise seven feet above the roof. Be careful taking vents through the roof when windows, doors, or ventilating openings are present. Vents must be ten feet away from these openings or three feet above them.

When plumbing the drain for a washing machine, keep this in mind. The standpipe from the trap must be at least 18 inches high, but it must not be more than 30 inches high The piping must be accessible for clearing stoppages.

FIXTURE PLACEMENT

Fixture placement is normally shown on all blueprints. There are code requirements that dictate elements of locating fixtures. Rough-in measurements vary from fixture to fixture and from manufacturer to manufacturer. To be safe, you should obtain rough-in books from your supplier for each type and brand of fixture you will be roughing in.

Standard toilets will rough-in with the center of the drain twelve inches from the back wall. This measurement of 12 inches is figured from the finished wall. If you are measuring from a stud wall, allow for the thickness of drywall or whatever the finished wall will be. Most plumbers rough-in the toilet twelve and a half inches from the finished wall. The extra ½ inch gives you a little breathing room if conditions are not exactly as you planned. From the front rim of the toilet, you must have a clear space of 18 inches between the toilet's rim and another fixture.

From the center of the toilet's drain, you must have 15 inches of clearance on both sides. This means you need a minimum width of 30 inches to install a toilet. If you are plumbing a half-bath, the room must have a minimum width of 30 inches and a minimum depth of 5 feet. These same measurements for clearance apply to bidets.

If you plan to install a tub waste using slip-nuts, you will have to have an access panel to gain access to the tub waste. If you do not want to have an access panel, use a tub waste with solvent-weld joints. Many people object to access panels in their hall or bedroom. Avoid slip-nut connections and you can avoid access panels.

If you take the time to learn your local plumbing code and are willing to think ahead in your planning, DWV and water systems will not give you many problems. Common sense will take you a long way in the installation process. Refer to the following illustrations for a visual conception of DWV systems and riser diagrams.

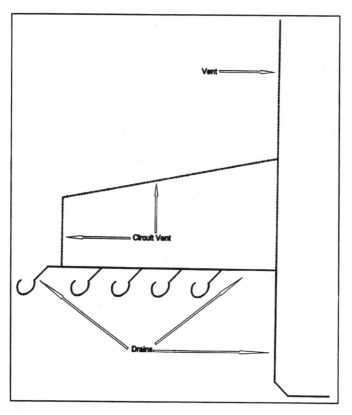

FIGURE 3.1 Circuit vent.

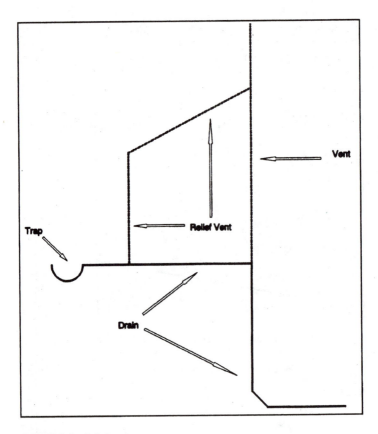

FIGURE 3.2 Relief vent.

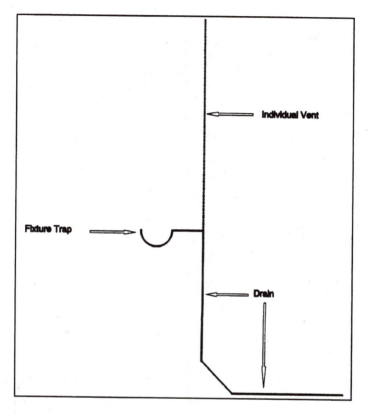

FIGURE 3.3   Individual vent.

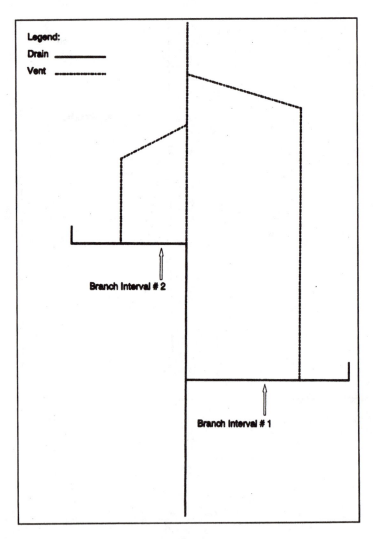

FIGURE 3.4 Stack with two branch intervals.

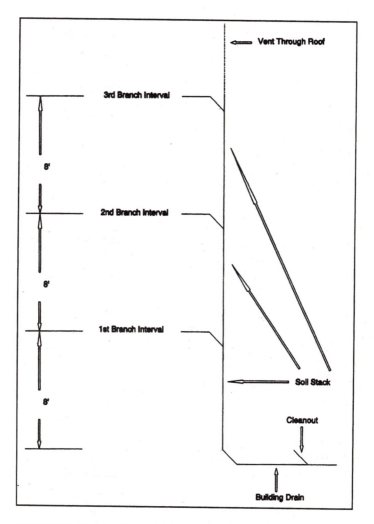

FIGURE 3.5 Branch-interval detail.

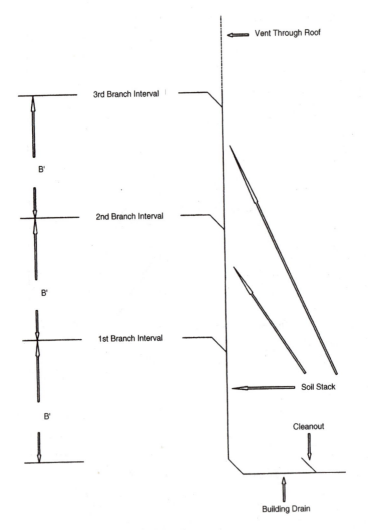

FIGURE 3.6 Branch-interval detail.

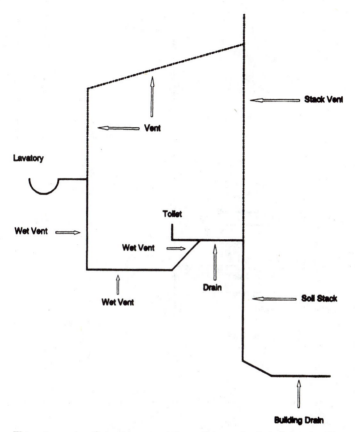

Wet vents are pipes that serve as a vent for one fixture and a drain for another. Wet vents, once you know how to use them, can save you a lot of money and time. By effectively using wet vents you can reduce the amount of pipe, fittings, and labor required to vent a bathroom group or two.

FIGURE 3.7 Wet vents.

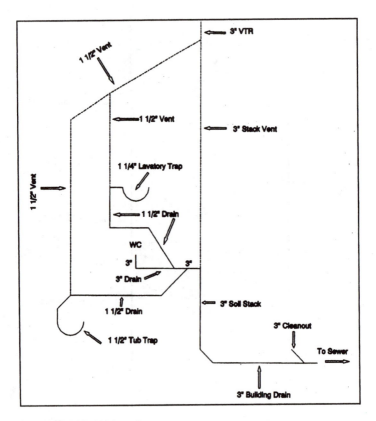

FIGURE 3.8 DWV riser diagram.

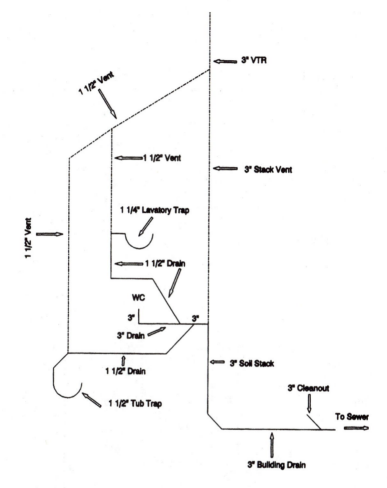

FIGURE 3.9 DWV riser diagram.

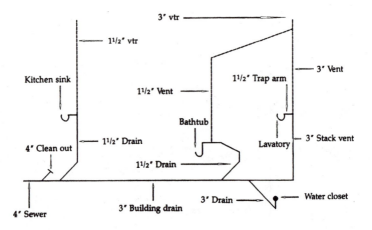

FIGURE 3.10 DW riser diagram.

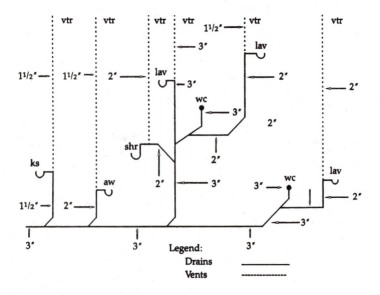

FIGURE 3.11 DWV layout.

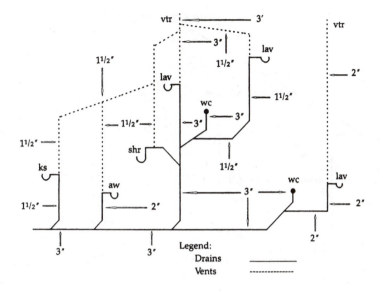

FIGURE 3.12 Efficient use of DWV pipes.

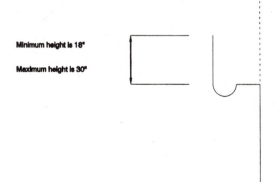

FIGURE 3.13 Washing-machine standpipe height.

VENT TERMINATION RULES

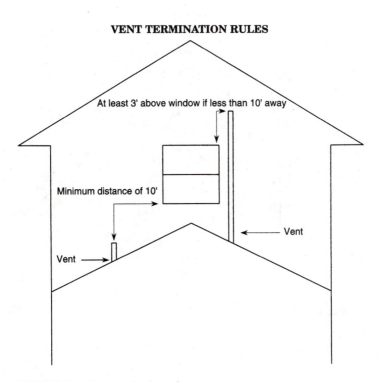

At least 3' above window if less than 10' away

Minimum distance of 10'

Vent

Vent

FIGURE 3.14 Vent termination rules.

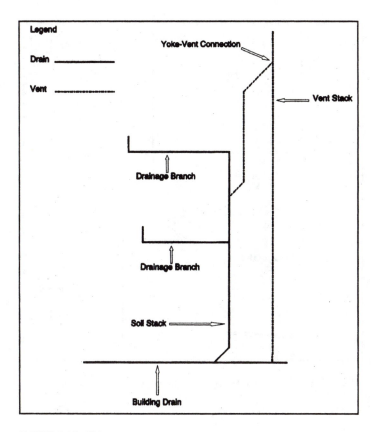

FIGURE 3.15 Yoke vent.

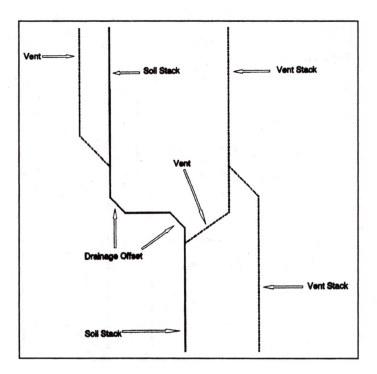

FIGURE 3.16 Example of venting drainage offsets.

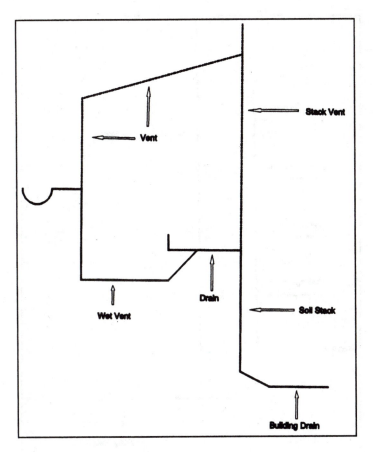

FIGURE 3.17 Stack vent.

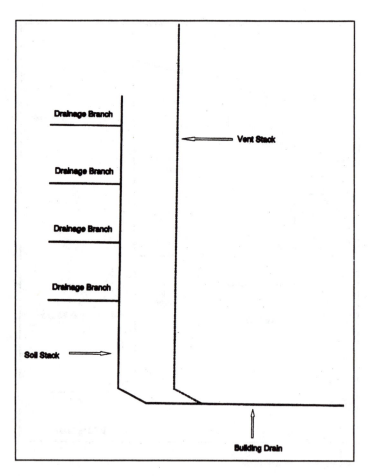

FIGURE 3.18 Vent stack.

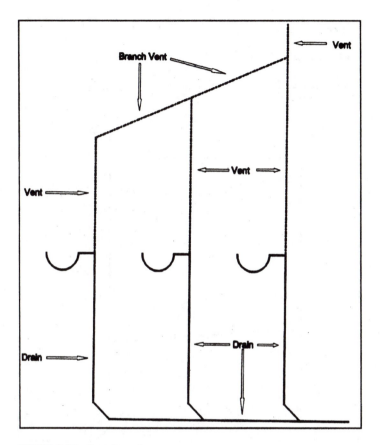

FIGURE 3.19 Branch vent.

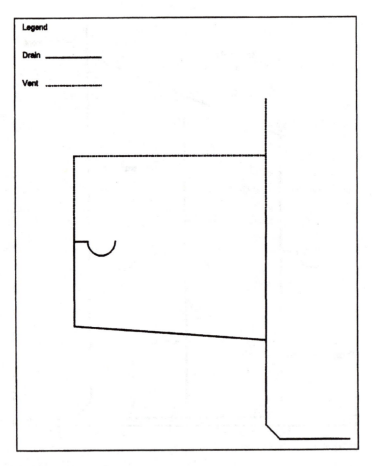

FIGURE 3.20 Level vent.

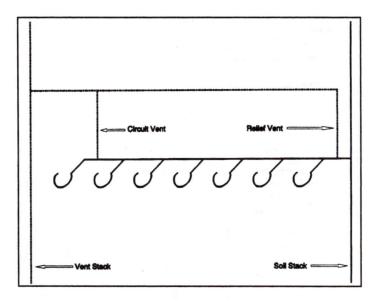

FIGURE 3.21 Circuit vent with a relief vent.

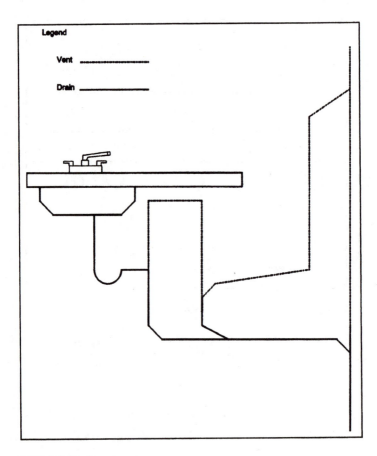

FIGURE 3.22   Island vent.

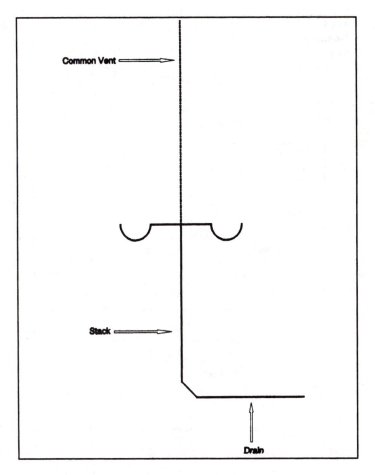

FIGURE 3.23 Common vent.

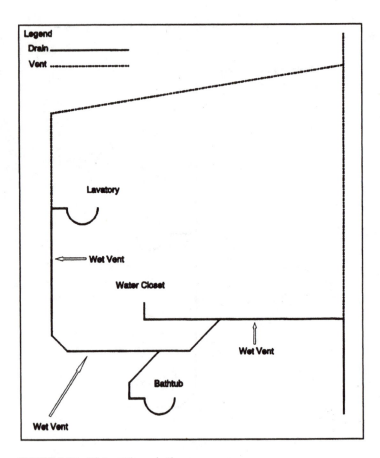

FIGURE 3.24 Wet-venting a bathroom group.

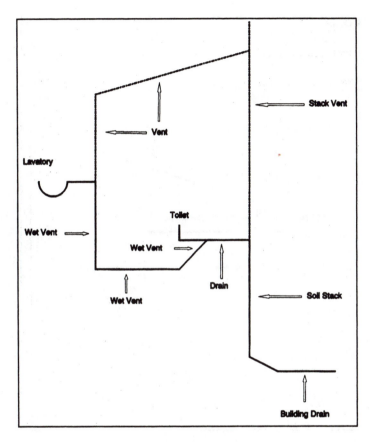

FIGURE 3.25 Wet-venting a toilet with a lavatory.

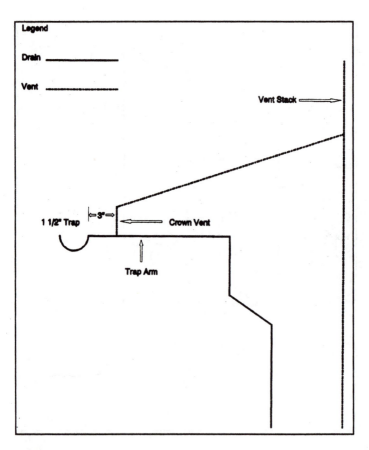

FIGURE 3.26 Crown venting.

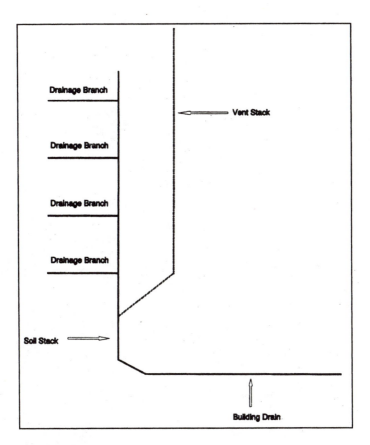

FIGURE 3.27 Vent stack.

4

EXISTING PLUMBING
SYSTEMS

Plumbers who are not accustomed to doing remodeling work are often caught between a rock and a hard place. They don't have the experience to know what to look out for when they get into remodeling, and the risks can be substantial. Remodeling work can be very different from new work or service work. For an unsuspecting plumber, a few remodeling jobs is all it takes to lose enough money to put a real strain on staying in business. Working with existing plumbing systems can be quite an adventure.

There is no doubt that remodeling can be hard work. It can also be dangerous work. Additionally, the work involved can be much more extensive than it first appears to be. This fact often tricks plumbers into giving bids and prices that are far lower than they should be. Remodeling is a good field to get into. It survives when new work dwindles, and remodeling can be done throughout the year. Snow and cold weather don't shut down remodelers. If you are thinking of getting into remodeling, don't do it before you read this chapter.

PIPE SIZES

One of the first considerations when adding to an existing system is the size of existing pipes. Before you can create your plumbing design, you must inspect the existing pipes you will be tying into. Check the size of the sewer where it enters a building. If it is a 3 inch pipe, you could be in for big trouble. A 3 inch sewer shall not carry the discharge of more than two water closets. If you are planning to install a third water closet, on a 3 inch sewer, you are out of luck. Before the job can be done, you will have to install a new sewer. It may be more cost effective to run a new sewer, independent of the old sewer, for the new plumbing. In most cases, it will make more sense to replace the existing sewer with a larger one. Before you become too involved in your remodeling job, make sure the sewer is large enough to carry the increased fixture-unit load. This may sound simple, but if you fail to confirm pipe sizes, you can run into extensive cost overruns.

As with the sewer, check the size of the incoming water service. Verify the ability of the water service to provide an adequate water supply. It will be rare that you will have to replace the water service with a larger one. Water distribution pipes must be inspected to assure adequate sizing. Take notice of the type of pipe the water distribution system is piped with. Since you will have to connect your new plumbing to these existing pipes, it helps to know the type of pipe used in the system.

A common problem with older homes is the size of the water pipes, especially the hot water pipes. It is not unusual to find homes where all the hot water distribution is made through ½ inch pipes. As you may know, ½ inch pipe is not large enough to supply the hot water for a house. If you cannot find ¾ inch water pipes to connect to, you may have to apply for a variance. Generally, if the only water pipes in the house are ½ inch, the code enforcement office will allow you to connect to it for the new installation. While this is a common practice, don't count on it. The code office can require you to run the proper size piping for your installation. If this happens, they may also make you bring the remainder of the home's plumbing up to current code requirements. If at any time you have questions about your responsibilities to the plumbing code, consult your local plumbing inspector.

THE WATER HEATER

With average remodeling jobs, the existing water heater will be sufficient. However, if you are installing a whirlpool tub or anything that will demand large amounts of hot water, consider the possibilities of adding a new water heater. It is your responsibility to install plumbing that will work well and that will conform with your local plumbing code.

OLD PLUMBING THAT MUST BE MOVED

It is easy to get caught up in figuring out how you will install new plumbing, without considering what may need to moved in the existing system. If an attic is being converted to living space, the existing vents may have to be relocated. Should existing walls be removed, there may be concealed plumbing to move. It could be water pipes, drains, or vents. This is a potential task that can be difficult to foresee.

When discussing the overall remodeling plans with a general contractor or owner, take wall removal into consideration. If a wall is going to be removed, any pipes in the wall will have to be relocated. When a wall is scheduled for demolition, look in the attic and in the basement or crawl space for pipes entering or exiting the wall. If you see pipes that will have to be moved, take note of the type of pipe you will be moving. Determine the type of material used in the pipe and the purpose of the pipe. This information will help you in designing a plan for the successful relocation of the pipe.

If you find the existing water distribution system to be piped with galvanized steel pipe, give serious consideration to replacing all of the galvanized pipe. Galvanized steel pipe is famous for rusting from the inside out. As the pipe rusts, two conditions develop that affect the water distribution system.

The first condition is restricted water flow within the pipe. As the pipe rusts, the rough surface collects minerals and other undesirable material to restrict the water flow. In time, the pipe can become closed to the point that only a dribble of water will come out of your faucets. The second affect of rusting galvanized water pipes is the development of leaks. Where the pipe has been threaded, the wall of the pipe is thinner. The threading process weakens the pipe and rust attacks these weak spots. As the rust continues, the pipe threads deteriorate and leak. Unless there are extenuating circumstances, replace all the galvanized water pipe you can.

There have been numerous occasions when I have found existing plumbing systems to be piped with illegal materials. The most common violation in the water distribution system is the use of polyethylene pipe for the entire water distribution system. Since this pipe is not approved for use with hot water, it cannot be used as a water distribution pipe in houses with hot water. The cold water pipes must be made from the same material used for the hot water pipes.

Remodeling Considerations

✔ Is the building drain large enough to handle increased fixture loads?

✔ Will the existing sewer be large enough to accept the addition of new fixtures?

✔ If a larger sewer is needed, should you add an additional sewer or increase the size of the existing sewer?

✔ Is the size of the existing water service adequate for your planned changes?

✔ Are the water mains in the building sized to accept additional loads?

✔ Is it possible to obtain a variance for undersized existing piping?

✔ Can the existing water heater handle additional demands?

✔ Will any existing plumbing have to be relocated to meet remodeling needs?

FIXTURES

When it is your job to replace existing fixtures, prepare for the unexpected. Working with old fixtures can turn your hair gray quickly. Old plumbing fixtures can present you with some undesirable working conditions. The list begins with rusted mounting nuts on faucets and continues to include everything from broken, or nonexistent closet flanges, to odd sized bathtubs.

If you have never tried to remove a double-bowl concrete laundry tub, you cannot imagine how heavy they are. Cast iron bathtubs are another heavy-weight fixture. Unless you have a small army to help you, plan on breaking these accidents waiting to happen up with a sledge hammer. The danger in moving such heavy fixtures outweighs the loss incurred by destroying the fixtures with a demo hammer. Having a 400-pound, cast iron tub chase you down a set of stairs will convince you that I am right in my assessment.

Galvanized Steel Drains

Galvanized drains have earned quite a reputation in the world of pipe blockages. Like galvanized water pipes, galvanized drains fall victims to rust. In the case of drains, the inside roughness, caused by rust, catches hair, grease, and other unidentified cruddy objects. As these items become snagged on the rough spots, the interior of the drain slowly closes. In time, the build-up will block the passageway of the pipe entirely. The result is a drain pipe that will not drain.

When this type of blockage is attacked with the average snake, the snake only punches a hole in the blockage. The drain will work for awhile, but it will not be long before the pipe is stopped up again. Unless the blockage is removed with a cutting head on an electric drain cleaner, the pipe will become restricted soon after the snaking.

When a garbage disposer is added to a kitchen sink with a galvanized drain, expect problems. Since garbage disposers send food particles down the drain, the rough spots in the drain will catch and hold the food. You can almost count on having to replace any old galvanized pipe used as a drain for garbage disposers. In addition to the rust problem, the fittings used with galvanized pipe are not as effective as modern fittings. New fittings utilize longer turns than those used with galvanized pipe. A galvanized quarter-bend takes a much sharper turn than a schedule 40 plastic quarter-bend. These tight turns are responsible for inducing stoppages within the pipe.

In addition to habitual stoppages, galvanized drains will develop leaks at their threads. This is caused by the rust working on the weak spots at the threads. In general, galvanized pipe should be replaced whenever feasible.

SEPTIC SYSTEMS

If you are adding plumbing to a house that uses a septic system, be careful. The septic system may not be adequate to handle the increased demand from new plumbing. Before you add plumbing to an old septic system, have the system checked out. You may be able to determine the system's capability by reviewing the original installation permit and plans. If this paperwork is not available, hire a professional to render an opinion on the ability of the system to handle the increased load.

Septic systems are expensive to install and expand. It would be a shame to add a new bathroom, only to find that the septic system is not adequate to process the waste. The cost of adding new chambers or lines in the septic field could shock you. While you may not be happy to discover the need to invest extra money in the septic system, it will be better to find it out before the new plumbing is installed.

ROTTED WALLS AND FLOORS

When you plan to replace old plumbing fixtures, you can never be sure what you will find during the process. It is not uncommon for toilets to leak and rot the floor beneath them. Bathing units are subject to the risks of rotten floors and walls. These structural problems will complicate your life as a plumber.

When you remove the old toilet and find a rotten floor, someone will have to pay to have the floor repaired. Toilets contribute to the rotting of floors through leaks around wax seals, tank-to-bowl bolts, and condensation. While the quantity of water running onto the floor may be minor, the damage can be major. As time passes, the water works its way into the underlayment, sub-floor, and floor joists. Small leaks may not stain the ceiling below a toilet until the damage is done.

If you are concerned about water damage around the base of the toilet, you can test for it before removing the old toilet. Take a knife or screwdriver and probe the floor around the base of the toilet. If the point of the probing instrument sinks into the floor, you have a problem. If the water damage is only in the underlayment or sub-floor, the cost of repairing the damage will be minimal. If you were not already planning to replace the floor covering, you will incur additional expense. To correct the problem, the finished floor covering will have to be replaced. Trying to make a patch in the floor covering will result in a floor that does not match.

Bathtubs and showers may have caused hidden damage to the floor or the walls. When people step out of the bathing unit, they often drip water on the floor. If there is not a good seal at the base of the tub or shower, this water will run under the bathing unit. Prolonged use under these conditions will result in severe water damage to the floor and floor joists.

If the walls surrounding the bathing unit leak, water will invade the wall cavity. The water may rot the studs or the base plate of the wall. Given enough time, the water will penetrate the wall's plate and enter the floor structure. This type of damage may go unnoticed until you remove the old fixtures. When the damage is found, it will have to be repaired before you can move ahead with your plumbing.

The structural damage caused by water can be substantial. I have seen floors rot to the point that the fixtures fall through the floor. It is not only embarrassing to have your toilet or tub fall through the floor—it's dangerous. If you will be responsible for the costs incurred to repair water damage, allow for the possibility in your remodeling budget.

CUTTING INTO CAST IRON STACKS

When remodeling, it is not unusual to be working with cast iron drains and vents. Cutting these pipes to install a new fitting can be very dangerous. Many of vertical pipes are not well secured. When you cut the pipe to install your new fitting, the pipe above you may come crashing down. This is a potentially fatal situation.

If you must cut into a vertical cast iron stack, be sure it will not fall when it is cut. Find a hub on the pipe and use perforated strap to secure the pipe. Wrapping the galvanized strapping around the pipe, under the hub, and nailing it to the studs or floor joist will usually protect you. Whenever possible, have a helper on hand when you cut the stack. The method used to secure the pipe will vary with individual circumstances, but be sure the pipe will not move when it is cut.

ADDING A BASEMENT BATHROOM

Putting a bathroom in a basement can cause a problem when it comes time to vent it. The bathroom must have a vent that goes up into the living space above the basement. You may either take the vent through the roof or tie it into another suitable vent. If you are tying into an existing vent, you must tie in at least six inches above the flood level rim of the fixture served by the existing vent.

You must make sure that you have a sewer installed low enough to accept the waste from the basement bath. If a sump and sewage pump is required, the cost will rise dramatically. These types of considerations must be factored in before you give a price for the job.

SIZE LIMITATIONS ON BATHING UNITS

When adding a bathroom within an existing house, you must be aware of access limitations. The size of your ingress areas will dictate the maximum size of your fixtures. For example, one-piece, fiberglass tub-shower combinations will not fit through the doors and stairways of most homes. One-piece units are used almost exclusively in new construction and add-on additions. Under new construction conditions, these bathing units can be set in place before walls and doors are installed.

Even if you are able to get one of these units in a front door, it probably will not go around turns or up stairways. To compensate for this problem, you will have to look to sectional bathing units. There are many types of sectional units available. There are two-piece, three-piece, four-piece, and five-piece units on the market. With so many to choose from, you should not have any difficulty in finding one to suit your needs.

CUT-OFFS THAT WON'T

Remodeling jobs are frequently complicated by valves that will not close. In these cases, you may have to cut the water off at the water meter. Never assume that an existing valve will function properly. Cut the valve off and confirm that the water is off by opening a faucet. Many old valves become stuck and will not close. Even when the handle turns, the valve may not be closing. If you fail to double check the effectiveness of the valve, you may flood the home when you cut pipes or loosen connections.

WORKING WITH OLD PIPES

Working with old pipes can complicate the life of a plumber. However, modern devices have made the job of coupling two different types of pipe together much easier. Let's take DWV copper pipe as an example. There are still a large number of plumbing systems in operation that were constructed from DWV copper. This thin-walled copper has a long life as a drain and vent material. When you find DWV copper, it is usually in good working condition. Since this pipe produces few problems, there is no need to replace it. However, it is rarely cost effective to use copper to extend a system in today's plumbing practices. When you want to add to a system configured from copper, you will probably want to connect to the copper with plastic pipe.

When your goal is to mate plastic pipe with copper pipe, you have a few options. The most common methods for converting copper to plastic is the use of threaded connections or rubber couplings. Either of these adapters will give satisfactory results without much effort or expense.

The first step in connecting to an existing copper pipe is cutting it. You will need to make room for your tee or wye. Roller-type cutters are the best choice for cutting copper, if you have room to operate them. When space is limited, you can cut the copper pipe with a hacksaw or a reciprocating saw, fitted with metal-cutting blades. The principals for soldering DWV copper are the same as those used for copper water pipe.

When you will be cutting in a copper fitting, you must consider how much pipe to remove for the fitting. Another consideration must be on of flexibility in the existing pipe. If the existing pipe cannot be moved forward, backward, or vertically, you have to use an additional fitting to make the connection. Once the pipe is cut, you prepare and install your new fitting. It is important to have all the pipes entering the new fitting to seat properly. This is not to say that the pipe must be jammed into the fitting to the hilt, but you must have enough pipe in the fitting to assure a solid joint. If your existing pipe has very little play in it, you may not be able to install the new fitting without a slip coupling.

Slip couplings do not have the ridges near their center points that standard couplings do. A slip coupling can slide down the length of a pipe, without being stopped by the ridges found in normal couplings. This sliding ability makes a slip coupling indispensable for some adaptation work. To install a normal fitting, you must have room to move exiting pipes back and forth. With slip couplings, this range of movement is not required.

If you are forced to use a slip coupling, you will cut additional pipe out of the existing system to install your new fitting. Your new fitting will be installed on one end of the existing pipe. The slip coupling will be slid back on the other existing pipe. A new piece of pipe will be placed in the vacant end of your new fitting. The pipe will be long enough to reach nearly to the end of the existing pipe. Then, the slip coupling will be slid back towards the new pipe. In doing so, you can connect the two pipes without a need for moving the existing pipes.

Once the branch fitting is installed, you are ready to install the threaded adapter. Your threaded adapter may have male or female threads. A piece of pipe will be installed between the branch fitting and the adapter. These joints will be soldered with normal practices. When the pipe and fittings have cooled, you are ready to switch pipe types.

When the copper has cooled, apply pipe dope to the threads of the male adapter. The male adapter may be copper or plastic, depending upon which order you have chosen for the adapters. Screw the plastic adapter into or onto the copper adapter. When the plastic adapter is tight, your conversion has been made. From this point, you can run your plastic pipe in the normal manner.

If you wish there was an easier way to convert DWV copper to schedule 40, plastic pipe, there is. Rubber couplings can make the conversion between the two pipe types simple. When you use rubber couplings, there is no need for soldering. The amount of pipe you cut out of the existing system is not critical when using rubber couplings. Your measurements are not critical, because you can bridge the span with your plastic pipe. Rubber couplings can slide up the existing pipes, like a slip coupling, so pipes without movement are no problem. All aspects considered, I can see no reason to use any other type of conversion method in normal applications.

The first step is to cut a section of pipe out of the existing plumbing system. Remove enough pipe to allow the installation of your branch fitting and two pieces of pipe. With the old pipe removed, slide a rubber coupling on each end of the existing pipes. Hold your branch fitting in place, and take measurements for the two pieces of pipe that will join the fitting to the existing pipes. Install the two pipes into the ends of your branch fitting.

Hold the branch fitting, and its pipe extensions, in place to check your measurements. The plastic pipes, protruding from the branch fitting, should extend to a close proximity of the copper pipes. With the pipes in place, slide the rubber couplings over the plastic pipes. Position the stainless steel clamps to the proper locations and tighten the clamps. When the clamps are tight, your conversion is complete. Now you can work directly from the plastic branch fitting to complete your installation.

Galvanized Drains

Many houses are still equipped with galvanized drains and vents. Due to galvanized pipe's potential for problems, it is a good idea to replace galvanized pipes whenever possible. On the occasions where you will not be replacing the pipe, you must use adapters to make the steel pipe compatible with new plumbing materials.

In many situations you will be forced to cut the old galvanized piping to install a branch fitting for your new plumbing. A hacksaw is the tool most often used to accomplish this task. If you have a reciprocating saw, you can use it with a metal-cutting blade to make the job easier. When using an electric saw, be careful not to get electrocuted. The saw must be insulated and designed for this type of work. Old drains often hold pockets of water; if you hit this water with a faulty saw, you will get a shock.

Rubber couplings are the easiest way to convert galvanized pipe to a different type of pipe. If you elect to use rubber couplings, the procedure will go about the same as described for copper pipe. You cut out a section of the galvanized pipe to accommodate your new branch fitting. Then, you install the branch fitting and its pipe sections with the rubber couplings. The couplings simply slide over the ends of each pipe and are held in place by the stainless steel clamps.

Since you cannot solder a fitting onto galvanized pipe, if you choose to use threaded adapters, you must have threads available to work with. When cutting in a branch fitting, you must cut the old pipe and remove it until you get to a threaded fitting or pipe end. This can be difficult. The old pipe may have concealed joints, or the pipe may be seized in its fittings. Pipe wrenches will be a must for disassembling old galvanized drains.

Once you have worked your way to a threaded fitting or an end of the pipe with threads on it, you can install your threaded adapters. Apply pipe dope to the male threads of the connection and screw the adapter into or onto the threads. Tighten the adapter with a pipe wrench. When the adapters are tight, you can proceed with your new piping material.

When cutting in a branch fitting on galvanized pipe, rubber couplings are the easiest route to take. The rubber couplings work fine and reduce the time and frustration involved in working with old galvanized pipes. If you are only converting a trap arm, or similar type of section, threaded adapters are a reasonable choice. With a trap arm, you will only be unscrewing one section of the steel pipe. Once it is removed, you will have the threads of the fitting to screw an adapter into. Under these conditions, threaded adapters are the best choice.

CAST IRON PIPE

Cast iron pipe has been used for many years to provide drains and vents in plumbing systems. Even today, cast iron is still used for these purposes. For the most part, cast iron has been replaced with plastic pipe in modern systems, but cast iron still does see use in some modern systems. There are two types of cast iron that you are likely to encounter. The first type is a service weight pipe with hubs. This pipe is often called service weight cast iron, or bell and spigot cast iron. The other type is a lighter weight pipe that does not have a hub.

Service weight cast iron is what you are most likely to encounter. This pipe has been around for a long time and is still used. It is the pipe used in most older homes. Service weight cast iron is usually joined with the use of caulked lead joints at the hub connections. While caulked lead joints were used to install this pipe, modern adapters can be used to avoid working with molten lead. The options for using adapters with bell and spigot cast iron will be discussed in the following paragraphs.

The lightweight, hubless cast iron pipe used in some jobs is a much newer style of cast iron. The connections used for this type of pipe are a type of rubber coupling. They are not the same rubber couplings referred to throughout this chapter. These special couplings are designed for joining hubless cast iron. The couplings have a rubber band that slides over the pipe and fitting. There is a stainless steel band that slides over the rubber coupling. The band is held in place by two clamps. When the clamps on the band are tight, they compress the rubber coupling and make the joint. These special couplings are easy to work with, but they are not your only option. The heavy rubber couplings discussed elsewhere in this chapter will work fine with this pipe.

Cast iron pipe can be cut with roller cutters, metal-cutting saw blades, and some people can even cut it with a hammer and a chisel. For the average job, ratchet-style snap cutters are the best tool for cutting cast iron pipe. These tools can be used to cut pipe when there is very little room to work with. If you can get the cutting chain around the pipe, and have a little room to work the handle back and forth, you can cut the pipe quickly and easily. These soil-pipe cutters are expensive, but you can rent them from most rental stores.

Rubber Couplings

The heavy rubber couplings that have been discussed earlier will work fine with either type of cast iron pipe. They will be used in the same method as described earlier. When working with hubless cast iron, you can use the special rubber couplings designed to work with the hubless pipe. If you use these special couplings to mate cast iron to plastic, you will need an adapter for the plastic pipe. It is possible to use the couplings without the plastic adapter, but you shouldn't. The plastic adapter will have one end formed to the proper size to accommodate the special coupling. The adapter is glued onto the plastic and then the special coupling slides over the factory-formed end of the adapter.

Cast Iron Doughnuts

When working with the fittings for service weight cast iron, there are special adapters available to allow plastic pipe to be installed in the

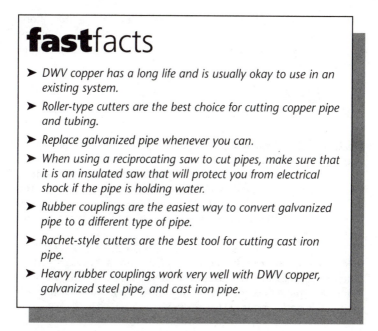

fastfacts

➤ *DWV copper has a long life and is usually okay to use in an existing system.*

➤ *Roller-type cutters are the best choice for cutting copper pipe and tubing.*

➤ *Replace galvanized pipe whenever you can.*

➤ *When using a reciprocating saw to cut pipes, make sure that it is an insulated saw that will protect you from electrical shock if the pipe is holding water.*

➤ *Rubber couplings are the easiest way to convert galvanized pipe to a different type of pipe.*

➤ *Rachet-style cutters are the best tool for cutting cast iron pipe.*

➤ *Heavy rubber couplings work very well with DWV copper, galvanized steel pipe, and cast iron pipe.*

hub. Normally, when pipe is placed in the hub of service weight cast iron, it is held in place by a hot lead joint. To avoid using molten lead, you can use a ring adapter. Most plumbers call these ring adapters doughnuts.

These ring adapters are not always easy to use. They fit in the hub of the cast iron and the new pipe is driven into the ring. As the new pipe goes into the ring, the ring grips the pipe and forms a water-tight joint. The problem comes in trying to get the pipe into the ring. You must lubricate the pipe and the inside of the ring gener-ously before attempting the installation. Special lubricant is made for the job and sold by plumbing suppliers.

Place the doughnut in the hub of the fitting. Apply plenty of lubrication to the inside of the ring and to the outside of the pipe. Place the end of the pipe in the ring and push. The pipe will proba-bly not go very far. When the pipe is as far in the ring as it will go, you will have to use some force. Take a block of wood and place it over the exposed end of the pipe. Hit the wood with a hammer to apply even pressure to the pipe. With some luck and a little effort, the pipe will be driven into the doughnut. Once the pipe is all the way into the adapter ring, you can proceed with your new piping in the normal manner.

COPPER WATER PIPES

A vast majority of today's homes have copper pipe installed for the distribution of potable water. If you do not wish to make your new installation in the same type of pipe, you will have to use adapters. When you prefer to work with plastic pipe, you must convert the copper to a suitable pipe type. This can be done with a variety of adapters. Of all the adapters available, threaded adapters are the most universal.

Cutting copper pipe is not difficult. The job is done easiest with roller-cutters, but can be accomplished with a hacksaw. When remodeling, miniature roller-cutters often come in handy. Measuring to determine how much pipe to remove for a branch fitting will be done as described for DWV copper.

Threaded Adapters

By using threaded adapters, you can mate copper water pipes with any other approved material. Cut in your tee and solder the threaded adapter to a piece of copper at the point you wish to make your conversion. After the soldered joint cools, apply pipe dope to the male threads to be used in the connection. Screw the male threads into the female threads until the fitting is tight. At this point, you conversion is complete. You can continue your new installation from the opposite end of the threaded adapter.

Compression Fittings

Compression fittings are another possibility for converting copper to a different type of pipe. Take a compression tee and cut it into the copper pipe. The same measurement techniques used for copper fittings will work with compression fittings. Since the branch fitting is of the compression type, there is no soldering to be done.

Saddle Valves

Self-piercing saddle valves are a good choice when adding new plumbing for an ice maker or similar appliance. The saddle will clamp to the copper pipe and allow the use of plastic or copper tubing between the saddle and the appliance. The saddle valve will connect to the tubing with a compression fitting. The saddle is held in place on the copper water pipe with a clamp and bolt device. With a self-piercing saddle valve, you do not have to drill a hole in the copper water pipe. When the handle on the saddle is turned clockwise, to its full extent, the copper water pipe is pierced. When the handle is turned back counterclockwise, the water can flow into the tubing.

PAINTED COPPER PIPES

The copper water pipes in older homes may be covered in paint. At one time, it was fashionable to paint exposed pipes to blend in with the decor of the home. The paint may have made the pipes more attractive, but it definitely makes soldering these pipes more difficult. If you are required to cut a branch fitting into painted pipes, the paint must be removed from the area to be soldered.

Most plumbers cut the painted pipes first, and then try to remove the paint. They usually use their regular sanding cloth and struggle to get the paint off. There is an easier way to remove the old paint. You should remove the paint before you cut the pipes. Before the pipes are cut, they are solid and easy to apply pressure to with the sandpaper. Once the pipes are cut, they wobble and bounce around, making it difficult to apply steady pressure with the sanding cloth. Another common problem with cutting the pipes before the paint is removed is the water in the pipes. When the pipes are cut, water often runs down the pipe, getting the paint wet, and the wet pipe gets the sanding cloth wet. With the pipe and the sandpaper wet, the paint is much harder to get off.

The standard sanding cloth used for copper pipe will remove most of the paint, but you will have to sand the pipe several times to accomplish the goal. If you know you will be working with painted pipes, put some steel wool in your tool box. Also invest in some sandpaper with a heavy grit. The rougher sandpaper will do a better job than the fine-grit paper usually used to sand copper. By using the steel wool and coarse sandpaper before the pipes are cut, the paint should come off with minimal effort.

If the paint refuses to come off, you can burn it off. If you have enough clear space to avoid an unintentional fire, use your torch to burn the paint off the pipes. When flammable materials are nearby, you must protect against an accidental fire. You can reduce the risk of unwanted fire in many ways. You can use a heat shield on your torch, if you have one. If you don't have a heat shield, you can make one from ordinary household goods. By folding aluminum foil into several layers, you can make a reasonable heat shield. A piece of sheet metal will also serve as a heat shield.

Spraying water on the flammable materials will reduce the risk of accidental fires. For best results, wet down all flammable materials. Place a heat shield between your flame and any flammable materials. Have an adequate fire extinguisher on hand, just in case. Proceed cautiously with your work, keeping an alert eye on your flame and the flammable materials. If you see the flammable material start to smoke, turn black, or bubble, cut your torch off and dowse the hot material with water. If the progression of an unwanted fire has gone beyond the early stage, quickly extinguish it with your fire extinguisher. If you use a heat shield and wet down the surrounding materials, you should not have a problem with unwanted fire.

After the paint is removed, cut the pipes. Prior to cutting the pipes, be sure the water is off. Drain as much water from the pipes as possible. If you can see water standing in the pipe, you know soldering will not be easy. It is not unusual for water to remain in pipes after they are cut. If your pipes are running horizontally, you may be able to pull down on the pipe to allow the water to escape. When working with a vertical pipe, you will not be able to tip it down to drain it. Instead, you will remove the water with a drinking straw. Don't worry, you will not have to suck the water out of the pipe. Place the straw in the vertical pipe. When the straw is in the pipe, blow through it. The pressure will force the standing water up and out of the pipe. Unless you have water leaking past a valve, this will clear enough water for you to do you soldering.

Apply a generous amount of flux to the pipe where the fitting will be placed. Before, installing the fitting, heat the flux with your torch. As the flux bubbles and burns, it will clean the pipe. After this cleaning, proceed with your normal soldering procedures to complete your job.

Remodeling work can be very different from new work. If you don't have experience in remodeling, be careful in giving prices to customers. A job that looks easy can turn into a major project. Assess all conditions carefully, and proceed with caution. If you do this, you should gain the experience you need with a minimal loss of profit.

5

SERVICE AND
REPAIR WORK

After all your plumbing is installed and tested, you may still have some work to do. It is not uncommon for new plumbing installations to develop leaks in the first few days of operation. Neither is it unusual for new fixtures and faucets to need some type of adjustment in the early weeks of their usage. Plumbers who do a lot of service and repair work are well aware of the many types of problems that can occur after plumbing is installed. The tips in this chapter will help you to troubleshoot and correct deficiencies in newly installed plumbing and older plumbing that has developed problems. This chapter is broken down into logical segments to help you solve your faucet and fixture failures.

RELIEF VALVES ON WATER HEATERS

The relief valve on a water heater is a very important safety device. It protects against a water heater that is developing excessive temperature or pressure. Without a dependable relief valve, a water heater can turn into a bomb.

Relief valves can develop drips and leaks. If the valve is leaking around the threads where it is screwed into the water heater, you must remove and reinstall the valve. The cause of this type of leak is usually a lack of pipe dope or Teflon tape. If the valve drips out of its discharge tube, you have to decide why it's leaking.

The leak could be coming from sediment in the relief valve. If debris gets between the valve's seat and washer, drips are going to happen. To test this, open the valve and allow it to blow some water out. Then, close the valve and see if it seals. Sometimes, letting the relief valve blow off some water will clear the debris. If the valve continues to drip, you will have to replace it.

When the relief valve is blowing off at full capacity, you may have a bad valve. The other possibility for this problem is thermal expansion. If the water distribution system is equipped with a backflow preventer, thermal expansion can force the relief valve to blow off. Change the relief valve and observe it for several days, or until it blows off again. If the valve operates normally, you have corrected your problem. If the new valve blows off, you either have thermal expansion or a problem with the water heater.

If you have a backflow preventer, thermal expansion is a good guess. To reduce the effects of thermal expansion, you can install air chambers on the water pipes at the water heater. In many cases, the installation of air chambers will resolve your problem. If the air chambers don't get the job done, you may have to install an expansion tank. The expansion tank will be similar to the type used on boilers. As the thermal expansion occurs, the expanding water can escape into the expansion tank.

The reason backflow preventers cause thermal expansion is their unwillingness to allow expanding water to run back down the water service pipe. Without a backflow preventer, expanding water can push back down the water service and into the water main. The backflow preventer blocks the path and does not allow water to flow back out of the building.

When the faucets are closed, there is no place for the expanding water to go. Because of this, it forces the relief valve to blow off due to excessive pressure. To correct the problem, you must provide some place for the expanding water to go.

WATER HEATER FAILURES

When a water heater is not up to par, there may be limited, or no, hot water. Working with water heaters is potentially dangerous. With electric water heaters, there is high voltage around all of the interior parts that may fail. These parts are the heating elements and thermostats. With gas water heaters, there is the risk of fire or explosion from the gas. Oil-fired water heaters also have the potential for fire-related risks.

Electric Water Heaters

If you have a limited supply of hot water that diminishes quickly, you probably have a bad heating element. Water heaters in most homes have two heating elements. Some heating elements bolt in, and others screw in. If both elements go bad, you will not have any hot water. When only one element is bad, you will have a limited supply of hot water. To determine if the elements are bad, they must be tested with a meter, while the electricity is turned on.

Electric water heaters also have thermostats that may go bad. On water heaters in average homes, there are two thermostats, an upper and a lower. Again, electrical power is present at the thermostats. The thermostats are set into a bed of insulation, behind the access panels of the water heater. The wires providing the power are also hidden in the insulation. Only trained technicians should remove the access covers on electric water heaters. The high voltage is sitting just behind the panels, waiting to zap untrained hands. If you get hit by the power from these wires, you could be killed.

There are reset buttons on electric water heaters. These buttons are located behind the access panels, near the electrical wires. Sometimes these buttons can be all it takes to get a water heater back on line. The simple step of pushing a reset button can be all that is needed.

There is a dial located behind the access cover of an electric water heater for setting the temperature of the water. This dial can be set with the electrical power turned off. The dial is normally set at the factory, and is sometimes adjusted when a water heater is installed. If you decide to change the setting on a water heater, cut the power off to the water heater first. The dial is simply turned right or left to raise and lower the water temperature. For additional troubleshooting information on water heaters, refer to Chapter 12.

FROST-FREE SILLCOCKS

Frost-free sillcocks do not require much attention. If sediment gets between the seat and the washer, the sillcock may drip. The packing nut on the sillcock could be loose and leaking, but these are the only two problems you are likely to experience. If the packing nut is leaking, all you have to do it tighten it. Unless the nut is cracked or the packing is defective, tightening it will be all that is necessary.

If water is dripping out of the sillcock, you will have to remove the stem. Cut the water off to the pipe supplying the sillcock. As you look at the handle of the sillcock, you will see a nut holding the stem into the body. Loosen the nut, and the stem will come out. Inspect the washer on the end of the stem. If the washer is good, leave it alone. If it is cut or worn, replace it. While the stem is out, cut the water back on to the pipe supplying the sillcock. Water will blow out of the end of the body, flushing any sediment and debris from the seat. After a few moments, cut the water off. Put the stem back into the body and tighten the retainer nut. Cut the water back on and inspect the results of your work. This procedure should cure your drip.

WASHING MACHINE HOOK-UPS

Washing machine hook-ups may be nothing more than two boiler drains, or they may consist of a faucet device. In either case, there is not much call for adjustments. The boiler drains have a packing nut that may leak. They may also drip from a bad washer or debris between the washer and the seat. To correct either of these problems, follow the directions given in the section on sillcocks.

If the hoses connect to a faucet, there will be little you can do if it is defective. This type of faucet is inexpensive and has very few repairable parts. In most cases, if the faucet is bad, it must be replaced. If you have problems with water pressure, check the screens in the ends of the hoses for the washing machine. There are cone-shaped screens in each hose. These screens filter sediment from the water pipes. In new installations, it is not uncommon for these screens to become clogged.

TUB WASTES

If a bathtub is not holding water or is not draining properly, you must adjust the tub waste. The most common style of tub waste is the trip-lever type. To adjust these tub wastes, remove the two large screws that hold the cover plate on the overflow fitting. When the screws are removed, grasp the plate and pull the assembly up out of the overflow tube. If the tub is not holding water, you must lengthen this assembly. If the tub is not draining, you must shorten the assembly.

To adjust the assembly, loosen the nuts on the threaded portion of the assembly. When the nuts are loose, you can turn the threaded rod to adjust the length of the assembly. Finding the proper setting is done on a trial and error basis. Make your adjustment, reinstall the assembly, and test the drain. Continue this process until you have perfected the setting.

If you have a tub waste that is controlled by the stopper in the tub, there will be no assembly in the overflow tube. If you operate the tub waste by pushing down on the stopper, the only adjustment you can make is to loosen or tighten the threads of the stopper. There is a rubber gasket on these stoppers. By turning the stopper, you can raise or lower the level of the rubber gasket. If you must lift and turn the stopper to operate the tub waste, there is no adjustment you can make to the tub waste.

TUB AND SHOWER FAUCETS

When you must work on the faucets for a tub or shower, you should refer to the instructions that are available from the manufacturers of the faucets. With the large number of different faucets available, it is difficult to give you precise instructions on working with them. Since I cannot cover all of the faucets, I will give you instructions for the most common types.

Two-Handle Faucets

Two-handle faucets may be used for either bathtubs, showers, or a combination of the two. When two-handle faucets are used for the combination of a tub and shower, the tub spout will have a diverter on it. This is the little rod that you pull up to divert water from the tub spout to the shower head. When the diverter rod is lifted, the bulk of the water should come out of the shower head. If you lift the rod and still have a substantial flow of water coming from the tub spout, you need a new diverter. This requires replacing the existing tub spout with a new one.

If water drips from the tub spout after the faucets are cut off, you either have a bad washer, a bad seat, or debris between the washer and the seat. To correct these problems, you will need a set of tub wrenches. These deep-set wrenches allow you to reach inside the wall to turn the nuts holding the stem into the faucet body. Before removing the faucet stems, cut the water off to the faucet body.

The handles of faucets will be attached with screws. The screws are hidden by caps that pop off. Using a knife, pry the caps out of the center of the handles. Loosen the screws, and remove the handles. There will be trim escutcheons to remove. Most of these can be removed by turning them counterclockwise. In some cases, the escutcheons will be held in place with a set screw. If this is the case, you will need a hex wrench to loosen the set screw. Slide the tub wrench over the stem and into the wall. When the wrench is on the retainer nut, turn the wrench counterclockwise. Once the nut is loose, the stem will come out.

Inspect the washer and replace it if it appears damaged. If you suspect that there is debris in the faucet body, flush it out. Put the faucet back together and test it. If it still leaks, you may have to replace the seat. Cut the water off and remove the stem. Insert a seat wrench into the seat and turn it counterclockwise. After a few turns, the seat will come out. Replace it and put the faucet back together. If all these efforts fail, replace the faucet.

If water is leaking around the handles of the faucet, you must either tighten the packing nuts or replace the "O" rings. To do either of these jobs, you must remove the handles from the stems. If you are only tightening the packing nuts, you do not have to cut the water off. If you plan to replace the "O" rings, you must cut the water off. The packing nuts are the nuts that are around the stem. With the water on, you will see the water leaking past the nut, if it needs to be tightened. Turning the packing nut clockwise should stop the leak in a new faucet. In an old faucet, you may have to replace the packing material.

If you must replace the packing, cut the water off. Loosen the packing nut and slide it back on the stem. Wrap the packing material—a greasy string will work,—around the stem and replace the packing nut. The nut will compress the packing and hold it in place. If you have to replace the "O" rings, they are located on the outside of the faucet stem. Once you have removed the stem from the faucet, you will see the "O" rings. If the rings are damaged, replace them and reinstall the stem.

Three-Handle Faucets

Three-handle faucets are the same as two-handle faucets, except for the diverter. Instead of having a diverter on the tub spout, the center handle diverts the water from the tub spout to the shower head. If the water is not diverting properly, you have to either replace the washer or the seat. In new installations, you may have to clear debris from within the body of the diverter. The techniques for this work are the same as described above.

Single-Handle Faucets

There are two basic types of single-handle faucets. They are cartridge-type and ball-type faucets. While the two types may look the same on a fixture, there are significant differences on the internal parts of the faucets. With either type of faucet, you must cut the water off before repairing them. In both types, you must remove the handle from the stem. This is where the similarities end.

Before proceeding, be sure the water is off. To work with a cartridge-type faucet, remove the trim collar that sits over the stem. This collar just pulls up for removal. You should see a thin clip holding the stem into the faucet body. Using a pair or pliers, remove this clip. By gripping the clip on the tab extending from the body, pull it straight out. Once the clip is out, grasp the stem with your pliers and pull it out. The cartridge will have "O" rings on the outside of it. These cartridges are made to be replaced, not repaired. The only exception is the "O" rings. If the "O" rings are bad, replace them.

With a ball-type faucet, there are many parts to contend with. The handle on this type of faucet is usually held in place with a screw. Once the screw is out, lift the handle off the stem. There should be a large, knurled nut holding the ball in place. You may have to use pliers to loosen this nut, but be careful not to scratch the finish.

When the large nut is removed, you will see a ball sitting in the faucet. It will have "O" rings, springs, and other loose parts fitted into it. Remove the ball carefully. You can purchase repair kits for these balls, or you can replace the entire assembly. It is generally best to rebuild the entire ball assembly.

Lavatory Faucets

Most lavatory faucets will have a single body with one or two handles. The repair and replacement procedures for the internal parts are essentially the same as those described above. The only major difference is that lavatory faucets are easier to work with. You will not need deep-set wrenches to reach the nuts on a lavatory faucet. Once the stem escutcheons are removed, you can loosen the retainer nuts with an adjustable wrench.

LAVATORIES

Pop-Up Adjustment

After a new installation, the pop-up assembly in a lavatory may need to be adjusted. This is done from under the lavatory bowl. Check the knurled nut that holds the ball and rod in the drain assembly. If it is not tight, tighten it. If you need more adjustment, adjust the lift rod. You can do this where the lift rod attaches to the pop-up rod or where it attaches to the lift assembly. By making the lift rod shorter, where it protrudes above the faucet, you can gain more leverage to seal the drain. This is the procedure to use if the bowl is not holding water. If the bowl is not draining fast enough, allow more rod to stick up above the faucet. The extra length will push the rod down farther and cause the bowl to drain faster.

If you fill a lavatory with water, only to have it slowly drain out, you may have a putty problem. When the pop-up assembly seems to be properly adjusted, but water still leaks out, there may not be a good seal around the drain. If the drain was installed without sufficient putty, water will slowly leak out of the bowl and into the drain. This type of leak can be hard to locate. Since the leaking water leaks into the drain, it doesn't show up under the bowl. To remedy the problem, you must remove the drain assembly and reinstall it.

Slip-Nuts

The drainage connections under the lavatory are often made with slip-nuts and washers. It is not uncommon for these connections to develop small leaks. If you notice water dripping from the slip-nut, tightening the nut should solve your problem. If it doesn't, remove the slip nut and replace the gasket that the nut compresses to create a leakproof seal.

Compression Fittings

Compression fittings often need to be tightened after the first few days of use. Until the joints have settled in, they may leak from pipe vibrations or movement. When a compression fitting is leaking, tightening the compression nut will usually correct the leak. If the leak is a bad one, you may have to replace the compression sleeve or nut. Sometimes the nuts crack and cause water to spray about. In either event, compression leaks are not difficult to fix.

TOILETS

It is very common for new toilets to require some adjustments after being installed. The adjustments needed will usually be in the toilet tank. Older toilets can suffer from several problems. Most are easy to correct.

Ballcocks

There are two common types of ballcocks. The first has a horizontal float rod with a float or ball on the end of it. The other type has a plastic float that moves up and down the shaft of the ballcock. The ballcock controls the amount of water that enters the toilet tank. Ballcocks are adjustable to increase or reduce the amount of water in the tank. Ideally, the tank's water level should be level with the fill-line etched into the interior of the tank.

The float-rod ballcock is very common and is easily adjusted. If you want more water in the tank, you bend the float rod down. By bending the rod downward, it requires more water to float the ball high enough to cut the water off. If you want less water, you bend the float rod up. This causes the ball to reach its cut-off point earlier, resulting in less water. The float rods are easy to bend with your hands.

To bend the rod, place the palm of your hand either over or under the rod, depending upon which direction you wish to bend it. With the rod cradled in the palm of one hand, use your other hand to bend it at the end where the float is. Be careful not the bend the rod sideways. If the float gets too close to the side of the tank, it may stick on the tank. When the float is hung up on the tank, the ballcock cannot work properly. If the ball sticks in the down position, the water will not cut off. If it sticks in the up position, the water will not cut on.

Vertical ballcocks have not been around as long as float-rod ballcocks, but they are very popular. One of the big advantages to a vertical ballcock is that there is no concern for the float sticking to the side of the tank. As float-type ballcocks age, the float rod often becomes loose. When the rod is not screwed into the ballcock securely, the float is apt to come into contact with the side of the tank. Vertical ballcocks remove this risk.

To control the water flow with a vertical ballcock, you move the float up and down the shaft of the ballcock. Moving it down produces more water and moving it up produces less. To move the float, you must squeeze the thin metal clip, located on a metal rod next to the float. While squeezing the clip, you can move the float up or down. When you like the placement, release the clip and the float will remain in the selected position.

Flappers

If you are working on a flush valve that has a flapper, it may need to be adjusted. If the chain between the flapper and the toilet handle is too short, the flapper will not seal the flush hole. This allows water to run out of the tank and into the bowl. As the water level drops, the ballcock refills the tank. This is an endless cycle that wastes a lot of water. If you notice the ballcock cutting on at odd times or see water running into the bowl long after a flush, investigate the flapper's seal. By moving the chain to the next hole in the handle, the seal should be improved. If not, adjust the length of the chain where it is connected to the handle. The point is, make sure the flapper sits firmly on the flush hole. Check the flapper to see if it appears worn. If adjustments don't work, replace the flapper.

If the chain between the flapper and the handle is too long, the flapper may become entangled in the chain. This causes the flapper to be held up, allowing water to constantly run down the flush hole. This problem can be corrected by shortening the length of the chain.

Tank Balls

Tank balls are sometimes used instead of flappers. If you have a tank ball, you also have lift wires and a lift-wire guide. The guide is attached to the refill tube in the toilet tank. The lift rods attach to the ball, run through the guide, and attach to the toilet handle. If the alignment of the tank ball is not set properly, water will leak past it, into the toilet bowl.

When you are having problems with the tank ball, inspect the location of the ball on the flush hole. If it is not properly seated, adjust the lift wires or the guide to gain a satisfactory seal. The guide can usually be turned with your fingers to make a better alignment. If the lift wires are too short and hold the ball up, you must bend them to add length.

If these attempts do not prove fruitful, you may have to replace the tank ball. This is simply a matter of unscrewing the ball from the lift rod and replacing it with a new ball. If a new ball doesn't do the trick, the flush valve may be pitted or defective. On a brass flush valve, you can use sandpaper to remove pits in the brass. By sanding the area where the tank ball sits, you smooth out the rough spots and encourage a better seal. The pits are voids that a tank ball cannot seal. They allow water to leak through to the toilet bowl. Unless you are working with an old toilet, the flush valve should not be pitted.

Tank-to-Bowl Bolts

As a new toilet is used, the tank-to-bowl bolts may work loose. This is especially true if the tank is not supported by a wall behind it. If you develop leaks at the tank-to-bowl bolts, tightening the bolts should stop the leaks. Be careful not to tighten the bolts too much. If you do, the china will break.

The Flush-Valve Gasket

If you have water flooding out between the tank and bowl when the toilet is flushed, you have a poor seal at the flush-valve gasket. This is the gray sponge or black rubber gasket that is fitted over the flush valve when installing the toilet. If the tank-to-bowl bolts are not tight enough to compress this gasket, you will have water leaking each time the toilet is flushed. Normally, tightening the tank-to-bowl bolts will stop this type of leak. If tightening the tank-to-bowl bolts doesn't correct the problem, you will have to disassemble the toilet to investigate the problem further.

   It is possible the flush-valve nut is cracked. It is also possible that the threaded portion of the flush valve has a hole in it. The gasket may have been shifted during the installation, rendering it ineffective. A good visual inspection should reveal the cause of your leak. After making the necessary corrections, reassemble the toilet and test it. If you have a continual problem, try replacing the flush valve.

KITCHEN FAUCETS

Kitchen faucets are about the same as lavatory faucets when it comes to repairing them. Normally, the only major difference is the spray hose assembly, if the kitchen faucet has one. If the spray is not working well, it may be clogged with sediment. If the water does not divert from the kitchen spout to the spray as it should, the handle of the spray may be restricted. You can unscrew the spray head and attempt to clean it, or you can replace the entire head.

KITCHEN SINKS

Kitchen sinks can offer a service plumber with several types of problems. Many of the problems are fairly easy to correct. One of the most difficult aspects of kitchen sinks is often that the devices being worked on may be seized so tightly that replacing them is very hard. I've had many occasions when I had to cut off nuts that should have been able to be removed with normal turning.

Basket Strainers

Normally, the only problem you may have with a basket strainer is a poor seal where the drain meets the sink. Removing the drain and reinstalling it with a good seal of putty will stop the leak. If the sink will not hold water, the seal on the basket may be defective. If you suspect the basket is bad, replace it with a new one.

Continuous Waste

The continuous waste for sinks with multiple bowls may develop leaks. The leaks will usually be small leaks at the slip-nut connections. To correct the leaks, tighten the slip-nuts. If this doesn't work, remove the slip-nuts, remove the washer and replace it. Then replace the slip-nut and tighten it.

Compression Fittings

The compression fittings under a kitchen sink are the same as those found under a lavatory. It is not unusual for these connections to begin leaking after the system is placed in use. Follow the same methods as those given for lavatories to correct deficiencies in the compression fittings under the kitchen sink.

Garbage Disposers

The most common problem with garbage disposers is normally when they become jammed. If a disposer hums, but does not turn, the cutting heads are stuck. To correct this problem, you should cut the power off to the disposer. Then, place a broom handle into the disposer and wedge it against the jammed cutting blades. With a little forceful leverage, you should be able to free the blades. Some disposers come with a special wrench designed to clear the cutting heads.

If the disposer fails to start, you can try pushing the reset button. Most disposers have a reset button on the bottom of the unit. If pushing the reset button doesn't correct the problem, you may have a blown fuse or circuit breaker. If there are no blown fuses or breakers in the electrical panel, you can either suggest that an electrician be called in, or suggest a full replacement of the unit. If you are comfortable with electrical meters and working with electricity, you can test the wiring at the disposer yourself for electrical power.

ADDITIONAL TROUBLESHOOTING

Additional troubleshooting can be found in Chapter 12. Experience is the best teacher when it comes to service and repair work. However, base knowledge is needed, and information like you have just read is the fastest way to build a good foundation for service work. Let's move now to the next chapter and learn about plumbing fixtures.

6

FIXTURES

Trim-out plumbing can appear deceivingly simple. When you watch an experienced plumber set fixtures, the job looks like work almost anyone can do. Well, almost anyone can do the job, but for most it will not be as easy as it looks. Without knowing the secrets for successfully setting fixtures, your attempts may be costly and frustrating. There is an art to installing plumbing fixtures. Without the proper knowledge, you are likely to break china fixtures and make many mistakes along the way. This chapter is going to show you how to avoid the most common problems encountered with finish plumbing.

Most plumbing fixtures are installed near the end of a job. The only fixtures installed during the rough-in are usually bathing units and hose bibbs. Since these are the first fixtures you will have to install, I will start my instructions with them.

INSTALLING SILLCOCKS

If you asked the average person to give you a list of the plumbing fixtures installed at their home, they would never think to include sillcocks. While you may not think of a sillcock as a fixture, the plumbing code does. Sillcocks are one of the few fixtures that may be installed during the rough-in stage of your job. In many cases the sillcocks will not be installed until the end of the job. The determin-

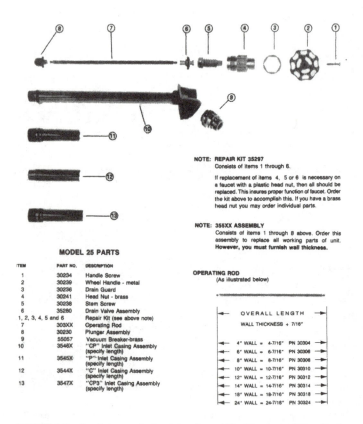

NOTE: REPAIR KIT 35297
Consists of items 1 through 6.

If replacement of items 4, 5 or 6 is necessary on a faucet with a plastic head nut, then all should be replaced. This insures proper function of faucet. Order the kit above to accomplish this. If you have a brass head nut you may order individual parts.

NOTE: 355XX ASSEMBLY
Consists of items 1 through 8 above. Order this assembly to replace all working parts of unit. However, you must furnish wall thickness.

MODEL 25 PARTS

ITEM	PART NO.	DESCRIPTION
1	30234	Handle Screw
2	30239	Wheel Handle - metal
3	30236	Drain Guard
4	30241	Head Nut - brass
5	30238	Stem Screw
6	35280	Drain Valve Assembly
1, 2, 3, 4, 5 and 6		Repair Kit (see above note)
7	303XX	Operating Rod
8	30230	Plunger Assembly
9	55057	Vacuum Breaker-brass
10	3546X	"CP" Inlet Casing Assembly (specify length)
11	3545X	"P" Inlet Casing Assembly (specify length)
12	3544X	"C" Inlet Casing Assembly (specify length)
13	3547X	"CP3" Inlet Casing Assembly (specify length)

OPERATING ROD
(As illustrated below)

OVERALL LENGTH
WALL THICKNESS + 7/16"

4" WALL = 4-7/16"	PN 30304
6" WALL = 6-7/16"	PN 30306
8" WALL = 8-7/16"	PN 30308
10" WALL = 10-7/16"	PN 30310
12" WALL = 12-7/16"	PN 30312
14" WALL = 14-7/16"	PN 30314
18" WALL = 18-7/16"	PN 30318
24" WALL = 24-7/16"	PN 30324

FIGURE 6.1 Breakdown of a frost-free hose bib. *(Courtesy of Woodford Mfg. Co.)*

ing factor for when they are installed will be accessibility. If you will not have access to install the sillcocks when the house is finished, you must install them during your rough-in.

Sillcocks are one of the easiest plumbing fixtures to install. The first step in the installation is choosing a suitable location. When the location is known, you will have to drill a hole for the sillcock. If the sillcock is being installed on the side of a frame house, make sure the siding has been installed on the home before putting in the sillcock. Drill your hole through the siding and wall sheathing. When the sillcock will be mounted to a masonry wall, use a hammer drill or star drill to make your hole in the masonry wall.

Sillcocks will have a mounting flange to accept two fasteners for the securing of the sillcock to the wall. Use screws to affix the sillcock to wood. Use plastic or lead anchors and screws to attach the sillcock to a masonry wall. If you are installing a frost-free sillcock, make sure the end of the sillcock extends past the exterior wall. Ideally, it should extend into an area that will be heated. In unheated areas, insulate the pipe to comply with the plumbing code.

Once the back end of the sillcock is in the desired location, connect the water supply to the sillcock. You should install a stop-and-waste valve in the pipe feeding the sillcock. Install the valve so that the arrow on the side of the valve points in the direction of the sillcock. Most sillcocks will allow you to sweat a piece of copper into the back end of them. If you prefer, you can make the connection with a female adapter. Sillcocks generally have external threads to allow the use of a female adapter in the connection. In cold climates, it is advisable to use frost-free sillcocks.

BATHTUBS

Bathtubs are installed during the rough-in stage of the job. How the tub is installed will depend on the type of tub being used. We will start with one-piece, tub and shower combination units.

One-piece, tub and shower combinations are equipped with a nailing flange, and are usually made from fiberglass or an acrylic material. These combination units are the most common of all bathing units installed in modern plumbing systems. Installing these units is not difficult.

Slide the unit into the opening that has been framed to receive the bathing unit. If you are using a standard five-foot unit, the rough opening should allow at least an additional quarter of an inch for the installation. A full ½ inch of extra space makes the installation much easier. One person can install these units alone, but the process is easier when two people work together on the installation.

Once the unit is in the cavity, you must level the unit. A four foot level should be used when setting the tub/shower combination. Lay the level on the flood level rim of the tub and check the bubble. If the bubble shows the tub to be level from front to back, half the battle is over. If the level shows the tub to be out of level, you will have to make a judgement call. If the tub is pitching slightly forward, you can proceed with the installation. If the tub is pitching backwards, you should correct the problem. If you ignore a backward pitch, the tub will not drain all of its water effectively.

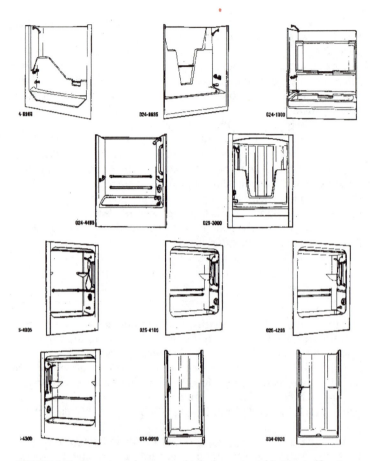

FIGURE 6.2 One-piece showers and tub-shower combinations. *(Courtesy of Eljer)*

If the tub is way out of level, contact the carpenters. It should be their responsibility to provide the plumber with a level floor to install the unit on. If the tub is only slightly out of plumb, you can correct the problem with sand. Place sand under the tub at the back of the bathing unit. The sand will raise the rear of the tub and bring it back into a level position. Most sub-floors are relatively level and I doubt you will have much problem with the front-to-back leveling of the tub.

Once you are level on the flood rim, put your level on the vertical edge of the unit. The odds are high you will have to make minor adjustments for this part of the installation. The adjustment is simple; move the base of the unit in or out to reach a level point. If the tub is difficult to move, be careful. Fiberglass and acrylic units can be broken in the blink of an eye. Don't force the unit. If the opening is properly sized, the resistance is due to a crooked approach. Make sure the unit is straight, and push or pull it as necessary. After a few attempts, you will reach the point where the unit is level on the vertical rise.

When the unit is level on the vertical rise, double check your horizontal leveling point on the flood level rim. If both portions of the unit are level, you are ready to secure it to the stud walls. To secure the unit, place nails or screws through the nailing flange and into the studs. Most plumbers use roofing nails for this part of the job. Place the nail or screw in the center of the nailing flange and insert it into the stud. The fasteners should be placed liberally around the nailing flange.

Once the nails or screws are in, the bathing unit is set. All that is left is to connect the drain. For tub/shower combos, you will use a tub waste and overflow as a drain. Personally, I like the type that is made of schedule 40 plastic that glues together. Some plumbers prefer brass drains with slip-nuts. If you use a tub waste with slip-nuts, you must provide a panel in the wall to allow access to the piping. With glue-together joints, the access panel is not required. For the few of you that will be using drum traps, you must leave access to the clean-out on the trap.

Tub wastes go together easily, but it helps to have a second set of hands available. The first step is installing the tub shoe, also known as a drain ell. The tub shoe is the part with female threads that sits under the drain in the bottom of the tub. The shoe has a flange that presses against the bottom of the tub. There will be a rubber gasket in your tub-waste kit that fits between the flange and the bottom of the tub. The drain for the tub is usually chrome, and it has male threads on the end of it.

Apply plumber's putty around the flange of the chrome drain. Take the putty in your hand and roll it into a line. Wrap the line around the drain's flange. Apply pipe dope or teflon tape to the threads of the drain. If you use teflon tape, apply it so it will be tightened as the drain is turned clockwise. Hold the tub shoe under the drain with the gasket in place. Put the chrome drain through the drain hole and screw it clockwise into the tub shoe. When the connection is snug, point the end of the shoe's pipe towards the head of the tub.

Next, place the tee included with your tub waste on the end of the shoe's pipe. From this tee, you will have a pipe that rises vertically to the overflow. The pipe should rise vertically in a nearly level manner. If the vertical pipe is cocked, cut the shoe's pipe to allow the tee to fit closer to the tub. This should allow the overflow pipe to rise upward without being cocked. In very rare circumstances, you might have to make the shoe's pipe longer to eliminate the cocked overflow pipe. In all my years as a plumber, I cannot remember a situation when this was the case. Normally, the shoe's pipe will need to be shortened, if it needs adjustment at all.

Once you have the tee set, you are ready to install the overflow fittings and pipe. The overflow portion of the tub drainage can come in two styles. The overflow fitting may be attached to the overflow pipe, or it may be a separate fitting. In schedule 40 tub wastes, the overflow fitting will not usually have the overflow pipe attached. Brass tub wastes will have the overflow pipe attached to the fitting. Hold the overflow fitting in place at the overflow hole in the tub. Take a measurement that will allow you to cut the overflow pipe to the correct size.

Cut the overflow pipe and install it in the tee. If you are working with separate pieces, install the overflow fitting on the pipe and point it into the overflow hole of the tub. There will be a thick sponge gasket that goes between the flange on the overflow fitting and the back of the tub. Install this gasket on the overflow fitting.

The next connection will be between the tee and the tub's trap. With a brass tub waste, you have two options. You can screw an 1½ inch female adapter onto the external threads of the tee, or you can screw the supplied tailpiece into the internal threads. Apply pipe dope or teflon tape to the threads of the connection you make. When you are using plastic drainage pipe, I recommend using a female adapter for the connection. If you use a tailpiece connection, you will have to have the tailpiece connected to the trap with a trap adapter and a slip-nut. This leaves the possibility of a leak in a concealed location.

By using the female adapter, you have a glued joint between the trap and the tee. Either method is acceptable, but you will need an access panel if you opt for the slip-nut connection. When all of the other connections are made, tighten the tub drain into the tub shoe. Since most tub drains are equipped with cross-bar strainers, two screwdrivers are excellent tools for this task. Cross the screwdriver between the strainer and turn the drain clockwise until the putty under the rim spreads out evenly. Once the tub waste is connected to the trap, you are ready to install the trim pieces on the interior of the tub.

The trim-out parts will vary with various tub wastes. Most glue-together wastes will use a twist-and-turn stopper, or a push-up stopper. When these stoppers are used, the overflow plate is generally a solid chrome plate with 1 or 2 screws attaching it to the overflow flange. There will be an opening between the surface of the tub and the face-plate to allow excess water to escape down the overflow pipe. When you install the face-plate, the screws should be tightened until they compress the sponge gasket on the back side of the tub.

Installing the drain stopper for these units is a matter of screwing it into the tub shoe. There is no need for pipe sealant on the threads of the stopper. All you do is screw it into the shoe. With the push-type stopper, it is pushed once to depress it and hold water in the tub. It is depressed a second time to release the water in the tub. With the lift-and-turn style stopper, you lift the stopper, turn it, and allow it to seat in the drain. This procedure holds water in the tub. Lifting and turning the stopper again will allow the tub's water to drain. I think these two types of tub wastes are the best available.

If you choose to use a trip-lever waste, the installation will be a little different. For these units, the opening and closing of the tub's drain is controlled with a lever. The lever is mounted in the faceplate of the overflow fitting. There will be two screws holding this plate in place. When you unpack your tub waste, you will see the chrome

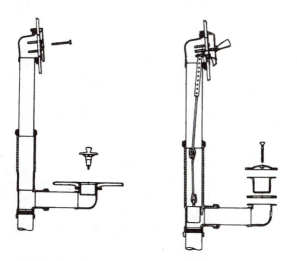

FIGURE 6.3 Typical tub wastes. (*Courtesy of Delta Faucet Company*)

face-plate with a long rod extending from it. These rods are adjustable for tubs with different heights. Most tub wastes will be pre-set for standard tubs.

If your tub is extra deep or unusually shallow, you may have to adjust the rod attached to the overflow's faceplate. This is done by loosening the retainer nut and turning the rod. Turn the rod counterclockwise to lengthen it and clockwise to shorten it. To install the faceplate on a trip-lever waste, feed the rod into the overflow pipe. When the rod is all the way into the pipe, screw the faceplate to the overflow flange.

Depending upon the style of trip-lever waste you are using, you may have either of two types of drains. The first time will be a cross-bar strainer with a mesh strainer that attaches to the tub shoe. The other type will have a stopper plug that has a crooked piece of flat metal attached to it. With the mesh-strainer type, all you have to do

fastfacts

➤ *One-piece, tub-shower units are the most common in new construction.*

➤ *The rough opening for the bathing unit should be sized so that it is at least one-quarter of an inch wider than the bathing unit.*

➤ *The bathing unit should be level when it is installed.*

➤ *If there is a slight forward pitch, towards the drain, this is acceptable.*

➤ *A backwards pitch is not acceptable.*

➤ *Bathing units must be leveled both horizontally and vertically.*

➤ *Never force a bathing unit into a rough opening. If the opening is too small, have the general contractor or carpenters enlarge it.*

➤ *A tub waste that connects with mechanical joints requires an access panel.*

➤ *A tub waste with solvent-welded or soldered joints does not require an access panel.*

is screw the mesh strainer to the tub shoe. For the other type, feed the flat metal section into the tub drain. When it is all the way into the shoe's drain, the stopper will be sitting in the proper position. If either of these types of tub wastes fail to hold water in the tub, adjust the length of the rod attached to the lever.

BATHTUBS WITHOUT A SHOWER SURROUND

The waste and water connections on these tubs will go the same as explained above. The only major difference is how the tub is installed. This type of tub does not have the same type of nailing flange to secure to the stud walls. Standard bathtubs require the installation of a support system. The support is usually a piece of wood, like a 2 x 4 stud.

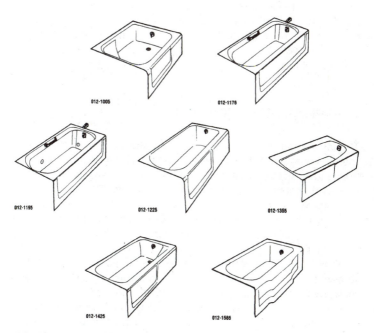

FIGURE 6.4 Typical bathtubs that are not equipped with shower surrounds. *(Courtesy of Eljer)*

The support should be about 58 inches long. It will be nailed to the studs so that the inside edge of the tub can sit on the support. To determine the proper height for the support you will have to measure your tub. With the tub sitting on the sub-floor, hold it up until it is level from the apron to the inside edge. This is the width of the tub, not the length. Measure from the bottom, inside edge of the tub to the sub-floor. Most tubs will measure between 13 and 14 inches. The height of your measurement will determine the top of your support.

If there are 13½ inches between the floor and the edge of the tub, the support will be installed so that the top of it is thirteen and a half inches above the floor. This way, when the tub is resting on the support, it will be level from the back wall to the apron. When you nail the support in place, be sure it is level along its length. When the tub is set in place it should be level from front to back and side to side.

Once the tub is sitting on the support, check to be sure the tub is level. Cast iron tubs are heavy enough to hold themselves in place. Steel tubs should be secured to the studs at the top of their drywall flange. Some steel tubs will be drilled with holes around the flange to allow you to nail or screw the tub to the wall. If there are no holes, nail roofing nails into the studs, just above the flange. The big head on the roofing nail will come down over the flange and secure the tub. Fiberglass tubs will be attached to the studs in the same manner as steel tubs.

ONE-PIECE SHOWER STALLS

The installation of a fiberglass or acrylic shower stall will go much the same way as a tub/shower unit. Set the unit in place and level it horizontally and vertically. Cut your holes for the faucets in the same way as described for tub/shower combinations. The faucet height for a shower is generally four feet above the sub-floor. The shower head height is 6½ feet above the sub-floor. The big difference between installing a shower stall and a tub/shower unit is the drainage piping.

Shower stalls do not use tub wastes; they use a shower drain. The shower drain is easy to install, but the job goes better if you have some help. Put a ring of putty around the bottom or the shower drain's rim. Push the drain through the drain opening in the shower base. From below the shower, slide the fiber gasket onto the threaded portion of the drain. Next, install the large nut, that came with the drain, onto the threads. Turn the nut clockwise until it is tight. This is

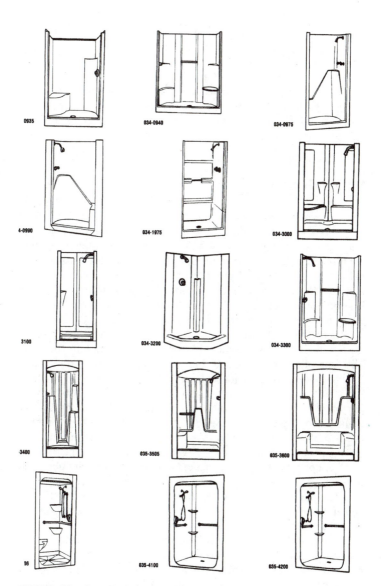

FIGURE 6.5 One-piece shower stalls. *(Courtesy of Eljer)*

where it is advantageous to have help. When possible, have someone in the shower, holding the drain, so it will not turn as you tighten the nut.

Once the drain is installed in the shower, you can connect it to the shower's trap. Most shower drains will be made from the same plastic pipe that you are using for your DWV system. The pipe size should be 2 inch. All you have to do to make the connection is install a piece of pipe between the trap and the drain.

SHOWER BASES

There may be a time when you will use only a shower base, instead of a one-piece stall. If you plan to tile the shower walls, you will use only a base. Shower bases are a little different to install. Once you have the base sitting in the proper position and level, use roofing nails to hold it in place. Don't drive the nails into the base; drive the nails into the studs just above the base. The head on the roofing nail is large enough to come down over the base and hold it in place when the nail is driven into the stud wall.

The drain of a shower base could be different from the type used on a stall. Some bases come with the drain molded into the base. This type of drain has a metal collar extending down from the base. You have two options for connecting your drain pipe to these shower drains. The most common method employs the use of a rubber gasket. The gasket is put into the metal collar of the shower drain. Then, the pipe is pressed up from below and into the rubber gasket. Another approach is to have the pipe protruding into the collar to allow the gasket to be driven down onto the pipe.

In either case, the rubber gasket will need to be well lubricated to receive the pipe. You will often have to drive the gasket onto the pipe using a hammer. Installing these rubber gaskets is easier when you have help. It can be difficult to get the pipe into the gasket. With a second set of hands, the job will go smoother.

The second option for this type of drain is to seal the shower drain with molten lead. This is not a job for a homeowner or inexperienced plumber. The molten lead is extremely hot and can cause severe damage if it comes into contact with the human body. Due to the risk of working with hot lead, I will not elaborate on the methods used in working with lead. Without professional training and experience, you should not work with hot lead.

TOILETS

Installing a toilet is not difficult. The job can be done by a professional in fifteen minutes if the rough-in work was done correctly. When you prepare for the toilet installation, you should see a pipe for the water supply and a closet flange. These items were installed during the rough-in. The first step is installing the closet bolts. These are the bolts that slide under the rim of the flange and protrude up through the holes in the base of the toilet. Closet bolts do not come with new toilets; you must purchase them separately. Place the wide, flat base of the bolts into the holes of the flange, on each side of the drain. Slide the bolts in the groove until they are an equal distance from the back wall and in line with the center of the drain.

Take a wax ring and place it on the flange. The wax ring is another item you will have to purchase separately. The ring should go directly over the drain, but it must not block the opening of the drain. Wax rings should be warm when they are installed. If your job

081-0215

081-0415

081-1475

081-1555

081-4805

081-7080

081-2400

FIGURE 6.6 One-piece toilets. *(Courtesy of Eljer)*

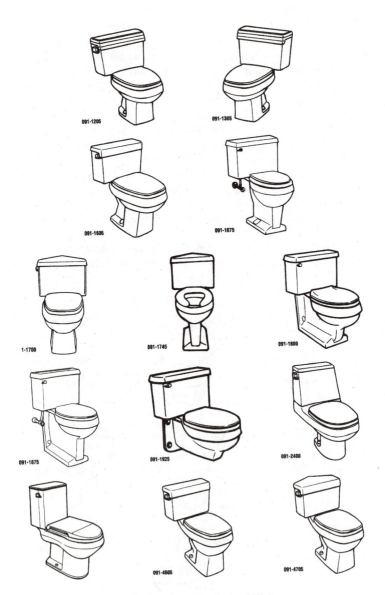

FIGURE 6.7 A variety of toilet types. *(Courtesy of Eljer)*

is unheated and cold, warm the wax ring with the heater in your vehicle before installing it. The next step is to place the toilet bowl over the bolts and onto the wax ring. Press down firmly on the bowl to compress the wax ring. When the bowl is sitting with its base on the floor, you must align the bowl with the back wall. Measure from the back wall to the holes in the bowl that will accept the bolts from the toilet seat. When these holes are an equal distance from the back wall, install the nuts on the closet bolts. With some brands of toilets, you will install a plastic disc on the bolts before putting the nuts on. These disks allow you to snap cover caps in place later.

Turn the nuts clockwise with your hand until they are snug. Use an adjustable wrench to tighten the nuts, but don't get carried away. If you tighten these bolts too much, the toilet will crack. Alternate from one nut to the other as you tighten them to avoid stress on either one of the nuts. When the toilet bowl will not easily twist to either side, the nuts are tight enough.

Normally, once the nuts are tight, the bolts will need to be cut off. Cut the bolts with a hacksaw. When the bolts are short enough, install the cover caps to hide the bolts and nuts. If you had to install plastic disks on the bolts, the cover caps should snap into place on the disks. If you did not have any disks, put putty in the cover caps and press them onto the nuts. The putty will hold the caps in place.

If you are installing a one-piece toilet, you are ready to connect the water supply. If you are installing a standard, two-piece toilet, your next step will be the installation of the tank. All the parts needed to install the tank should have been provided with the tank. Take the large sponge or rubber gasket and place it over the threaded portion of the flush valve on the bottom of the tank. The bolts that hold the tank onto the bowl are called tank-to-bowl bolts. Normally, there will be two tank-to-bowl bolts, but some toilets require three.

The tank-to-bowl bolts will be packaged with the bolts, nuts, metal washers, and rubber washers. First, slide the rubber washers onto the bolts until they rest against the head of the bolt. Set the toilet tank onto the bowl, with the flush-valve gasket sitting on the flush hole of the tank. It helps if you have someone to hold the tank while you connect it to the bowl. If you are working alone, be careful not to let the tank fall. The china will break if it hits a hard surface. When the tank is in place, insert the tank-to-bowl gaskets into the holes at the base of the tank.

The bolts will go through the holes in the tank and bowl. Place the metal washers on the bolts and follow them with the nuts. Run these nuts up hand-tight. Gradually tighten the nuts. Alternate between the nuts during tightening to avoid stress on any single area. Don't over-tighten these nuts or the tank will crack from the pressure. When the nuts are reasonably tight, move on to the next step of the installation. After the installation is complete, the nuts can be tightened if they leak.

All that is left is the seat and the water connection. To install the toilet seat, put it in place, with the seat bolts going through the holes behind the seat. Install the nuts on the bolts, and the seat is installed. Seats do not come with new toilets; they must be purchased separately.

For the water connection, make sure the water is turned off to the supply pipe for the toilet. At this stage of plumbing, the main cut-off valve is usually the only way to cut the water off to the supply pipe. Once the water is off, cut the supply pipe to install a cut-off on it. If your supply pipe is coming out of the wall, you will use an angle stop. If the supply is coming out of the floor, you will use a straight stop. Attaching the cut-off will be done using the same methods used in installing the pipe. Before you install the stop, place an escutcheon on the supply pipe. When copper pipe is used for the supply, it is not uncommon to use a compression stop. Using compression stops eliminates the need to solder the stops on the supply pipes. This is advantageous in keeping an open flame away from finished walls and cabinets.

To install a compression stop, slide the large compression nut onto the pipe. Next, slide the large compression sleeve on the pipe.

fastfacts

➤ *Wax rings should be warm prior to installation.*

➤ *Toilet bowls and tanks should be aligned evenly with the wall behind them.*

➤ *Don't overtighten bolts on china fixtures.*

➤ *When tightening bolts for china fixtures, alternate the tightening process to avoid too much stress on one point of the fixture.*

➤ *When bending metallic supply tubes, be careful to avoid crimping them.*

Then, put the stop on the pipe. Push the sleeve up against the stop and screw the compression nut onto the stop. Hold the stop with an adjustable wrench as you tighten the nut with another wrench. There is no need to apply pipe dope to the threads of a compression fitting. The joint is made when the compression sleeve is compressed and held in place by the compression nut. When I use copper pipe, I use compression stops.

After the stop is installed, you will install the closet supply. This supply tube will have a flat head with a washer on the head. Most modern, metallic closet supplies come with a plastic washer built onto the supply. The stop and closet supply do not come with the new toilet. They must be purchased separately. You will have to cut the supply tube to the desired length. Metallic supply tubes are best cut with roller-cutters.

Hold the head of the closet supply to the inlet of the ballcock. Bend the tube as needed to make it line up with the stop. When bending metallic supplies, be careful not to crimp the tubing. You may want to buy a spring bender to bend your tubing. Professional plumbers can bend the tubing with their hands, but inexperienced people often crimp the tubing. Spring benders reduce the risk of damaging the tubing during the bending process. When the supply is in the right shape, cut it to the proper length.

Slide the ballcock nut onto the supply tube with the threads pointing up, towards the threads of the ballcock inlet. Slide the small compression nut and sleeve onto the supply tube, with the threads facing the stop. You do not have to apply pipe dope to any of these threads. Hold the closet supply against the ballcock inlet, and tighten the ballcock nut onto the threads. When it is hand-tight, put the supply tube into the stop and tighten the compression nut. When everything is properly aligned, tighten both connections with an adjustable wrench. At this point, you have completed the toilet installation.

LAVATORIES

While there are many types of lavatories, the plumbing for the various types is essentially the same. Most professional plumbers trim out the lavatory bowl before installing it. This saves time and the working conditions are better. There are two primary types of lavatory faucets. The first type has both handles and the spout built into a single unit. The second type has the three pieces as separate units that are installed independently and then connected together. One-piece faucets are the most common.

FIGURE 6.8 Wall-hung lavatory with legs. *(Courtesy of Eljer)*

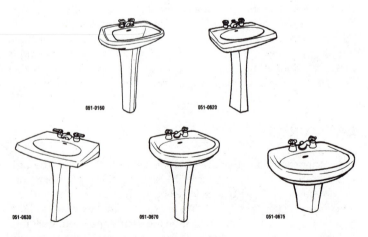

051-0160

051-0620

051-0630

051-0670

051-0675

FIGURE 6.9 Pedestal lavatories. *(Courtesy of Eljer)*

052-0298

052-0301

052-0308

052-0358

052-0368

052-0378

052-0389

052-7048

053-0364

053-0374

053-0384

056-1000

FIGURE 6.10 Self-rimming lavatories. *(Courtesy of Eljer)*

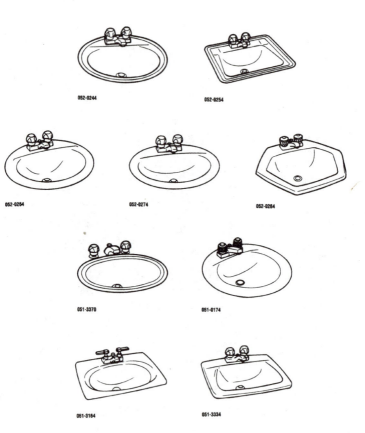

FIGURE 6.11   Self-rimming lavatories. *(Courtesy of Eljer)*

One-Piece Lavatory Faucets

These faucets are the easiest type to install. Take your lavatory bowl and set it on a counter or on the floor. Your faucet will have a base with two threaded inlets extending below the base. If your faucet has a gasket to fit the base of the faucet, place it over the threaded inlets and against the faucet base. If you don't have a gasket, make one from plumber's putty. Roll the putty into a long string and wrap it around the edges of the faucet base. Place the faucet on the lavatory with the inlets going through the holes in the bowl. There will be three holes in the bowl, the center hole will be used later.

Press the faucets onto the bowl to compress the putty, if your faucet did not come with a gasket. Take the ridged washers and place them on the threaded inlets from below the lavatory. Follow the washers with the mounting nuts. Tighten these nuts until the faucet is held firmly in place. If you plan to install faucets that do not have a common body, the procedure is different.

Three-Piece Faucets

Three-piece faucets require you to install each handle and the spout separately. After they are all mounted, you connect them with small tubing. There are many variations of three-piece faucets. Most of them require you to place a nut and metal washer on the body of each unit. Then, you push the units up, one at a time, through the holes in the lavatory and screw a mounting flange onto the body of the faucet. The flange is followed by a collar and the handle for the faucet. The spout will usually be held in place by a mounting nut below the bowl.

The spout will have an inlet on each side for hot and cold water. You will connect a tube from each faucet body to the spout. This tubing is very sensitive and crimps easily. This type of faucet is more time consuming to install than a single-body faucet. In most cases, the tubing connection is made with compression fittings. Some models have the tubing soldered to the spout. Since there are so many possibilities, refer to your installation instructions for exact details on these faucets.

The Pop-Up Assembly

Now you are ready to install the drain. The drain for a lavatory is frequently called a pop-up. When you look at the pop-up, there will be a number of pieces. The first thing you should do is unscrew the finished, trim piece of the pop-up. Place a ring of putty around the bottom of this piece. The body of the pop-up will have threads running down its length for a few inches. There should be a large nut on these threads. Spin it down to the end of the threads, near the middle of the pop-up body. Place the metal washer over these threads. Then, slide the beveled rubber washer onto the threads. The bevel should be pointing up, towards the bottom of the lavatory bowl. Apply pipe dope to the first few threads on the end of the body. Put the threads through the drain hole in the bowl, from underneath the bowl. Screw the finished part of the drain onto the threads. Don't use a wrench, a hand-tight installation is all that is required.

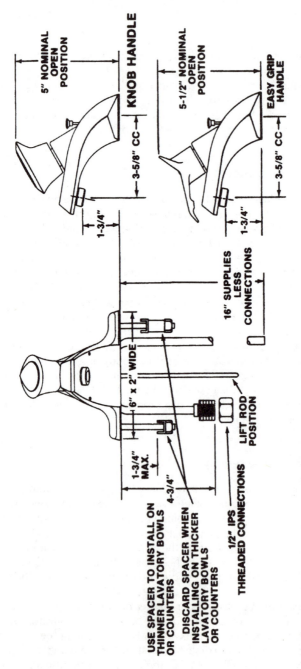

KNOB HANDLE

5" NOMINAL OPEN POSITION

3-5/8" CC

1-3/4"

EASY GRIP HANDLE

5-1/2" NOMINAL OPEN POSITION

3-5/8" CC

1-3/4"

16" SUPPLIES LESS CONNECTIONS

6" x 2" WIDE

LIFT ROD POSITION

1-3/4" MAX.

4-3/4"

USE SPACER TO INSTALL ON THINNER LAVATORY BOWLS OR COUNTERS

DISCARD SPACER WHEN INSTALLING ON THICKER LAVATORY BOWLS OR COUNTERS

1/2" IPS THREADED CONNECTIONS

FIGURE 6.12 Single-handle, single-body lavatory faucet. (*Courtesy of Moen, Inc.*)

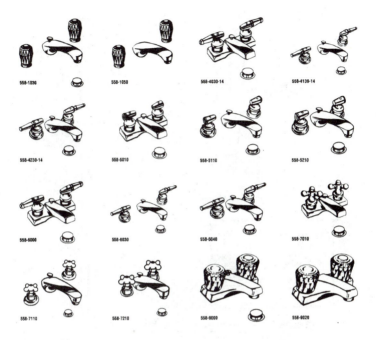

FIGURE 6.13 Three-piece lavatory faucets. *(Courtesy of Eljer)*

Pull the body down to seat the trim piece and putty against the bowl. Slide the beveled washer up to the drain hole. Tighten the large nut until the metal washer comes into firm contact with the beveled washer. Check the alignment of the finished drain and complete the tightening process. Tighten the nut until the putty spreads out and the beveled washer compresses. There will be a tailpiece with the pop-up assembly. The tailpiece will be about four inches long and will have thin threads on one end. Apply pipe dope to the threads and screw the tailpiece into the bottom of the pop-up body. A hand-tight installation is all that is required.

There should be a rod extending from the back of the pop-up body. If this rod has not been installed, install it now. Place the pop-up plug in the drain from inside the lavatory bowl. The pop-up rod will have a nylon ball on the end of it. If there is a knurled nut on the threads where the rod will go, remove it. Slide the nut down the rod

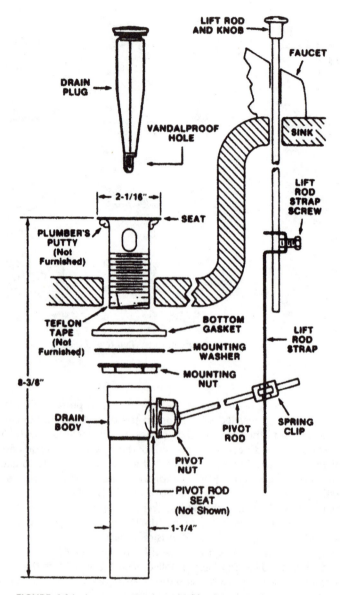

FIGURE 6.14 Lavatory pop-up assembly. *(Moen, Inc.)*

in a way that it can screw onto the drain assembly and hold the rod in place. Insert the rod into the assembly and tighten the knurled nut. You may have to use a pair of pliers to make a tight connection, but the fitting usually will not leak if it is hand-tight.

The last step in the pop-up installation is the lift-rod. Take the round rod and screw the head on it. Slide the rod through the hole in the faucet, between the handles. Take the perforated metal strip and allow the rod to penetrate the hole in the top of the strip, where the setscrew is. Take the thin metal clip and slide one end of it over the pop-up rod. The clip should remain near the end of this rod. Now, place the perforated metal strip so that one of its holes slides over the pop-up rod. Bend and place the remaining end of the clip on the pop-up rod to hold the metal strip in place. Tighten the setscrew to secure the lift rod to the metal strip. As you pull up on the lift-rod, the pop-up plug should go down to seal the drain. When you push the rod back down, the pop-up plug should come up to allow the drain to open. You may have to try various settings to obtain the proper adjustment between these connections.

Supply Tubes

Before you set the lavatory, attach the lavatory supplies to the faucets. Place the supplied nuts on the supply tubes. Put the head of each supply tube into an inlet of the faucet body. Screw the supply-tube nut to the threaded portion of the supply inlet. If you do not install the supply tubes now, you will need a basin wrench to tighten the nuts after the lavatory is set.

Wall-Hung Lavatories

With all of this done, you are ready to install the lavatory bowl. Each type of lavatory will install a little differently. For a wall-hung lavatory, the first step is installing the wall bracket. The bracket is screwed to the wall. There should be a piece of wood backing in the wall that was installed during the rough-in. Refer to your rough-in book for the proper height of the bracket. When the bracket is level and secure, place the lavatory on the bracket. Press down on the back rim of the bowl to seat it on the bracket. Some models will have flanges and holes below the lavatory to allow the installation of extra screws. These screws are a safety precaution to prevent the lavatory from being knocked off the bracket. That is all there is to installing a wall-hung lavatory.

Rimmed lavatories are not used much anymore. If you are using a rimmed lavatory, the first step is cutting a hole in the countertop. The lavatory should have a template supplied with it to show you how to cut the hole. Once the hole is cut, place the metal ring in the hole. To mount the bowl, push it up from below the counter, until it touches the ring. There will be sink clips provided to mount the sink. These clips attach to the ring and hold pressure on the bowl as they are tightened. There will be instructions with the clips on how to use them.

Self-rimming lavatories are still used. For these units, use the supplied template to cut a hole in the countertop. Apply a caulking sealant around the edge of the hole and set the bowl into the hole. The weight of the bowl and the plumbing connections are all that hold this type of lavatory in place.

When using a top with the bowl molded into it, all you have to do is place the top on the cabinet. The lavatory bowl is an integral part of the countertop and requires no additional installation.

Pedestal lavatories are the most difficult of all for the average person to install. The bowl is mounted to a wall bracket, just like a wall-hung lavatory. The pedestal is then placed under the bowl to hide piping and to help support the bowl. The complicated part of a pedestal sink is making the trap and supply connections. There is very little room to work with and a minor miscalculation will ruin the effect of the pedestal. Some pedestals have a hole in the base to allow you to screw them to the floor. Others depend on the weight of the bowl to hold the pedestal in place. With any of these specialty fixtures, refer to your installation instructions and rough-in book.

The trap will be attached to the trap arm and the lavatory's tailpiece. How these connections are made will depend on the type of trap you use. I will assume you are using a schedule 40 plastic drainage system. Your trap can be glued to the trap arm, if you use a schedule 40 plastic trap. If you prefer to use a chrome trap, you will need to make the trap-arm connection with a trap adapter. The trap adapter looks like a male adapter with a slip-nut and washer on it. The trap adapter is glued to the trap arm and allows the connection between the adapter and the trap to be made with a slip-nut and washer.

The connection between the trap and the tailpiece will always be made with a slip-nut and washer. A lavatory tailpiece has a diameter of 1¼ inches. If your trap is designed for an 1½ inch tailpiece, you can still use it. To modify the trap, all you need is a reducing washer for the slip-nut. These reducing washers convert the 1½ inch trap opening to an 1¼ inch opening.

To make the slip-nut connection, slide the nut up onto the tailpiece. Slide the washer onto the tailpiece and place the tailpiece in the trap. Slide the washer and nut down to the trap and tighten the nut. When the washer is compressed, the joint is made. With the use of trap adapters, you can use any standard trap to make your connection. If your trap is too low to connect to the tailpiece, you can use a tailpiece extension. These handy items come in various lengths and allow you to extend the length of the tailpiece. They connect to the tailpiece with a slip-nut and washer.

Make sure the water to the water pipes is turned off. Follow the instructions given under the toilet section to install your cut-off stops. Use the same instructions given for the closet supply to install the lavatory supplies. The supply tubes will have a different type of head, but the installation methods are about the same. The lavatory supply will have a head tapered to fit into the inlets of the faucet body. There will be a nut that slides up the supply tube and screws onto the threaded portion of the inlets. These nuts hold the supply tubes in place. You do not need to apply pipe dope to the inlet connections for the supply tubes. Also, slip-nut connections do not require the use of pipe dope.

THE KITCHEN SINK

As was the case with lavatories, kitchen sinks can be trimmed out before they are placed in the countertop. The faucets for kitchen sinks are usually of the single-body type. They are installed with the same procedures used to install lavatory faucets. If your faucet has a spray-hose attachment, the hose will connect to a tapped opening in the center of the faucet's base. To install the sprayer, start with the housing for the unit.

The housing will be inserted in the hole next to the faucet base. Some faucets are designed for the hose to retract directly through the faucet base, but most have a separate housing for the hose. Once you have put the housing in the hole, screw the mounting nut on the threads from below the sink. Feed the hose through the housing, with the brass threads going to the base of the faucet. Apply pipe dope to the threads and screw the hose connection into the female threads on the base of the faucet. If the spray head is not already attached, screw it on the end of the hose above the sink.

When you are not mounting a garbage disposer to the sink, a basket strainer will be used as the drain assembly. There are two

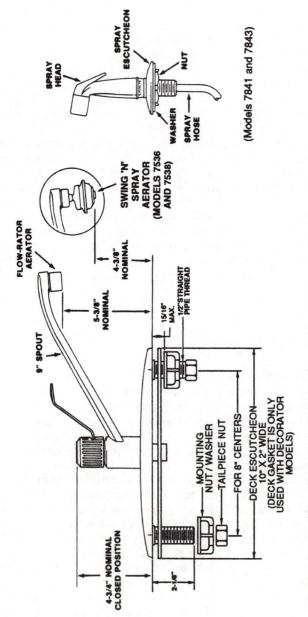

SPRAY
HEAD

SPRAY
ESCUTCHEON

NUT

WASHER

SPRAY
HOSE

(Models 7841 and 7843)

SWING 'N'
SPRAY
AERATOR
(MODELS 7536
AND 7538)

FLOW-RATOR
AERATOR

5-3/8"
NOMINAL

4-3/8"
NOMINAL

9" SPOUT

15/16"
MAX.

1/2" STRAIGHT
PIPE THREAD

MOUNTING
NUT / WASHER

TAILPIECE NUT

FOR 8" CENTERS

DECK ESCUTCHEON
10" X 2" WIDE
(DECK GASKET IS ONLY
USED WITH DECORATOR
MODELS)

4-3/4" NOMINAL
CLOSED POSITION

2-1/8"

FIGURE 6.15 Single-handle, single-body kitchen faucet. (Courtesy Moen, Inc.)

common types of basket strainers. The first uses a large nut to secure the strainer to the sink. The second uses a flange with threaded rods to secure the drain; this is the easiest type to work with.

With either type, take the drain apart to install it. Apply putty under the rim of the finished piece of the drain and insert it in the drain hole. Press the drain down to spread the putty. If you are using the type with a large nut, slip the fiber washer over the threads of the drain, from below the sink. Follow the washer with the big nut. Tighten the nut until the putty is spread out and the drain is tight. If you have trouble with the drain turning as you attempt to tighten the nut, don't be surprised. If this happens, cross two screwdrivers through the bar grids in the drain. This will allow someone to hold the drain in place while you tighten the nut.

When you use the type of drain with a flange, you don't have to worry about the drain turning during the tightening. For this type, the flange is placed over the threaded part of the drain and the threaded rods are tightened. As you tighten the rods, the flange assembly puts pressure on the drain to force a good seal. At this point, you are ready to install the sink in the countertop.

Use the template supplied with your sink to cut a hole in the countertop. Apply a caulking sealant to the top edge of the hole and set the sink in place. If you are using a cast iron sink, the weight of the sink will hold it in place. If you are using a stainless steel sink, you will have to secure the sink with clips. The clips will come with the sink; refer to their directions for installation.

In most instances, the clips will slide into a channel on the bottom of the sink. Then, the shaft that the metal clip is attached to will be turned clockwise. As you tighten the shaft, the clip bites into the bottom of the countertop. The pressure from the clip pressing against the countertop pulls the sink down tight.You will use the same steps to connect the water supplies to the kitchen faucet that you employed for the lavatory faucet.

Most basket strainers will come with a flanged tailpiece. Unlike the lavatory tailpiece, these tailpieces do not screw into the drain. A tailpiece washer is placed on top of the tailpiece's flange. A slip-nut is slid up the tailpiece from the bottom. The tailpiece is positioned under the drain and the slip-nut is tightened onto the drain's threads. Flanged tailpieces come in different lengths to adapt to your drainage system. Tailpiece extensions can be added if needed to lengthen the tailpiece.

On a single-bowl sink, the remainder of the connection between the trap arm and the drain is the same as described for a lavatory.

The only difference is the size of the drainage. Kitchen sinks use a 1½ inch drain. If you have a sink with more than one bowl, you can use a continuous waste to connect the multiple bowls to a single trap. There are two styles of continuous wastes. The first is an end-outlet and the second is a center-outlet. The end-outlet brings the waste of one bowl to a tee, under the other bowl. A center-outlet brings the waste of both bowls to a tee in the the center, between the two bowls.

The placement of your trap arm will dictate the best type of continuous waste to use. Continuous wastes go

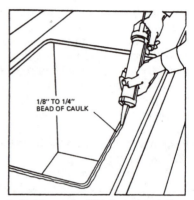

1/8" TO 1/4"
BEAD OF CAULK

Apply a light type of caulking compound completely around all edges of sink opening. Stay approximately 1/4" from cutout. Use only a 1/8" to 1/4" bead of compound. More will make sink difficult to seat.

FIGURE 6.16 Caulking a sink opening for a kitchen sink. *(Courtesy of Moen, Inc.)*

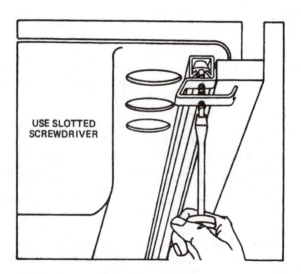

USE SLOTTED
SCREWDRIVER

FIGURE 6.17 Installation of sink clips. *(Courtesy of Moen, Inc.)*

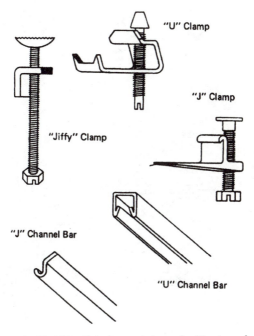

"U" Clamp

"J" Clamp

"Jiffy" Clamp

"J" Channel Bar

"U" Channel Bar

FIGURE 6.18 Sink clips and channels. *(Courtesy of Moen, Inc.)*

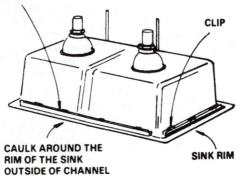

CHANNEL

CLIP

CAULK AROUND THE
RIM OF THE SINK
OUTSIDE OF CHANNEL

SINK RIM

FIGURE 6.19 Caulking locations on a kitchen sink.
(Courtesy of Republic sinks by UNR Home Products)

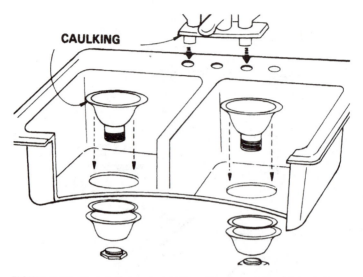

FIGURE 6.20 Basket-strainer installation. *(Courtesy of Republic sinks by UNR Home Products)*

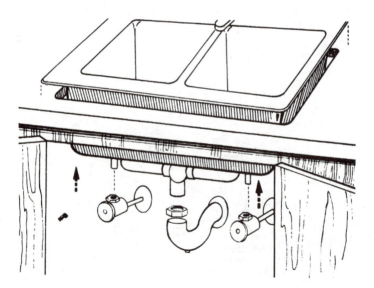

FIGURE 6.21 Detail of a center-outlet continuous waste for a kitchen sink. *(Courtesy of Republic sinks by UNR Home Products)*

together with slip-nuts and washers. Once the bowls are connected by the continuous waste, the connection between the trap arm and the tailpiece of the waste will go together just like the waste for a lavatory.

If you are installing a garbage disposer, it will be attached to the drain of the sink bowl. The disposer eliminates the need for a basket strainer. The finished part of the disposer's drain will be installed like that of a basket strainer. From below, you will slide a flange over the drain. The flange will be held in place with a snap-ring that fits on the drain collar. You will tighten threaded rods to secure the drain to the sink. When the drain is tight, the disposer will be mounted to the drain collar.

The mounting is usually done by holding the disposer on the collar and turning it clockwise. There will be a small plastic elbow to install on the side of the disposer. You will see a hole in the side of the disposer with a metal ring attached to it by screws. Remove the metal ring and slide it over the elbow until it reaches the flange on the elbow. Place the supplied rubber gasket on the face of the elbow. Hold the elbow in place and attach the metal bracket to the side of the disposer with the supplied screws. Your trap or continuous waste will attach to the elbow with slip-nuts and washers.

DISHWASHERS

The dishwasher will have a ⅝ inch drain hose attached to it. This drain should enter the cabinet under the kitchen sink and be attached to an airgap. An airgap is a device that mounts on the sink or countertop and has a wye connection. To install the air gap, remove the chrome cover by pulling on it. Remove the mounting nut from the threads. Push the airgap up through a hole in the sink or counter from below. Install the mounting nut and replace the chrome cover.

There will be two ridged connection points on the airgap. The first will accept the ⅝ inch hose from the dishwasher. Before putting the hose on the air gap, slide a stainless steel clamp over the hose. Next, push the hose onto the ridged connection and tighten the clamp around the hose and insert connection. Attach a piece of ⅞ inch hose to the other connection point, using the same procedure.

When the ⅞ inch hose leaves the airgap, it will attach to a garbage disposer or a wye tailpiece. The wye tailpiece attaches to the sink's tailpiece, like a tailpiece extension. The hose will attach to the wye portion of the extended tailpiece and will be held in

place with a stainless steel clamp. If you are connecting to a disposer, knock out the plug in the disposer first. There will be a thin
metal disk seated in the side of the disposer that must be removed.
This is normally done with a hammer and a screwdriver. Once the
plug is removed, attach the hose to the connection with a stainless
steel clamp.

There is a small box under the dishwasher where the water supply will be connected. Buy a dishwasher ell to make your connection. One end of this elbow has male threads and the other end is a
compression fitting. Apply pipe dope to the threads and screw it
into the dishwasher. Under the kitchen sink, you will need to tap
into the hot water.

The hot water connection can be made with a tee fitting or a special type of cut-off stop. The special stops are designed to feed the
sink and a dishwasher. The connection point for the dishwasher tubing will be the correct size, without any type of adapter. If you install
a tee fitting, the dishwasher must have its own cut-off valve. A stop-
and-waste valve is the type normally used. If you use a ½ inch valve,
you will need a reducing coupling. The tubing going to the dishwasher will have an inside diameter of ⅜ inch. Your reducing coupling will be a half-by-three-eighths reducer. Connect your tubing to
the reducer and run it to the dishwasher elbow. Make your connection at the compression fitting, and you're done.

ICE MAKERS

When you have an ice maker to hook up, use a self-piercing saddle
valve and ¼ inch tubing. The tubing will connect at the back of the
refrigerator with a ¼ inch compression fitting. The tubing may run
to the water supply at the kitchen sink or to another more accessible cold-water pipe. The saddle valve will clamp around the cold-
water pipe and be held in place by two bolts.

Before bolting the saddle on, make sure the rubber gasket that
came with the saddle is in place. It should be in the hollow of the
saddle, where the piercing will take place on the pipe. Secure the
saddle and connect the tubing to it. This connection will be made
with a compression fitting. Turn the handle clockwise until you cannot turn it any further. Then, turn the handle counterclockwise to
open the valve and allow water to run through the tubing.

WATER HEATERS

Water heater inlets may have male or female threads. The inlets will be marked to identify the hot and cold water connections. To connect a water heater, you install the appropriate adapters on the inlets and connect the inlets to the water piping. You must have a gate valve on the incoming water pipe and there should be no valves on the outgoing water pipe. Most places require a vacuum breaker to be installed on the inlet pipe. A temperature and pressure-relief (T&P) valve will be needed.

The T&P valve must be rated to safely operate with the water heater you are installing. The T&P valve will screw into the top or side of the water heater. A discharge pipe must run from the T&P valve to within 6 inches of the floor. Use pipe dope on all the threaded connections made in setting a water heater. This covers the range of fixtures most often installed, so let's move to the next chapter and discuss water heaters in greater depth.

7

WATER HEATERS

Water heaters are only a small portion of a plumbing system, but they are important, and they can be dangerous. Many plumbers take water heaters and the codes pertaining to them lightly. This is a mistake. Water heaters can become lethal if they are not installed properly. I feel very strongly about the safety issues associated with water heaters. My position is based largely on decades of watching plumbers and plumbing contractors take short-cuts with water heaters that could be disastrous.

GENERAL PROVISIONS

Water heaters are sometimes used as a part of a space heating system. When this is the case, the maximum outlet water temperature for the water heater is 140 degrees F., unless a tempering valve is used to maintain an acceptable temperature in the potable water system. It is essential that all potable water in the water heater be maintained throughout the entire system. Potability of water must be maintained at all times. Every water heater is required to be equipped with a drain valve near the bottom of the water heater. This is true, too, for hot water storage tanks. All drain valves must conform to ASSE 1005.

The location of water heaters and hot water storage tanks is important. Code requires both water heaters and hot water storage tanks to be accessible for observations, maintenance, servicing, and replacement. Every water heater is required to bear a label of an approved agency.

The temperature of water delivered from a tankless water heater may not exceed 140 degrees F. when used for domestic purposes. This portion of the code does not supersede the requirement for protective shower valves, as detailed in the code.

All water heaters shall be third-party certified. Water heaters must be installed in accordance with manufacturer's requirements. Oil-fired water heaters must be installation requirements of the *International Mechanical Code.* Electric water heaters must meet the requirements of the *ICC Electrical Code.* Gas-fired water heaters are required to meet the criteria of the *International Fuel Gas Code.*

All storage tanks and water heaters installed for domestic hot water must have the maximum allowable working pressure clearly and indelibly stamped in the metal, or marked on a plate welded thereto, or otherwise permanently attached. All markings of this type must be in accessible positions outside of the tanks. Inspection or reinspection of these markings must be readily possible.

Every hot water supply system is required to be fitted with an automatic temperature control. The control must be capable of being

fastfacts

➤ *Water heaters may be used as part of a space heating system.*

➤ *The location of water heaters is important.*

➤ *Water heaters must be accessible for observations, maintenance, servicing, and replacement.*

➤ *Water temperature from water heaters for domestic use may not exceed 140 degrees F.*

➤ *All water heaters must be third-party certified.*

➤ *All hot water supply systems must be fitted with automatic temperature controls.*

adjusted from a minimum temperature to the highest acceptable temperature setting for the intended temperature operating range.

INSTALLING WATER HEATERS

All water heaters are required to be installed in accordance with the manufacturer's recommendations, and within the confines of the plumbing code. Water heaters that are fueled by gas or oil must conform to both the plumbing code and the mechanical or gas code. Electric water heaters must conform to the requirements of the plumbing code and the provisions of NFPA 70, as listed in the plumbing code.

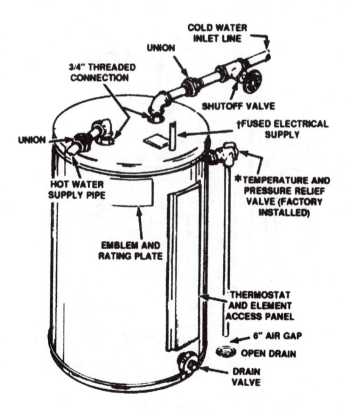

FIGURE 7.1 Electric water heater setup. *(Courtesy of A. O. Smith Water Products Co.)*

The installation of a water heater that has an ignition source when installed in a garage, requires that the water heater be installed on an elevated base that keeps the source of the ignition at least 18 inches above the garage floor.

Rooms are sometimes used as plenums for heating systems. When this is the case, water heaters using solid, liquid, or gas fuel are not allowed to be installed in rooms that contain air handling machinery when the rooms are being used as plenums. Additionally, fuel-fired water heaters are not allowed to be installed in sleeping rooms, bathrooms, or closets that can be accessed from either bathrooms or sleeping rooms, unless the water heater is equipped with a direct-vent system. There is one exception to this rule. If a water heater has a sealed combustion chamber or is directly vented to the outside, said water heater can be installed in a sleeping room, bathroom, or closet that is accessible from either of these rooms.

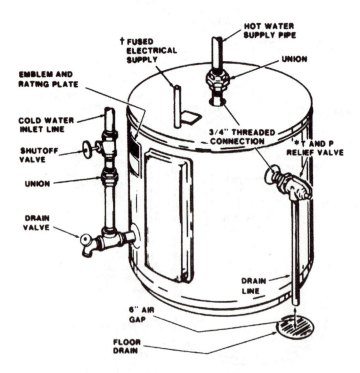

FIGURE 7.2 Electric water heater setup. *(Courtesy of A. O. Smith Water Products Co.)*

When earthquake loads are applicable, water heater supports must be designed and installed for the seismic forces in accordance with the *International Building Code*.

When water heaters are installed in attics, special provisions must be made for the water heaters. For example, an attic that houses a water heater must be provided with an opening and unobstructed passageway large enough to allow for the removal of the water heater. This should be common sense, but it is also part of the plumbing code. There are many measurements that come into play when planning the exit route for an attic water heater. They are as follows:

- *Minimum height: 30 inches*
- *Minimum width: 22 inches*
- *Maximum length: 20 feet*

A continuous solid floor is required in the exit area, and the flooring must be at least 24 inches wide. Another requirement calls for a level service area with minimum dimensions of 30 inches deep and 30 inches wide. This service area must be made in front of the water heater, or where the service area of the water heater is located. A clear access opening with minimum dimensions of 20 inches by 30 inches where the dimensions are large enough to allow removal of the water heater.

Location Considerations for Water Heaters that have an Ignition Source:

✔ If the water heater is being installed in a garage, it must be installed on a platform that is at least 18 inches above the garage floor.

✔ Will the room where the water heater is to be installed be used as a plenum for a heating system? If so, the water heater must not be installed in the room.

✔ Are you about to install a water heater in a sleeping room, bathroom or closet that can be accessed from either bathrooms of sleeping rooms? If so, the water heater must be equipped with a direct-vent system or has a sealed combustion chamber.

✔ Will the water heater be installed in an attic? If so, be sure to comply with the code requirements for accessibility.

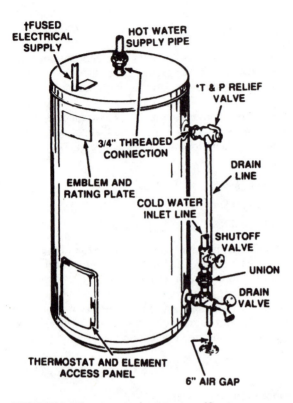

FIGURE 7.3   Electric water heater setup. *(Courtesy of A. O. Smith Water Products Co.)*

MAKING CONNECTIONS

Making connections to water heaters is not difficult, but the manner in which the connections are made must conform to code requirements. The first consideration is the installation of cutoff valves. A cold water branch line from a main water supply to a hot water storage tank or water heater must be provided with a cutoff valve that is accessible on the same floor, located near the equipment, and only serving the hot water storage tank or water heater. The valve used must not interfere with or cause a disruption of the cold water supply to the remainder of the cold water system.

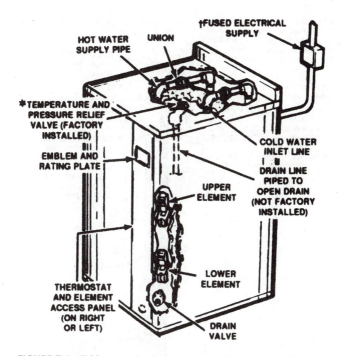

FIGURE 7.4 Tabletop-style of water heater. *(Courtesy of A. O. Smith Water Products Co.)*

Any means of connecting a circulating water heater to a tank must provide for proper circulation of water through the water heater. All piping that is required for the installation of appliances that will draw from the water heater or storage tank must comply with all provisions of the plumbing and mechanical codes.

SAFETY REQUIREMENTS

Safety requirements are essential to comply with when installing or replacing a water heater or hot water storage tank. One major concern is the siphoning of water from a water heater or storage tank. An antisiphon device, of a required type, is required to prevent siphoning. A cold water dip tube with a hole at the top, or a vac-

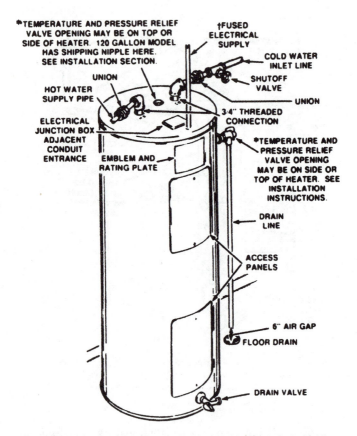

FIGURE 7.5 Electric water heater setup. *(Courtesy of A. O. Smith Water Products Co.)*

uum relief valve installed in the cold water supply line above the top of the water heater or storage tank are acceptable means of protection. Some water heaters and storage tanks receive their incoming water from the bottom of the unit. These types of heater and tanks must be supplied with an approved vacuum relief valve that complies with ANSI Z21.22.

Energy cutoff valves are required on all water heaters that are automatically controlled. The energy cutoff valve is designed to cut off the supply of heat energy to the water tank before the temperature of the water in the tank exceeds 210 degrees F. The installation

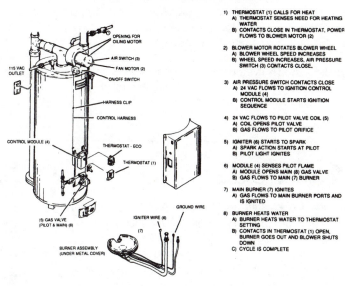

WATER HEATING CYCLE
(GAS AND ELECTRIC POWER ARE ON, "OFF/ON" SWITCH IS ON)

1) THERMOSTAT (1) CALLS FOR HEAT
   A) THERMOSTAT SENSES NEED FOR HEATING WATER
   B) CONTACTS CLOSE IN THERMOSTAT, POWER FLOWS TO BLOWER MOTOR (2)

2) BLOWER MOTOR ROTATES BLOWER WHEEL
   A) BLOWER WHEEL SPEED INCREASES
   B) WHEEL SPEED INCREASES, AIR PRESSURE SWITCH (3) CONTACTS CLOSE.

3) AIR PRESSURE SWITCH CONTACTS CLOSE
   A) 24 VAC FLOWS TO IGNITION CONTROL MODULE (4)
   B) CONTROL MODULE STARTS IGNITION SEQUENCE

4) 24 VAC FLOWS TO PILOT VALVE COIL (5)
   A) COIL OPENS PILOT VALVE
   B) GAS FLOWS TO PILOT ORIFICE

5) IGNITER (6) STARTS TO SPARK
   A) SPARK ACTION STARTS AT PILOT
   B) PILOT LIGHT IGNITES

6) MODULE (4) SENSES PILOT FLAME
   A) MODULE OPENS MAIN (8) GAS VALVE
   B) GAS FLOWS TO MAIN (7) BURNER

7) MAIN BURNER (7) IGNITES
   A) GAS FLOWS TO MAIN BURNER PORTS AND IS IGNITED

8) BURNER HEATS WATER
   A) BURNER HEATS WATER TO THERMOSTAT SETTING
   B) CONTACTS IN THERMOSTAT (1) OPEN, BURNER GOES OUT AND BLOWER SHUTS DOWN
   C) CYCLE IS COMPLETE

FIGURE 7.6 Heating cycle for a water heater. *(Courtesy of A. O. Smith Water Products Co.)*

of an energy cutoff valve does not relieve the need for a temperature-and-pressure relief valve. Both are required.

Every electric water heater must be provided with its own electrical disconnect switch that is in close proximity to the water heater. In the case of gas-fired or oil-fired water heaters, cutoff valves must be installed close to the water heaters to stop the fuel flow when needed.

RELIEF VALVES

Pressure relief valves and temperature relief valves, or a temperature-and-pressure relief valve (the one most commonly used) are required on all water heaters and storage tanks that are operating above atmospheric pressure. The valves used must be approved and conform to ANSI Z21.22 ratings. Relief valves must be of a self-closing (levered) type. In no case shall the relief valve be used as a means of controlling thermal expansion.

Relief valves must be installed in the shell of a water heater tank. Any temperature relief valve must be installed so that it is actuated by the water in the top 6 inches of the tank being served by the valve. When separate tanks are used, the valves must be installed on the tank and there must not be any type of valve installed between the water heater and the storage tank. It is prohibited to install a cutoff valve or check valve between a relief valve and the water heater or tank being serviced by the relief valve. Never omit the installation of required relief valves. The result of doing so can be catastrophic.

fastfacts

➤ *Relief valves must be of a self-closing (levered) type.*

➤ *Relief valves must be installed in the shell of a water heater tank.*

➤ *Temperature relief valves must be installed so that it is actuated by the water in the top 6 inches of the tank being served by the valve.*

➤ *Cut-off valves and check valves must not be installed between a relief valve and the water heater or tank being serviced by the relief valve.*

➤ *Temperature relief valves must have a maximum setting of 210 degrees F.*

➤ *Pressure relief valves must have a maximum setting of 150 PSI or the maximum rating as stated by the manufacturer of the water heater, whichever is less.*

➤ *Discharge tubes from relief valves must not be connected to a sanitary drainage system.*

➤ *Discharge tubes must run full size from the outlet of the relief valve to the discharge point.*

➤ *Traps are prohibited on discharge tubes and pipes.*

➤ *Discharge piping must drain by gravity.*

➤ *Discharge tubing must terminate atmospherically not more than 6 inches above the finished floor.*

➤ *Discharge tubing must not be threaded on the terminal end.*

All relief valves, whether temperature, pressure, or a combination of the two, and all energy cutoff devices must bear a label of an approved agency. The valves and devices must have a temperature setting of not more than 210 degrees F. and a pressure setting that does not exceed the tank or water heater manufacturer's rated working pressure, or 150 PSI, whichever is less. The relieving capacity of each relief valve must equal or exceed the heat input to the water heater or storage tank.

Since relief valves may create a discharge, the disposal of the discharge must be dealt with. In no case is it allowable for the discharge tube from a relief valve to be connected directly to a drainage system. The discharge tube must be provided in a full-size tube or pipe that is the same size as the outlet of the relief valve. You have two choices on the termination point of the discharge tube. It can be piped to the outside of a building, or it can terminate over an indirect waste receptor that is located inside of a building.

When freezing conditions may exist, the discharge tubing or piping for a relief valve must be protected from freezing. This is done by having the discharge tubing discharge through an air gap and into an indirect waste receptor that is located in a heated space. Local regulations may allow some other form of installation, so check your local code requirements.

Any risk of personal injury or property damage must be avoided when piping a discharge tube from a relief valve. The discharge piping must be installed so that it is readily observable by building occupants. Traps on discharge tubes and pipes are prohibited. All discharge piping must drain by gravity. The tubing must terminate atmospherically not more than 6 inches above the floor, and the end of the discharge tubing or piping is not allowed to be threaded.

On occasions when discharge piping is installed so that it leaves the room or enclosure that contains a water heater and relief valve that discharges into an indirect waste receptor, there must be an air gap installed before, or at the point of leaving the room or enclosure. Discharge tubes from relief valves must not discharge into a safety pan. Materials used for discharge piping shall be made to the standards listed in the plumbing code or shall be tested, rated, and approved for such use in accordance with ASME A112.4.1.

SAFETY PANS

Safety pans are required for water heaters and storage tanks that are installed in locations where leakage may cause property damage.

Water heaters and storage tanks shall be placed in safety pans that are constructed of galvanized steel or other approved metal materials. The minimum thickness of the metallic pan shall be 24 gauge. Electric water heaters must be installed in pans when leakage may cause property damage, but the pan may be made of a 24 gauge metal pan or a high-impact plastic pan that has a minimum thickness of 0.0625 inches. All piping from safety pan drains must be made with materials approved by the plumbing code.

Safety pans must have a minimum depth of 1½ inches and be of sufficient size and shape to receive all dripping or condensate from the tank or water heater contained in the pan. A safety pan must drain by an indirect waste. The drainage pipe or tube from the pan must have a minimum diameter of 1 inch or the outlet diameter of the relief valve, whichever is larger.

The drain tube or pipe from a safety pan must run full-size for its entire length and terminate over a suitably located indirect waste receptor or floor drain, or extend to the exterior of the building and terminate not less than 6 inches or more than 24 inches above the adjacent ground level.

Unfired hot water storage tanks must be insulated so that heat loss is limited to a maximum of 15 British thermal units (Btus) per hour, per square foot of external tank surface area. For purposes of determining this heat loss, the design ambient temperature shall not be higher than 65 degrees F.

VENTING WATER HEATERS

The venting of water heaters which require venting is regulated by the plumbing code. All venting materials used must be in compliance with all code requirements. Venting systems might consist of approved chimneys, Type B vents, Type L vents, or plastic pipe. The recommendations of the equipment manufacturer must be observed in selecting the proper venting material and installation procedure.

Vents must be designed and installed to develop a positive flow adequate to convey all products of combustion to the outside atmosphere. Condensing appliances which cool flue gases nearly to the dewpoint within the appliance, resulting in low vent gas temperatures, may use plastic venting materials and vent configurations unsuitable for non-condensing appliances. All unused openings in a venting system must be closed or capped to the satisfaction of the local code enforcement officer.

| Diameter of Connector | | Galvanized Sheet |
Inches	mm	Gauge No.
5 or less	125 or less	28
Over 5 to 9	Over 125 to 225	26
Over 9 to 12	Over 225 to 300	22
Over 12 to 16	Over 300 to 400	20
Over 16	Over 400	16

FIGURE 7.8 Chimney connector made of single-wall pipe and serving low heat appliances must be no less than the gauges listed above. *(Reprinted from the 2000 Uniform Plumbing Code with the permission of the International Association of Plumbing and Mechanical Officials)*

Type B vents are not allowed for use with water heaters which may be converted readily to the use of solid or liquid fuels. Water heaters listed for use with chimneys only may not be vented with Type B vents. Manually operated dampers must not be installed in chimneys, vents, or chimney or vent connectors of fuel burning water heaters. Fixed baffles on the water heater side of draft hoods and draft regulators are not to be considered as dampers.

Vent Connectors

Vent connectors used for gas water heaters with draft hoods may be constructed of noncombustible materials having resistance to corrosion not less than that of galvanized sheet steel and be of a thickness not less than that specified in the code. Or, they may be of Type B or Type L vent material. When single-wall metal vent connectors are used, they must be securely supported with joints fastened with sheet metal screws, rivets, or another approved means. Such connectors must not originate in an unoccupied attic or concealed space and shall not pass through any attic, inside wall, floor, or concealed space, and shall be located in the same room or space as the fuel burning water heater.

Supporting Vent Systems

Combustion products, vents, vent connectors, exhaust ducts from ventilating hoods, chimneys, and chimney connectors must not extend into or through any air duct or plenum, except for when the venting system may pass through a combustion air duct. The base of all vents supported from the ground must rest on a solid masonry or concrete base extending at least 2 inches above adjoining ground level. If the base of a vent is not supported from the ground and is not self-supporting, it must rest on a firm metal or masonry support. All venting systems must be supported adequately for its weight and design. No water heater is allowed to be vented into a fireplace or into a chimney that serves a fireplace.

Vent Offsets

With minor exceptions, gravity vents must extend in a generally vertical direction with offsets not exceeding 45 degrees. These vents are allowed to have one horizontal offset of not more than 60 degrees. All offsets must be supported properly for their weight and must be installed to maintain proper clearance to prevent physical damage and to prevent the separation of joints.

Offsets with angles of more than 45 degrees are considered to be horizontal offsets. Horizontal vent connectors must not be greater than 75 percent of the vertical height of the vent and must comply with all code regulations.

Vent connectors in a gravity-type venting system must have a continuous rise of not less than ¼ inch per foot of developed length. This is measured from the appliance vent collar to the vent. If a single-wall metal vent connector is allowed and installed, it must have a minimum clearance of 6 inches from any combustible material.

Termination

Vents must terminate above the roof surface of the building being served. The pipe must pass through a flashing and terminate in an approved or listed vent cap that is installed in accordance with the manufacturer's recommendations. There is an exception to this. A direct vent or mechanical draft appliance will be acceptable when installed according to its listing and manufacturer's instructions.

Gravity-type venting systems, with the exception of venting systems which are integral parts of a listed water heater, must terminate at least 5 feet above the highest vent collar being served. Type B gas vents with listed vent caps 12 inches in size or small are permitted to be terminated in accordance with the code requirements so long as they are at least 8 feet from a vertical wall or similar obstruction. All other Type B vents must terminate not less than 2 feet above the highest point where they pass through the roof and at least 2 feet higher than any portion of a building within 10 feet.

Type L vents shall not terminate less than 2 feet above the roof through which it passes, nor less than 4 feet from any portion of the building which extends at an angle of more than 45 degrees upward from the horizontal. No vent system is allowed to terminate less than 4 feet below or 4 feet horizontally from, nor less than 1 foot above any door, openable window, or gravity air inlet into any building. As usual, there are exceptions.

Vent terminal of direct vent appliances with inputs of 50,000 Btu/H or less shall be located at least 9 inches from an opening through which combustion products could enter a building. Appliances with inputs in excess of 50,000 Btu/H but not exceeding 65,000 Btu/H shall require 12 inch vent termination clearances. The bottom of the vent terminal and the air intake shall be located at least 12 inches above grade.

Area

The internal cross-sectional area of a venting system must not be less than the area of the vent collar on the water heater, unless the venting system has been designed in accordance with other code requirements. In no case shall the area be less than 7 square inches, unless the venting system is an integral part of a listed water heater.

Venting Multiple Appliances

It is acceptable to connect multiple oil or listed gas-burning appliances to a common gravity-type venting system provided the appliances are equipped with an approved primary safety control capable of shutting off the burners and the venting system is designed in compliance with the code requirements.

Multiple appliances connected to a common vent system must be located within the same story of the building, unless an engineered system is being used. The inlets for multiple connections must be offset in a way that no inlet is opposite of another inlet. Oval vents may be used for multiple appliance venting, but the venting system must be not less than the area of the largest vent connector plus 50 percent of the areas of the additional vent connectors.

Btu	watts
1000	293
2000	586
4000	1172
5000	1465
100,000	29,300

Size of Combustion Air Openings or Ducts[1] for Gas- or Liquid-Burning Water Heaters

Column 1 Buildings of Ordinary Tightness		Column 2 Buildings of Unusually Tight Construction	
Condition	Size of Opening or Duct	Condition	Size of Opening or Duct
Appliance in unconfined[2] space	May rely on infiltration alone.	Appliance in unconfined[2] space: Obtain combustion air from outdoors or from space freely communicating with outdoors.	Provide 2 openings, each having 1 sq. in. (645 mm²) per 5,000 Btu/h input.
Appliance in confined[4] space 1. All air from inside building	Provide two openings into enclosure each having 1sq. in. (645 mm²) per 1,000 Btu/h input freely communicating with other unconfined interior spaces. Minimum 100 sq. in. (0.06 m²) each opening.	Appliance in confined[4] space: Obtain combustion air from outdoors or from space freely communicating with outdoors.	1. Provide two vertical ducts or plenums: 1 sq. in. (645 mm²) per 4,000 Btu/h input each duct or plenum. 2. Provide two horizontal ducts or plenums: 1 sq. in. (645 mm²) per 2,000 Btu/h input each duct or plenum.
2. Part of air from inside building	Provide 2 openings into enclosure[3] from other freely communicating unconfined[2] interior spaces, each having an area of 100 sq. in. (0.06 m²) plus one duct or plenum opening to outdoors having an area of 1 sq. in. (645 mm²) per 5,000 Btu/h input rating.		3. Provide two openings in an exterior wall of the enclosure: each opening 1 sq. in. (645 mm²) per 4,000 Btu/h input. 4. Provide 1 ceiling opening to ventilated attic and 1 vertical duct to attic: each opening 1 sq. in. (645 mm²) per 4,000 Btu/h input.
3. All air from outdoors: Obtain from outdoors or from space freely communicating with outdoors.	Use any of the methods listed for confined space in unusually tight construction as indicated in Column 2.		5. Provide 1 opening in enclosure ceiling to ventilated attic and 1 opening in enclosure floor to ventilated crawl space: each opening 1 sq. in. (645 mm²) per 4,000 Btu/h input.

1 For location of opening, see Section 507.3.
2 As defined in Section 223.0.
3 When the total input rating of appliances in enclosure exceeds 100,000 Btu/h, the area of each opening into the enclosure shall be increased 1 sq. in. (645 mm²) for each 1,000 Btu/h over 100,000 Bth/h.
4 As defined in Section 205.0.

FIGURE 7.9 Gas or liquid-burning water heaters must meet certain standards in the size of combustion air openings or ducts. *(Reprinted from the 2000 Uniform Plumbing Code with the permission of the International Association of Plumbing and Mechanical Officials)*

Existing Systems

New water heaters installed as replacements must meet code criteria before they can be connected to existing venting systems. The existing system must have been installed lawfully at the time of its installation. Code compliance with the internal area of the venting system must exist. Any connection made must be made in a safe manner.

Draft Hoods

Draft hoods for water heaters must be located in the same room or space as the combustion air opening for the water heater. The draft hood must be installed in the position for which it is designed and must be located so that the relief opening is not less than 6 inches from any surface other than the water heater being served, measured in a direction 90 degrees to the plane of the relief opening. Exceptions could exist if manufacturer's recommendations vary.

Existing Masonry Chimneys

Existing masonry chimneys with not more than one side exposed to the outside can be used to vent a gas water heater. There are, however, some conditions which must apply for this to be the case. The local code may require unlined chimneys to be lined with approved materials. Effective cross-sectional area of the chimney must not be more than four times the cross-sectional area of the vent and chimney connectors entering the chimney. The effective area of the chimney when connected to multiple connectors must not be less than the area of the largest vent or connector plus 50 percent of the area of the additional vent or connector.

Automatically controlled gas water heaters connected to a chimney which also serve equipment burning liquid fuel must be equipped with an automatic pilot. A gas water heater connector and a connector from an appliance burning liquid fuel may be connected to the same chimney through separate opening, providing the gas water heater is vented above the other fuel-burning appliance, or both may be connected through a single opening if joined by a suitable fitting located at the chimney. Multiple connections must not be made at the same horizontal plane of another inlet. Any chimney used must be clear of obstructions and cleaned if previously used for venting solid or liquid fuel-burning appliances.

Connectors

Chimney connectors must comply with code requirements as set forth in tables in the local code book. When multiple connections are made, the connector, the manifold, and the chimney must be sized properly. Gravity vents must not be connected to vent systems served by power venters, unless the connection is made on the negative pressure side of the power exhauster. Single-wall metal chimneys require a minimum clearance of 6 inches from combustible materials.

Connectors must be kept as short and as straight as possible. Water heaters are required to be installed as close as possible to the venting system. Connectors must not be longer than 75 percent of the portion of the venting system above the inlet connection unless a part of an approved engineered system.

A connector to a masonry chimney must extend through the wall to the inner face of the liner, but not beyond. The connector must be cemented to the masonry. A thimble may be used to facilitate removal of the connector for cleaning, in which case the thimble shall be permanently cemented in place. Connectors shall not pass through any floor or ceiling.

Draft regulators are required in connectors serving liquid fuel burning water heaters, unless the water heater is approved for use without a draft regulator. When used, draft regulators must be installed in the same room or enclosure as the water heater in such a manner that no difference in pressure between air in the vicinity of the regulator and the combustion air supply will be permitted.

Mechanical Draft Systems

It is acceptable to vent water heaters with mechanical draft systems of either forced or induced draft design. Forced draft systems must be designed and installed to be gastight so as to prevent leakage of combustion products into a building. Connectors vented by natural draft must not be connected to mechanical draft systems operating under positive pressure. Systems using a mechanical draft system must be made to prevent the flow of gas to the main burners when the draft system is not performing so as to satisfy the operating requirements of the water heater for safe performance.

Exit terminals of mechanical draft systems must be located not less than 12 inches from any opening through which combustion products could enter the building, nor less than 7 feet above grade when located adjacent to public walkways.

Ventilating Hoods

Ventilating hoods can be used to vent gas burning water heaters installed in commercial applications. Dampers are not allowed when automatically operated water heaters are vented through natural draft ventilating hoods. If a power venter is used, the water heater control system must be interlocked so that he water heater will operate only when the power venter is in operation.

SIZING WATER HEATERS

Here are some tables that can be helpful when sizing water heaters:

Number of Bathrooms	1 to 1.5			2 to 2.5				3 to 3.5			
Number of Bedrooms	1	2	3	2	3	4	5	3	4	5	6
First Hour Rating², Gallons	42	54	54	54	67	67	80	67	80	80	80

Notes:
¹The first hour rating is found on the "Energy Guide" label.
²Non-storage and solar water heaters shall be sized to meet the appropriate first hour rating as shown in the table.

FIGURE 7.10 The minimum capacity for water heaters must be as shown above. *(Reprinted from the 2000 Uniform Plumbing Code with the permission of the International Association of Plumbing and Mechanical Officials)*

Number of bedrooms	1	2	3
Storage capacity (gallons)	20	30	30
Input in Btuh	2.5 KW	3.5 KW	4.5 KW
Draw (gallons per hour)	30	44	58
Recovery (gallons per hour)	10	14	18

FIGURE 7.11 Water heating minimum sizing table for electric heaters assuming that less than two full bathrooms exist.

Number of bedrooms	1	2	3
Storage capacity (gallons)	30	30	30
Input in Btuh	70,000	70,000	70,000
Draw (gallons per hour)	89	89	89
Recovery (gallons per hour)	59	59	59

FIGURE 7.12 Water heating minimum sizing table for oil-fired heaters assuming that less than two full bathrooms exist.

Number of bedrooms	2	3	4	5
Storage capacity (gallons)	30	40	40	50
Input in Btuh	36,000	36,000	38,000	47,000
Draw (gallons per hour)	60	70	72	90
Recovery (gallons per hour)	30	30	32	59

FIGURE 7.13 Water heating minimum sizing table for gas-fired heaters assuming that less than two full bathrooms exist.

Number of bedrooms	2	3	4	5
Storage capacity (gallons)	40	50	50	66
Input in Btuh	4.5 KW	5.5 KW	5.5 KW	5.5 KW
Draw (gallons per hour)	58	70	72	88
Recovery (gallons per hour)	18	22	22	22

FIGURE 7.14 Water heating minimum sizing table for electric heaters assuming that 2 to 2½ bathrooms exist.

Number of bedrooms	2	3	4	5
Storage capacity (gallons)	30	30	30	30
Input in Btuh	70,000	70,000	70,000	70,000
Draw (gallons per hour)	89	89	89	89
Recovery (gallons per hour)	59	59	59	59

FIGURE 7.15 Water heating minimum sizing table for oil-fired heaters assuming 2 to 2½ bathrooms exist.

Number of bedrooms	3	4	5	6
Storage capacity (gallons)	40	50	50	50
Input in Btuh	38,000	38,000	47,000	50,000
Draw (gallons per hour)	72	82	90	92
Recovery (gallons per hour)	32	32	40	42

FIGURE 7.16 Water heating minimum sizing table for gas-fired heaters assuming 3 to 3½ bathrooms exist.

Number of bedrooms	3	4	5	6
Storage capacity (gallons)	50	66	66	80
Input in Btuh	5.5 KW	5.5 KW	5.5 KW	5.5 KW
Draw (gallons per hour)	72	88	88	102
Recovery (gallons per hour)	22	22	22	22

FIGURE 7.17 Water heating minimum sizing table for electric heaters assuming 3 to 3½ bathrooms exist.

Number of bedrooms	3	4	5	6
Storage capacity (gallons)	59	59	59	59
Input in Btuh	70,000	70,000	70,000	70,000
Draw (gallons per hour)	89	89	89	99
Recovery (gallons per hour)	59	59	59	59

FIGURE 7.18 Water heating minimum sizing table for oil-fired heaters assuming 3 to 3½ bathrooms exist.

SAFETY COMMENTARY

I've been plumbing for over 25 years. During these years I've done just about every type of plumbing that there is. My career has involved working with all sorts of plumbers and plumbing contractors. As a plumbing contractor for the last 20 years, I've used a lot of other plumbing contractors as subcontractors. My work as even extended into teaching code classes and apprenticeship classes at Central Maine Technical College. I'm telling you this to give you an idea of my background and experience in the industry. This is because I want to tell you something that I feel is extremely important about water heaters.

During my time in the field, I've run into many occasions when plumbing contractors failed to obtain permits for the installation of water heaters. A plumbing permit is required for every water heater installation and replacement. Yet, a good number of contractors feel that they can get by without a permit. I've heard dozens of contractors say that it takes more time to get a permit and inspection than it does to install a water heater. This can be true, but the point is that a permit and inspection is required by code.

Some code enforcement offices are more active than others in enforcing the need for permits and inspections when water heaters are being replaced. But, all codes that I know of do require a permit and inspection for the installation or replacement of a water heater. This shouldn't be a big deal, but it is, and it can be a very big deal.

About two years ago I lost the service contract on over 180 apartment units because I refused to replace a water heater without a permit and inspection. I didn't like losing the account, but I was not going to violate the plumbing code and risk myself and my business to a lawsuit for any cost. It can be hard for reputable plumbing contractors to compete with bootleggers who are willing to work without permits. The time and money spend on permits and inspections does drive up the cost of a job. Still, you should never install or replace a water heater without adhering to the plumbing code.

Why do I feel so strongly about permits for water heaters? There are several reasons. First, code requires a permit and inspection, so this should be reason enough. Secondly, installing a water heater illegally can open up a huge risk for a lawsuit. If you install a water heater in accordance with code requirements and a problem erupts later, your insurance should cover your losses. This probably would not be the case if you had installed the water heater in violation of the plumbing code. Leaving ethics, morals, and lawsuits out of it, there is still the issue of personal safety.

Believe it or not, I have found water heaters where a plug had been installed in place of a relief valve. Any plumber with even minimal experience knows the risk to this. I've found cutoff valves installed in illegal locations around water heaters. If the proper safety precautions are not taken, a water heater can become a large bomb. I remember watching a video from a manufacturer that showed water heaters blowing up. In one case, the water heater went from a basement location right the roof of a home. The explosions can be very violent.

I won't try to beat you over the head with my vision, but seriously consider the risk you will be taking if you don't install water heaters properly and in accordance with all code provisions. Protect yourself by getting permits and inspections.

chapter 8

GAS PIPING

The installation of fuel-gas piping is a substantial responsibility. If a plumber has a leak in a water pipe, the result is something getting wet. A similar leak in a gas pipe can result in a serious explosion. All code requirements have importance and should be followed, and this is especially true when working with gas piping and gas connections. The installation, modification, and maintenance of fuel-gas systems are all addressed in the code.

Starting with the point of delivery, the code covers all piping from that point to the connections with each utilization device. Code coverage includes design, materials, components, fabrication, assembly, installation, testing, inspection, operation, and maintenance of such piping systems. What is the point of delivery? It is the outlet of the service meter assembly or the outlet of the service regulator or service shutoff valve where a meter is not provided, or where the service meter assembly is located within a building, at the entrance of the supply pipe into the building. When working with undiluted liquefied petroleum (LP) gas systems the point of delivery is the outlet of the first stage pressure regulator. Any LP gas storage system must be designed and installed in accordance with the fire prevention code and NFPA 58.

When a gas system is modified or added to all pipe sizing must conform with the sizing requirements of the code. If an additional gas appliance is being added, the existing piping must be checked to determine if it has the required capacity for all appliances being served. When it is determined that an existing gas system is not large enough for additional appliances, the existing system must be upgraded to meet demand capacities. This can be done by enlarging the existing system or by running a new system for the appliances to be served.

All exposed gas piping, except for black steel pipe, must be marked and identified by a yellow label that shows the word "Gas" in black letters. This identification marker must be spaced at intervals which do not exceed 5 feet. No marking is required on piping that is in the same room as the equipment being served. Any tubing that carries medium-pressure gas must me marked with a label at the beginning and end of each section of tubing used to convey gas.

It is a violation of code to make interconnections between multiple meters. When multiple meters are installed on the same premises to serve multiple users, each meter must be served by an independent gas system. It is not acceptable to connect gas piping from one user's system to the system of another user. To avoid mistaken interconnections, piping from multiple meter installations must be marked with an approved permanent identification by the installer. The markings must be clear and readily identifiable.

PIPE SIZING

The sizing of gas pipe depends on many factors. Code books contain tables that are used to determine proper pipe sizing. You can see many such tables at the back of this chapter. All pipe used for an installation, extension, or alteration must be sized to supply the full number of outlets for the intended purposes. When a gas supply has a pressure of 0.5 psig, or less, and the gas meter is located withing 3 feet of the building exterior, all building piping from the meter outlet downstream, including the pipe outlets, must have a minimum diameter of ½ inch. The hourly volume of gas demand at each gas outlet must not be less than the maximum hourly demand, as specified by the manufacturer of the appliance being served.

Calculating gas demand can seem intimidating, but it's really not all that difficult to do. The goal is to determine the cubic feet per hour of gas required. In order to do this, you must divide the maximum Btu/h input of an appliance by the average Btu/h heating value per cubic foot of gas being served. This is a simple formula, so long as you know what the Btu/h rating is for the appliances being served. Many appliances will be labeled with a Btu/h rating, but some will not bear the labeling. If you are faced with appliances that

Typical Minimum Demands in BTUs Per Hour	
Appliance	Demand
Water Heater (30 gallon)	30,000
Water heater (40–50 gallon)	50,000
Refrigerator	3,000
Range	65,000
Clothes Dryer	35,000
Residential Fireplace Log Lighter	25,000
Residential Barbecue	50,000

FIGURE 8.1 Gas appliance table.

are not labeled with a rating, you can refer to a table in your code book that will allow you to estimate the rating needed to perform the gas demand calculation. Whether you are working with known ratings or estimates from a table, you must size the gas piping to maintain and supply gas in a capacity that is not less than the actual demand of the installed appliances. The sizing of all gas piping must be in keeping with code requirements.

Number of dwelling units	Ranges only	Ranges and water heaters
2	.85	.77
4	.65	.59
6	.54	.49
8	.46	.44
10	.42	.40
15	.36	.34
20	.31	.30
30	.25	.24
40	.23	.22
50	.21	.20

FIGURE 8.2 Multiplier of total connected load.

MATERIALS FOR GAS PIPING

Materials for gas piping are available in various types. Some of the piping options available include:

- *Aluminum-alloy pipe*
- *Aluminum-alloy tubing*
- *Brass pipe*
- *Copper pipe*
- *Copper-alloy pipe*
- *Copper tubing (Type K or L)*
- *Copper-alloy tubing (Type K or L)*
- *Copper tube (Type ACR)*
- *Corrugated stainless steel tubing*
- *Ductile iron pipe*
- *Plastic pipe*
- *Plastic tubing*
- *Steel pipe*
- *Steel tubing*

All corrugated stainless steel tubing used for gas piping must be tested, listed, and installed in accordance with ANSI LC-1. Plastic pipe, or tubing and compatible fittings must be installed only undergound outside of buildings. Only polyethylene pipe shall be used with LP gases and such applications must comply with NFPA 58. Plastic pipe is allowed to terminate above ground outside of buildings where installed in premanufactured anodeless risers or service head adapter risers that are installed in accordance with the manufacturer's installation instructions.

Copper tubing has some restrictions on its use. For example, copper tubing cannot be used to convey gas that contains more than an average of 0.3 grain of hydrogen sulfide for 100 standard cubic feet of gas. Copper tubing systems must be identified with an appropriate label, with black letters on a yellow field, to indicate the piping system conveys fuel gas. The labels must be permanently attached to the tubing within one foot of the penetration of a wall, floor, or partition and at maximum intervals throughout the length of the tubing run. The labels must be visible to facilitate inspections.

Aluminum-alloy pipe and tubing is not approved for underground use or for use outside of a structure. When it comes to fittings used in gas applications, the fittings must be approved for use with fuel-gas systems. All fittings must be compatible with, or shall be of the same material as, the pipe or tubing being used. At no time are bushings allowed to be used. All flange fittings must conform to ASME B16.1 or ASME B15.5. Any gasket material used for flanged fittings must be approved and compatible with the fuel gas being conveyed.

If used pipe is to be reused for gas piping, the used pipe must not have been used previously for any purpose other than conveying gas. Reused pipe must be clean, in good condition, free from internal obstructions, and burred ends must be reamed to the full bore of the pipe. Before any pipe, tubing, fittings, valves, or other devices can be reused they must be cleaned thoroughly, inspected, and determined to be equivalent to new materials.

All joints and connections must be of an approved type and tested to be gas tight at required test pressures. When connections are made between different types of materials, the joints must be made with approved adapter fittings. Any connections made between different metallic piping materials must be made with dielectric fittings. This is to isolate electrically above-ground piping from underground piping or to isolate electrically different metallic piping materials joined underground.

fastfacts

➤ *Plastic pipe and tubing must be installed only underground outside of buildings.*

➤ *Copper tubing has some restrictions on its use.*

➤ *Aluminum-alloy pipe and tubing is not approved for underground use.*

➤ *Aluminum-alloy pipe is not approved for use outside of a building.*

➤ *Bushings are not allowed in gas piping and fittings.*

➤ *Connections made between different metallic piping materials must be made with dielectric fittings.*

PREPARATION AND INSTALLATION

Preparation and installation of gas pipe materials varies, depending upon the type of material being used. All pipe has to be prepared in a similar fashion. Ends of piping must be cut squarely. The pipe must be reamed and chamfered. Any burrs or obstructions in or on the pipe ends must be removed. Every pipe end must have a full-bore opening and must not be undercut. In addition to pipe preparation, all joints must be prepared properly. This is true of all types of joints.

When working with brazed joints, you must ensure that all joint surfaces are cleaned. An approved flux must be applied where it is required. Filler used to make a brazed joint must conform to AWS A5.8. Brazing materials must have a melting point that is in excess of 1,000 degrees F., and alloys used for brazing must not contain more than 0.05 phosphorous. All brazed joints must be made by certified brazers.

Flared joints must be created with tools designed for the purpose of flaring joints. When mechanical joints are needed, they must be made with fittings that are specified by the manufacturer for the gas service and must be installed according to the manufacturer's recommendations. Pipe-joint compound or sealant tape must be used on threaded joints. All threads used with threaded joints must conform to ASME B1.20.1. Joint surfaces must be cleaned by an approved procedure when joints are to be welded. Any filler material used to weld a joint must be of an approved type.

Joints made between aluminum-alloy pipe and tubing or fittings must be flared or made as mechanical joints. Brass piping allows for connections to be made by brazing, mechanical joints, threaded connections, or welded joints. When copper or copper-alloy pipe is used, joints must be brazed, mechanical, threaded, or welded. Joints made between copper or copper-alloy tubing or fittings must be brazed, flared, or mechanical. Flanged joints are required when ductile iron pipe is used for gas piping. All connections must conform to the manufacturer's instructions.

JOINING PLASTIC PIPE AND FITTINGS

The installation of plastic pipe and fittings requires joints as described in the code. When plastic pipe and fittings are used, joints between them may be made with a solvent-cement method, a heat-fusion method, or with compression couplings or flanges. The method of joining the pipe and fittings must be compatible with the materials and the manufacturer's recommendations. Threading plastic pipe or tubing is prohibited. When polyethylene pipe, tubing, or fittings are used, joints shall be made by means of either heat-fusion or mechanical joints. Joints of this type must be made to sustain effectively the longitudinal pull-out forces caused by contraction of the piping or by external loading.

Heat-fusion joints are allowed only on polyolefin pipe or tubing. This includes polyethylene and polybutylene pipe and tubing. All joint surfaces must be clean and dry. The joint surfaces have to be heated to melt temperature and joined. Joints made this way must not be disturbed until they cool to a tolerance allowed by the manufacturer.

Plastic gas piping can be joined with the use of mechanical compression fittings. These fittings must be designed and approved for use with plastic pipe that carries natural gas or LP-gas vapor. When mechanical fittings are used with polyethylene pipe the fittings must conform with ASTM D 2513 category 1, full-restraint, full-seal joints, and must be so marked.

The gasket material in a compression-type mechanical fitting must be compatible with the plastic piping and the gas being distributed. An internal tubular rigid stiffener must be used in conjunction with the fitting, and the stiffener must be flush with the end of the pipe or tubing. The stiffener must also extend at least to the outside end of the compression fitting when it is installed. No stiffener that has rough or sharp edges may be used. Split tubular stiffeners are prohibited, and no stiffener shall be force fitted into the plastic.

PVC pipe, tubing, and fittings used for gas piping can be joined with a solvent-cement joint. The procedures for making this type of joint are essentially the same as those used for making joints in drain, waste, and vent piping where PVC pipe is used. All services to be joined must be clean and dry. An application of an approved primer to both a pipe end and a fitting hub is required prior to cementing a joint. Any solvent cement used to make a connection must be of an approved type. All solvent-cement joints must be made while the cement is wet.

Polyolefin plastic pipe (polybutylene and polybutylene), tubing, and fittings are not allowed to be joined with solvent cement. Heat fusion and mechanical joints are acceptable ways of joining polyolefin plastic pipe, tubing, and fittings. It is not acceptable to use solvent cement or heat fusion to join different types of plastic.

Flanges and special joints must be qualified by the manufacturer for the intended use with plastic pipe and tubing. Any connection made with a flange or other special joint must be approved by a code official. Polyethylene pipe and tubing must not be flared at anytime. PVC pipe and tubing is usually not allowed to be flared. However, there can be times when a manufacturer specifies that PVC pipe can be flared for use in underground installations.

JOINING STEEL PIPE AND TUBING

Steel pipe and fittings can be joined with threaded joints, welded joints, or mechanical joints. When mechanical joints are used, they must be made with an approved elastomeric seal. All mechanical joints must be installed within the guidelines of the manufacturer's

fastfacts

➤ *Solvent-cement joints, heat-fusion joints, compression couplings, and flanges are acceptable means of making connections with various types of gas piping.*

➤ *Threading plastic pipe and tubing is prohibited.*

➤ *Polyethylene pipe, tubing, and fittings can be joined with heat-fusion or mechanical joints.*

➤ *Heat-fusion joints are allowed only on polyolefin pipe and tubing, which includes polyethylene and polybutylene pipe and tubing.*

➤ *Plastic gas piping can be joined with the use of mechanical compression fittings.*

➤ *PVC pipe, tubing, and fittings used for gas piping can be joined with a solvent-cement joint.*

➤ *Polyolefin pipe, tubing, and fittings must not be joined with solvent-cement joints.*

➤ *Solvent-cement joints are not allowed to join different types of plastic pipe or tubing.*

➤ *Heat-fusion joints are not allowed to join different types of plastic pipe or tubing.*

➤ *Polyethylene pipe and tubing must not be flared.*

recommendations. Normally, mechanical joints for steel gas piping are allowed only outside and underground. Steel tubing can be joined with welded joints or mechanical joints, but not with threaded joints.

Connecting metal pipe to plastic pipe is allowed only when the connection is made outside and underground. There is one exception to this rule: Plastic pipe is permitted to terminate above ground outside of buildings where installed in premanufactured anodeless risers or service head adapter risers that are installed in accordance with the manufacturer's installation instructions. Underground connections between metallic piping and plastic piping are to be made with mechanical fittings or factory-assembled, plastic-to-steel transition fittings when they are specified by the manufacturer for use in gas piping applications.

When installing a compression-end riser transition you must be sure that there is at least 12 inches of horizontal length of metallic piping underground at the end of any plastic piping installed. Metallic pipe installed below ground for the purpose of conveying gas must be protected from corrosion. The length and size of metallic pipe joined with plastic pipe must be adequate to prevent stress or strain on the plastic piping.

Anodeless risers and service head adapter risers are the only types of risers allowable when polyethylene gas pipe is terminated above ground. Any above-ground portion of polyethylene must be centered in the metallic casing to ensure that the temperature of the polyethylene pipe does not exceed a temperature of 150 degrees F.

All metallic pipe and tubing exposed to corrosive action must be protected from corrosion. An example of such a situation would be steel pipe installed below ground. Zinc coatings, the type used to galvanize pipe, is not acceptable protection for underground gas piping. Consult a local code official for advice in choosing a suitable means of protection.

Protective coatings and wrapping for gas piping must be of an approved type and installed, for the most part, with a machine. When wrapping is done in the field, the wrapping is to be limited to fittings, short sections of piping, and piping where the factory wrap has been damaged or stripped for threading or welding.

When joint compound is used to seal threaded joints, the compound must be applied only to male threads. Any joint compound used must be resistant to the action of liquefied petroleum gases. All threads on pipe and fittings must comply with ASME B1.20.1 and all code requirements. Any threads that are damaged must not be used. Welds that open or that are defective must be cut out and replaced by a new section of piping and connection.

INSTALLING GAS PIPING SYSTEMS

There are a number of rules to follow when installing a gas piping system. There are interior locations where gas piping may not be installed. It is a code violation to install gas piping inside any of the following:

- *Air ducts*
- *Clothes chutes*
- *Chimneys*
- *Vents*
- *Ventilating ducts*
- *Dumbwaiters*
- *Elevator shafts*
- *Concealed plenums*

Gas pipe that is installed in a concealed location must not be fitted with unions, tubing fittings, or running threads. Concealed piping is not allowed in solid partitions and walls, unless the gas piping is installed in a chase or casing. It is never acceptable to allow a gas pipe to be used as a grounding electrode.

Gas pipe that penetrates masonry work must be protected by being encased in a sleeve. The sleeve opening on the outside of a building must be sealed so that water will not run through the sleeve and into a building. Penetrating a building foundation below grade is not allowed when installing gas piping.

Gas piping that is going to be concealed should be run with black or galvanized steel pipe. When some other type of pipe or tubing is used, and the pipe or tubing is not protected by at least $1\frac{1}{4}$ inches of wood or framing member, the installation must be protected from damage with the use of shield plates. The plates must be made of steel and have a minimum thickness of $\frac{1}{16}$ inch. Shield plates must cover the area of the gas pipe or tubing where it is not protected adequately by the framing member. When shield plates are needed, they must extend a minimum of 4 inches above sole plates, below top plates, and to each side of a stud, joist, or rafter.

Unless a code official determines that no other option exists, you must not install gas piping in solid floor slabs, such as concrete. In the case of solid floors, house piping is to be installed above the floor. The piping can be installed in open or furred spaces, hollow partitions, hollow walls, attic space, or pipe chases.

On occasions when a code official determines that gas piping in a solid floor cannot be avoided, there are special rules to be followed. There are two basic options to consider. One way to install such piping is to sleeve it in Schedule 40 steel pipe that has tightly sealed ends and joints. Both ends of the casing must extend at least 2 inches beyond the point where the pipe emerges from the solid floor. The other option is to install piping in a channel in the floor. The channel in the floor must be covered so that access to the gas piping is available. The covering must prevent the entrance of corrosive materials, or the channel must be filled with a noncorrosive material that will cause a minimum of damage to a floor when the material is removed.

When gas pipe penetrates solid floors or solid walls, the pipe must be protected, unless otherwise approved by a code official. The proper type of protection is a sleeve. Gas pipe installed through a solid wall or solid floor must either be installed in a casing or through an opening of adequate size. Piping must be encased in a 1:3 mixture of cement and sand. The coating must not be less than ¾ inch thick.

Except where otherwise approved, gas piping is not allowed to be installed in a common trench with water, sewer, or drainage piping. All gas piping that is installed in contact with earth of other corrosive material must be protected from corrosion. Any dissimilar metal that is joined together underground requires the use of an insulated coupling. It is prohibited to place metallic piping in contact with cinders. The minimum allowable depth for burying gas pipe is 18 inches, unless otherwise specified by a code official. When individual gas lines are installed for outside lights, grills, or other appliances, the gas piping must be buried at least 8 inches deep. Additional depth may be required if it is believe that the piping will be subject to physical damage at an 8 inch depth. Buried gas piping must not be installed so as to be in contact with the ground or fill under a building or floor slab.

Pipe hangers and supports shall be in compliance with standard code requirements. Copper tubing conveying gas and running parallel to joists must be secured to the center of the joist at a maximum interval of 6 feet. If the copper tubing is running at an angle to joists, it must be installed either through holes in the joists that are at least 1½ times the outside diameter of the tubing or secured to and supported at maximum intervals of 6 feet. If the tubing is installed closer than 1¾ inches to the face of a joist, a steel nail plate must be installed on the face of the joist to prevent the tubing from being damaged by nails or staples. Copper tubing installed vertically in walls must be supported at the floor and ceiling level and protected by nail plates. The tubing must not be supported at any other point in the wall. Nail plates must have a minimum thickness of 0.0508 inch (16 gauge) and a minimum length of 4 inches beyond concealed penetrations.

Changes in direction for gas piping must be made by the use of fittings, or by the use of bends that are approved by the standard code. Bends used with metallic pipe must be smooth and free from buckling, cracks, or other evidence of mechanical damage. The bends must be made only with bending equipment and procedures that are intended for the purpose of making bends. Longitudinal welds of a pipe are required to be near the neutral axis of the bend. No bend shall exceed a 90 degree angle. An inside radius of a bend must not be less than 6 times the outside diameter of the pipe. If plastic piping is bent, it must be bent in accordance with the manufacturer's recommendations.

Gas outlets may not be installed behind doors. Any unthreaded portion of gas piping outlets must extend at least 1 inch through finished ceilings and walls. When outlets are extending through floors, slabs, or outdoor patios, the outlets must extend at least 2 inches above the floor, slab, or patio. Outlet fittings or piping must be fastened securely. All outlet locations must allow for the use of proper wrenches without any straining, bending, or damage to the piping. Flush-mounted-type-quick-disconnect devices are an exception. These devices are to be installed in compliance with the manufacturer's installation instructions. Any gas outlet that is not connected to an appliance must be capped to a gas-tight condition.

When plastic piping is used as a gas pipe, the pipe must not be used within or under any building or slab. The pipe is not allowed to operate at pressures greater than 100 PSIG for natural gas or 30 PSIG for LP gas. When plastic pipe is buried as a gas pipe it must be accompanied by a yellow insulated copper tracer wire or other approved conductor. The tracer wire must be installed adjacent to the underground plastic pipe and access must be provided to the tracer wire or the wire must terminate above ground at each end of the plastic gas piping. Sizing for the tracer wire shall not be smaller than 18 AWG and the insulation type must be suitable for direct burial.

TESTING INSTALLATIONS

The testing of installations is required before any system of gas piping is put into use. All components of the gas system must be tested to ensure that the system is gas tight. System components that will be concealed, must be tested prior to concealment. The testing of a gas system is to be done with either air or an inert gas. No other type of gas or liquid is allowed for testing purposes. The test pressure must be measured with a device that is acceptable to local code officials.

SHUTOFF VALVES

Shutoff valves must not be installed in concealed locations or any space that is used as a plenum. All gas outlets are required to have individual shutoff valves. The valves must be in the same room and within 6 feet of any appliance being served by a gas outlet. Access to the shutoff valve is required. With the exception of buildings meant to be used as single-family or two-family dwellings, cutoff valves are required for every gas system that serves a separate tenant. All cutoff valves must be accessible to the tenant who is served by the gas system. When a common gas system is used to service a number of individual buildings, shutoff valves are required outside of each building being served. Every gas meter must be equipped with a shut-off valve on the supply side of the meter.

Any shutoff valve used in a gas system must be of an approved type. The valves must be constructed of materials that are compatible with gas piping. All shutoff valves must be accessible. When

fastfacts

➤ *Shutoff valves must not be installed in concealed locations.*

➤ *Shutoff valves must not be installed in any space used as a plenum.*

➤ *Shutoff valves must be in the same room as the equipment being served by the valve.*

➤ *Shutoff valves must be within 6 feet of the equipment being served by the valve.*

➤ *Shutoff valves must be accessible.*

➤ *Shutoff valves are required for every gas meter.*

shutoff valves control separate piping systems, they must be placed an adequate distance from each other so that they will be readily accessible for operation, and they must be installed in a way that protects them from damage. All shutoff valves must be plainly marked with an identification tag that is attached by the installer of the valves. The tagging is required to ensure that each valve is readily identifiable.

REGULATIONS PERTAINING TO TWO-PSI AND HIGHER GAS PIPING

What is a two-psi gas piping system? It is a gas piping system that is equipped with a service regulator that is set to deliver gas at 2 psi. Sizing gas pipe that runs from the point of delivery to a medium pressure regulator requires considering many factors.

The first step is to determine the distance between the point of delivery and the most remote medium pressure regulator. This distance will be used in conjunction with a sizing table provided in your local code book. You will look at the table and find the length of piping needed to reach the medium pressure regulator. If your required distance falls between two numbers, choose the larger number to indicate your length.

You will have to decide what the outside diameter of your piping will be. The gas demand figures in the table will help you to establish the pipe size. If the demand that you need is not listed, choose the next higher rating. By determining pipe length and gas demand, you can see easily what the minimum pipe diameter must be. This type of sizing procedure works for pipe and tubing. You simply use the proper table and your sizing work is simple.

SPECIAL CONDITIONS

Special conditions sometime apply to pipe sizing for gas systems. There can be times when systems call for a more sophisticated sizing solution. The maximum design operating pressure for piping systems located within buildings is 5 psi, with the possible exception of a variance from a code official. There are also some conditions which must be met to allow such a high pressure. Generally, one or more of the following conditions must be met:

- *Piping system is located in an industrial processing or heating structure.*
- *Piping system is located in a research structure.*
- *Piping system is located in an area that contains boilers or mechanical equipment only.*
- *Piping system is welded steel pipe.*
- *Piping system is temporary for structures under construction.*

GAS FLOW CONTROLS

The gas flow controls on a piping system are very important. Gas pressure regulators, or gas equipment pressure regulators, are required when a gas appliance is designed to operate at a lower pressure than that of the main gas system. Access to the regulators is required. All regulators must be protected from physical damage. any regulator installed on the exterior of a building must be rated for such an installation. Second stage regulators for undiluted LP gas systems must be listed and labeled in accordance with UL 144.

Medium pressure regulators installed in 2 psi portions of systems must meet certain specifications. All of these regulators must be approved and suitable for the inlet and outlet gas pressure required by the application. The device must have a reduced outlet pressure under lockup conditions. Any medium pressure regulator used must have a rated capacity capable of serving the appliances installed. Access must be provided for all regulators. If a medium pressure regulator is installed indoors, it must be vented to the outdoors or equipped with a leak-limiting device.

Medium pressure regulators must not be installed in locations where they will be concealed. Any piping system served by one gas meter and containing multiple medium pressure regulators must have listed shutoff valves installed immediately ahead of each medium pressure regulator.

All regulators that require venting must be equipped with an independent vent to the outside air. This vent must be designed to prevent the entry of water, foreign objects, or any other form of obstruction. One exception to this rule is that second stage regulators equipped with and labeled for utilization with approved vent-limiting devices do not have to be vented to outside air.

UNDILUTED LIQUEFIED PETROLEUM GASES

Undiluted liquefied petroleum gases are butane and propane. The sizing procedure for gas systems conveying undiluted liquefied petroleum gases are about the same as those already discussed. You will simply use the tables in your local code book to determine the proper pipe or tubing size. Regulators used for these systems must be listed for the use and must be installed in accordance with their listing.

There are various options in the ways that appliance connections may be made. Rigid metallic pipe and fittings are suitable for such connections. Unless there is an official code variance, semi-rigid metallic tubing and metallic fittings can be used so long as they don't exceed 6 feet. These materials, when used, must be located entirely in the same room as the appliance. It is not acceptable to have semi-rigid metallic tubing enter a motor-operated appliance through an unprotected knockout opening.

Listed and labeled gas appliance connectors can be used to make connections. However, they must be installed in accordance with the manufacturer's instructions and located entirely in the same room as the appliance that they are connected to. Quick-disconnect devices that are listed and labeled, used with listed and labeled gas appliance connectors, can be used for making connections. Any listed and labeled gas convenience outlet used in conjunction with a listed and labeled gas appliance connector can also be used to make connections to appliances. Another option is a listed and labeled gas appliance connector that complies with ANSI Z21.69 that is designed for use with food service equipment that has casters, or that is otherwise subject to movement for cleaning. This same type of device can be used for connections to large movable gas utilization equipment. All connectors, tubing, and piping must be protected from physical damage.

The diameter of fuel connectors must not be smaller than the inlet connection to the appliance be connected to. Most connectors cannot be more than three feet in length. Connectors for ranges and domestic clothes dryers can be up to 6 feet long. At no time can a connector be concealed in a wall, nor can they be extended through a wall. This same ruling applies to floors, partitions, ceilings, or appliance housings. An approved shutoff valve that is not less than the nominal size of the connector must be accessible at the gas piping outlet immediately ahead of the connector. Any connector used must be sized to provide the total demand of the connected appliance, as described in a table in your local code book.

Gas appliances that are intended to be moved for routine cleaning, maintenance, or other reasons must be connected to a gas system with a flexible connector. The connector must be labeled for the application and protected from physical damage.

GAS-DISPENSING SYSTEMS

All gas dispensers are required to have an emergency switch to shut off all power to the dispenser. An approved backflow device that prevents the reverse flow of gas must be installed on the gas supply pipe or in the gas dispenser. Any gas-dispensing system installed inside of a building must be ventilated by mechanical means in accordance with the *International Mechanical Code*. When compressed natural gas fuel-dispensing systems for vehicles fueled by the gas are installed, they must be installed in accordance with NFPA 52 and the fire prevention code.

SUPPLEMENTAL GAS

Anytime that air, oxygen, or other special supplementary gas is introduced into a gas system, an approved backflow preventer must be installed. This device must be on the gas line to the equipment supplied by the special gas, and located between the source of the special gas and the gas meter.

Sometimes there is a need for standby gas. When LP gas or some other standby gas is interconnected with primary gas piping systems, an approved three-way, two-port valve, or approved backcheck safeguard shall be installed to prevent backflow into any system.

Gas piping should not be intimidating. But, working with gas is serious business. Learn and understand the gas code completely. Follow it. Don't cut corners. A gas leak has the potential to be much more devastating than a water leak. I promised you sizing tables, and here they are, on the following pages.

Size of Gas Piping

Maximum Delivery Capacity of Cubic Feet of Gas Per Hour of IPS Pipe Carrying
Natural Gas of 0.60 Specific Gravity Based on Pressure Drop of 0.5 Inch Water Column

Length in Feet

Pipe Size, Inches	10	20	30	40	50	60	70	80	90	100	125
1/2	174	119	96	82	73	66	61	56	53	50	44
3/4	363	249	200	171	152	138	127	118	111	104	93
1	684	470	377	323	286	259	239	222	208	197	174
1-1/4	1404	965	775	663	588	532	490	456	428	404	358
1-1/2	2103	1445	1161	993	880	798	734	683	641	605	536
2	4050	2784	2235	1913	1696	1536	1413	1315	1234	1165	1033
2-1/2	6455	4437	3563	3049	2703	2449	2253	2096	1966	1857	1646
3	11,412	7843	6299	5391	4778	4329	3983	3705	3476	3284	2910
3-1/2	16,709	11,484	9222	7893	6995	6338	5831	5425	5090	4808	4261
4	23,277	15,998	12,847	10,995	9745	8830	8123	7557	7091	6698	5936

Pipe Size, Inches	150	200	250	300	350	400	450	500	550	600
1/2	40	34	30	28	25	24	22	21	20	19
3/4	84	72	64	58	53	49	46	44	42	40
1	158	135	120	109	100	93	87	82	78	75
1-1/4	324	278	246	223	205	191	179	169	161	153
1-1/2	486	416	369	334	307	286	268	253	241	230
2	936	801	710	643	592	551	517	488	463	442
2-1/2	1492	1277	1131	1025	943	877	823	778	739	705
3	2637	2257	2000	1812	1667	1551	1455	1375	1306	1246
3-1/2	3861	3304	2929	2654	2441	2271	2131	2013	1912	1824
4	5378	4603	4080	3697	3401	3164	2968	2804	2663	2541

FIGURE 8.3 Here, the relationship between pipe size and natural gas carrying capacity is explained. *(Reprinted from the 2000 Uniform Plumbing Code with the permission of the International Association of Plumbing and Mechanical Officials)*

Size of Gas Piping

Maximum Delivery Capacity in Liters Per Second of IPS Pipe Carrying
Natural Gas of 0.60 Specific Gravity Based on Pressure Drop of 12.7 mm Water Column

Pipe Size, mm	Length in mm										
	914	1829	2743	3719	4633	5547	6492	7315	8352	9266	11582
15	1.4	0.9	0.8	0.6	0.6	0.5	0.5	0.4	0.4	0.4	0.3
20	2.9	2.0	1.6	1.3	1.2	1.1	1.0	0.9	0.9	0.8	0.7
25	5.4	3.7	3.0	2.5	2.3	2.0	1.9	1.7	1.6	1.5	1.4
32	11.0	7.6	6.1	5.2	4.6	4.2	3.9	3.6	3.4	3.2	2.8
40	16.5	11.4	9.1	7.8	6.9	6.3	5.8	5.4	5.0	4.8	4.2
50	31.9	21.9	17.6	15.1	13.3	12.1	11.1	10.3	9.7	9.2	8.1
65	50.8	34.9	28.0	24.0	21.3	19.3	17.7	16.5	15.5	14.6	13.0
80	89.8	61.7	49.6	42.4	37.6	34.1	31.3	29.1	27.3	25.8	22.9
90	131.4	90.3	72.5	62.1	55.0	49.9	45.9	42.7	40.0	37.8	33.5
100	183.1	125.9	101.1	86.5	76.7	69.5	63.9	59.5	55.8	52.7	46.7

Pipe Size, mm	Length in mm									
	13899	18532	23165	27798	32431	37064	41697	46330	50963	55596
15	0.3	0.3	0.2	0.2	0.2	0.2	0.2	0.2	0.2	0.1
20	0.7	0.6	0.6	0.5	0.4	0.4	0.4	0.3	0.3	0.3
25	1.2	1.1	0.9	0.9	0.8	0.7	0.7	0.6	0.6	0.6
32	2.6	2.2	1.9	1.8	1.6	1.5	1.4	1.3	1.3	1.2
40	3.8	3.3	2.9	2.6	2.4	2.2	2.1	2.0	1.9	1.8
50	7.4	6.3	5.6	5.1	4.7	4.3	4.1	3.8	3.6	3.5
65	11.7	10.0	8.9	8.1	7.4	6.9	6.5	6.1	5.8	5.5
80	20.7	17.8	15.7	14.3	13.1	12.2	11.4	10.8	10.3	9.8
90	30.4	26.0	23.0	20.9	19.2	17.9	16.8	15.8	15.0	14.3
100	42.3	36.2	32.1	29.1	26.8	24.9	23.4	22.1	21.0	20.0

FIGURE 8.4 Here, the metric relationship between pipe size and natural gas carrying capacity is explained. (*Reprinted from the 2000 Uniform Plumbing Code with the permission of the International Association of Plumbing and Mechanical Officials*)

Medium Pressure Natural Gas Systems for Sizing Gas Piping Systems
Carrying Gas of 0.60 Specific Gravity

Capacity of Pipes of Different Diameters and Lengths in Cubic Feet Per Hour for
Gas Pressure for 2.0 psi with a Drop to 1.5 psi

Pipe Size, Inches	Length in Feet											
	50	100	150	200	250	300	350	400	450	500	550	600
1/2	466	320	257	220	195	177	163	151	142	134	127	121
3/4	974	669	537	460	408	369	340	316	297	280	266	254
1	1834	1261	1012	866	768	696	640	595	559	528	501	478
1-1/4	3766	2588	2078	1799	1577	1429	1314	1223	1147	1084	1029	982
1-1/2	5642	3878	3114	2685	2362	2140	1969	1832	1719	1624	1542	1471
2	10,867	7469	5998	5133	4549	4122	3792	3528	3310	3127	2970	2833
2-1/2	17,320	11,904	9559	8181	7251	6570	6044	5623	5276	4984	4733	4515
3	30,818	21,044	16,899	14,463	12,818	11,614	10,685	9940	9327	8810	8367	7983

	650	700	750	800	850	900	950	1000	1100	1200	1300	1400
1/2	116	112	108	104	101	97	95	92	87	83	80	77
3/4	243	234	225	217	210	204	198	193	183	174	167	161
1	458	440	424	409	396	384	373	363	345	329	315	302
1-1/4	940	903	870	840	813	788	766	745	707	675	646	621
1-1/2	1409	1353	1304	1259	1218	1181	1147	1116	1060	1011	968	930
2	2713	2606	2511	2425	2347	2275	2209	2149	2041	1947	1865	1791
2-1/2	4324	4154	4002	3865	3740	3626	3522	3425	3253	3103	2972	2855
3	7644	7344	7075	6832	6612	6410	6225	6055	5751	5486	5254	5047
4	15,592	14,979	14,430	13,935	13,486	13,075	12,698	12,351	11,730	11,190	10,716	10,295

	1500	1600	1700	1800	1900	2000	2100	2200	2300	2400	2500	2600
1/2	74	71	69	67	65	63	62	60	59	57	56	55
3/4	155	149	145	140	136	132	129	126	123	120	117	115
1	291	281	272	264	256	249	243	237	231	226	221	216
1-1/4	598	578	559	542	526	512	499	486	475	464	454	444
1-1/2	896	865	837	812	788	767	747	728	711	695	680	665
2	1726	1667	1613	1564	1519	1477	1439	1403	1369	1338	1309	1282
2-1/2	2751	2656	2570	2492	2420	2354	2293	2236	2183	2133	2086	2043
3	4862	4696	4544	4406	4279	4162	4053	3953	3859	3771	3688	3611
4	9918	9578	9269	8986	8727	8488	8267	8062	7870	7691	7523	7365
5	17,943	17,327	16,768	16,258	15,789	15,357	14,957	14,585	14,238	13,914	13,610	13,325
6	29,054	28,057	27,151	26,325	25,566	24,866	24,218	23,616	23,055	22,531	22,038	21,576

FIGURE 8.5 The delivery capacity per hour in cubic feet is examined here. *(Reprinted from the 2000 Uniform Plumbing Code with the permission of the International Association of Plumbing and Mechanical Officials)*

Medium Pressure Natural Gas Systems for Sizing Gas Piping Systems Carrying Gas of 0.60 Specific Gravity

Capacity of Pipes of Different Diameters and Lengths in Liters Per Second for Gas Pressure of 13.8 kPa with a Drop to 10.3 kPa

Pipe Size, mm	Length in Meters											
	15.2	30.4	45.6	60.8	76.0	91.2	106.4	121.6	136.8	152.0	167.2	182.4
15	3.7	2.5	2.0	1.7	1.5	1.4	1.3	1.2	1.1	1.1	1.0	1.0
20	7.7	5.3	4.2	3.6	3.2	2.9	2.7	2.5	2.3	2.2	2.1	2.0
25	14.4	9.9	8.0	6.8	6.0	5.5	5.0	4.7	4.4	4.2	3.9	3.8
32	29.6	20.4	16.4	14.0	12.4	11.2	10.3	9.6	9.0	8.5	8.1	7.7
40	44.4	30.5	24.5	21.0	18.6	16.8	15.5	14.4	13.5	12.8	12.1	11.6
50	85.5	58.8	47.2	40.4	35.8	32.4	29.8	27.8	26.0	24.6	23.4	22.3
65	136.3	93.6	75.2	64.4	57.0	51.7	47.6	44.2	41.5	39.2	37.2	35.5
80	240.6	165.6	132.9	113.8	100.8	91.4	84.1	78.2	73.4	69.3	65.8	62.8
	197.6	212.8	228.0	243.2	258.4	273.6	288.8	304.0	334.4	364.8	395.2	425.6
15	0.9	0.9	0.8	0.8	0.8	0.8	0.7	0.7	0.7	0.7	0.6	0.6
20	1.9	1.8	1.8	1.7	1.7	1.6	1.6	1.5	1.4	1.4	1.3	1.3
25	3.6	3.5	3.3	3.2	3.1	3.0	2.9	2.9	2.7	2.6	2.5	2.4
32	7.4	7.1	6.8	6.6	6.4	6.2	6.0	5.9	5.6	5.3	5.1	4.9
40	11.1	10.6	10.3	9.9	9.6	9.3	9.0	8.8	8.3	8.0	7.6	7.3
50	21.3	20.5	19.8	19.1	18.5	17.9	17.4	16.9	16.1	15.3	14.7	14.1
65	34.0	32.7	31.5	30.4	29.4	28.5	27.7	26.9	25.6	24.4	23.4	22.5
80	60.1	57.8	55.7	53.7	52.0	50.4	49.0	47.6	45.2	43.2	41.3	39.7
100	122.7	117.8	113.5	109.6	106.1	102.9	99.9	97.2	92.3	88.0	84.3	81.0
	456.0	486.4	516.8	547.2	577.6	608.0	638.4	668.8	669.2	729.6	760.0	790.4
15	0.6	0.6	0.5	0.5	0.5	0.5	0.5	0.5	0.5	0.5	0.4	0.4
20	1.2	1.2	1.1	1.1	1.1	1.0	1.0	1.0	1.0	0.9	0.9	0.9
25	2.3	2.2	2.1	2.1	2.0	2.0	1.9	1.9	1.8	1.8	1.7	1.7
32	4.7	4.5	4.4	4.3	4.1	4.0	3.9	3.8	3.7	3.6	3.6	3.5
40	7.0	6.8	6.6	6.4	6.2	6.0	5.9	5.7	5.6	5.5	5.3	5.2
50	13.6	13.1	12.7	12.3	11.9	11.6	11.3	11.0	10.8	10.5	10.3	10.1
65	21.6	20.9	20.2	19.6	19.0	18.5	18.0	17.6	17.2	16.8	16.4	16.1
80	38.3	36.9	35.7	34.7	33.7	32.7	31.9	31.1	30.4	29.7	29.0	28.4
100	78.0	75.3	72.9	70.7	68.7	66.8	65.0	63.4	61.9	60.5	59.2	57.9
125	141.2	136.3	131.9	127.9	124.2	120.8	117.7	114.7	112.0	109.5	107.1	104.8
150	228.6	220.7	213.6	207.1	201.1	195.6	190.5	185.8	181.4	177.2	173.4	169.7

FIGURE 8.6 This is a metric version of the previous table. (*Reprinted from the 2000 Uniform Plumbing Code with the permission of the International Association of Plumbing and Mechanical Officials*)

Medium Pressure Natural Gas Systems for Sizing Gas Piping Systems Carrying Gas of 0.60 Specific Gravity

Capacity of Pipes of Different Diameters and Lengths in Cubic Feet Per Hour for Gas Pressure of 3.0 psi with a Drop to 1.5 psi

Pipe Size, Inches	50	100	150	200	250	300	350	400	450	500	550	600
						Length In Feet						
1/2	857	589	473	405	359	325	299	278	261	247	234	224
3/4	1793	1232	990	847	751	680	626	582	546	516	490	467
1	3377	2321	1864	1595	1414	1281	1179	1096	1029	972	923	881
1-1/4	6934	4766	3827	3275	2903	2630	2420	2251	2112	1995	1895	1808
1-1/2	10,369	7140	5734	4908	4349	3941	3626	3373	3165	2989	2839	2709
2	20,008	13,752	11,043	9451	8377	7590	6983	6496	6095	5757	5468	5216
2-1/2	31,890	21,918	17,601	15,064	13,351	12,097	11,129	10,353	9714	9176	8715	8314
3	56,376	38,747	31,115	26,631	23,602	21,385	19,674	18,303	17173	16,222	15,406	14,698

Pipe Size	650	700	750	800	850	900	950	1000	1100	1200	1300	1400
1/2	214	206	198	191	185	180	174	170	161	154	147	141
3/4	448	430	414	400	387	375	365	355	337	321	308	296
1	843	810	780	754	729	707	687	668	634	605	580	557
1-1/4	1731	1663	1602	1547	1497	1452	1410	1371	1302	1242	1190	1143
1-1/2	2594	2492	2401	2318	2243	2175	2112	2055	1951	1862	1783	1713
2	4995	4799	4623	4465	4321	4189	4068	3957	3758	3585	3433	3298
2-1/2	7962	7649	7369	7116	6886	6677	6484	6307	5990	5714	5472	5257
3	14,075	13,522	13,027	12,580	12,174	11,803	11,463	11,149	10,589	10,102	9674	9294
4	28,709	27,581	26,570	25,658	24,831	24,074	23,380	22,741	21,598	20,605	19,731	18,956

Pipe Size	1500	1600	1700	1800	1900	2000	2100	2200	2300	2400	2500	2600
1/2	136	131	127	123	120	117	114	111	108	106	103	101
3/4	285	275	266	258	251	244	237	231	226	221	216	211
1	536	518	501	486	472	459	447	436	426	416	407	398
1-1/4	1101	1063	1029	998	969	942	918	895	874	854	835	818
1-1/2	1650	1593	1542	1495	1452	1412	1375	1341	1309	1279	1252	1225
2	3178	3069	2970	2879	2796	2720	2649	2583	2522	2464	2410	2360
2-1/2	5064	4891	4733	4589	4457	4335	4222	4117	4019	3927	3842	3761
3	8953	8846	8367	8112	7878	7663	7463	7278	7105	6943	6791	6649
4	18,282	17,635	17,066	16,546	16,069	15,629	15,222	14,844	14,491	14,161	13,852	13,561
5	33,038	31,904	30,875	29,935	29,072	28,276	27,539	26,855	26,217	25,620	25,060	24,534
6	53,496	51,660	49,993	48,471	47,074	45,785	44,593	43,484	42,451	41,485	40,579	39,727

FIGURE 8.7 This chart shows the figures for a higher gas pressure of 3.0 psi. *(Reprinted from the 2000 Uniform Plumbing Code with the permission of the International Association of Plumbing and Mechanical Officials)*

**Medium Pressure Natural Gas Systems for Sizing Gas Piping Systems
Carrying Gas of 0.60 Specific Gravity**

Capacity of Pipes of Different Diameters and Lengths in Liters Per Second for
Gas Pressure of 20.7 kPa with a Drop to 10.3 kPa

Pipe Size, mm	Length in Meters											
	15.2	30.4	45.6	60.8	76.0	91.2	106.4	121.6	136.8	152.0	167.2	182.4
15	6.7	4.6	3.7	3.2	2.8	2.6	2.4	2.2	2.1	1.9	1.8	1.8
20	14.1	9.7	7.8	6.7	5.9	5.4	4.9	4.6	4.3	4.1	3.9	3.7
25	26.6	18.3	14.7	12.6	11.1	10.1	9.3	8.6	8.1	7.6	7.3	6.9
32	54.5	37.5	30.1	25.8	22.8	20.7	19.0	17.7	16.6	15.7	14.9	14.2
40	81.7	56.2	45.1	38.6	34.2	31.0	28.5	26.5	24.9	23.5	22.3	21.3
50	157.4	108.2	86.9	74.4	65.9	59.7	54.9	51.1	47.9	45.3	43.0	41.0
65	250.9	172.4	138.5	118.5	105.0	95.2	87.6	81.5	76.4	72.2	68.6	65.4
80	443.5	304.8	244.8	209.5	185.7	168.2	154.8	144.0	135.1	127.6	121.2	115.6
	197.6	212.8	228.0	243.2	258.4	273.6	288.8	304.0	334.4	364.8	395.2	425.6
15	1.7	1.6	1.6	1.5	1.5	1.4	1.4	1.3	1.3	1.2	1.2	1.1
20	3.5	3.4	3.3	3.1	3.0	3.0	2.9	2.8	2.6	2.5	2.4	2.3
25	6.6	6.4	6.1	5.9	5.7	5.6	5.4	5.3	5.0	4.8	4.6	4.4
32	13.6	13.1	12.6	12.2	11.8	11.4	11.1	10.8	10.2	9.8	9.4	9.0
40	20.4	19.6	18.9	18.2	17.6	17.1	16.6	16.2	15.4	14.6	14.0	13.5
50	39.3	37.8	36.4	35.1	34.0	33.0	32.0	31.1	29.6	28.2	27.0	25.9
65	62.6	60.2	58.0	56.0	54.2	52.5	51.0	49.6	47.1	45.0	43.0	41.4
80	110.7	106.4	102.5	99.0	95.8	92.9	90.2	87.7	83.3	79.5	76.1	73.1
100	225.9	217.0	209.0	201.9	195.3	189.4	183.9	178.9	169.9	162.1	155.2	149.1
	456.0	486.4	516.8	547.2	577.6	608.0	638.4	668.8	699.2	729.6	760.0	790.4
15	1.1	1.0	1.0	1.0	0.9	0.9	0.9	0.9	0.9	0.8	0.8	0.8
20	2.2	2.2	2.1	2.0	2.0	1.9	1.9	1.8	1.8	1.7	1.7	1.7
25	4.2	4.1	3.9	3.8	3.7	3.6	3.5	3.4	3.3	3.3	3.2	3.1
32	8.7	8.4	8.1	7.8	7.6	7.4	7.2	7.0	6.9	6.7	6.6	6.4
40	13.0	12.5	12.1	11.8	11.4	11.1	10.8	10.6	10.3	10.1	9.8	9.6
50	25.0	24.1	23.4	22.6	22.0	21.4	20.8	20.3	19.8	19.4	19.0	18.6
65	39.8	38.5	37.2	36.1	35.1	34.1	33.2	32.4	31.6	30.9	30.2	29.6
80	70.4	68.0	65.8	63.8	62.0	60.3	58.7	57.3	55.9	54.6	53.4	52.3
100	143.7	138.7	134.3	130.2	126.4	123.0	119.8	116.8	114.0	111.4	109.0	106.7
125	259.9	251.0	242.9	235.5	228.7	222.4	216.7	211.3	206.2	201.6	197.2	193.0
150	420.8	406.4	393.3	381.5	370.3	360.2	350.8	342.1	334.0	326.4	319.2	312.5

FIGURE 8.8 This table shows the metric figures for a gas pressure of 3.0 psi. *(Reprinted from the 2000 Uniform Plumbing Code with the permission of the International Association of Plumbing and Mechanical Officials)*

Medium Pressure Natural Gas Systems for Sizing Gas Piping Systems Carrying Gas of 0.60 Specific Gravity

Capacity of Pipes of Different Diameters and Lengths in Cubic Feet Per Hour for Gas Pressure of 5.0 psi with a Drop to 1.5 psi

Pipe Size, Inches	50	100	150	200	250	300	350	400	450	500	550	600
					Length in Feet							
1/2	1399	961	772	661	586	531	488	454	426	402	382	365
3/4	2925	2010	1614	1381	1224	1109	1021	949	891	842	799	762
1	5509	3786	3041	2602	2306	2090	1923	1789	1678	1585	1506	1436
1-1/4	11,311	7774	6243	5343	4735	4291	3947	3672	3445	3255	3091	2949
1-1/2	16,947	11,648	9353	8005	7095	6429	5914	5502	5162	4876	4631	4418
2	32,638	22,432	18,014	15,417	13,664	12,381	11,390	10,596	9942	9391	8919	8509
2-1/2	52,020	35,753	28,711	24,573	21,779	19,733	18,154	16,889	15,846	14,968	14,216	13,562
3	91,962	63,205	50,756	43,441	38,501	34,884	32,093	29,865	28,013	26,461	25,131	23,976

Pipe Size	650	700	750	800	850	900	950	1000	1100	1200	1300	1400
1/2	349	335	323	312	302	293	284	277	263	251	240	231
3/4	730	701	676	653	632	612	595	578	549	524	502	482
1	1375	1321	1273	1229	1190	1153	1120	1089	1035	987	945	908
1-1/4	2824	2713	2614	2524	2442	2368	2300	2237	2124	2027	1941	1865
1-1/2	4231	4065	3916	3781	3659	3548	3446	3351	3183	3037	2908	2794
2	8149	7828	7542	7283	7048	6833	6636	6455	6130	5848	5600	5380
2-1/2	12,988	12,477	12,020	11,608	11,233	10,891	10,577	10,288	9771	9321	8926	8575
3	22,960	22,057	21,249	20,520	19,858	19,253	18,698	18,187	17,273	16,478	15,780	15,160
4	46,830	44,990	43,342	41,855	40,504	39,271	38,139	37,095	35,231	33,611	32,186	30,921

Pipe Size	1500	1600	1700	1800	1900	2000	2100	2200	2300	2400	2500	2600
1/2	222	214	208	201	195	190	185	181	176	172	168	165
3/4	464	449	434	421	409	398	387	378	369	360	352	345
1	875	845	818	798	770	749	729	711	694	678	664	650
1-1/4	1796	1735	1679	1628	1581	1537	1497	1460	1425	1393	1363	1334
1-1/2	2691	2599	2515	2439	2368	2303	2243	2188	2136	2087	2042	1999
2	5183	5005	4844	4696	4561	4436	4321	4213	4113	4020	3932	3849
2-1/2	8261	7978	7720	7485	7270	7071	6886	6715	6556	6406	6267	6135
3	14,605	14,103	13,648	13,233	12,851	12,500	12,174	11,871	11,589	11,326	11,078	10,846
4	29,789	28,766	27,838	26,991	26,213	25,495	24,831	24,214	23,639	23,100	22,596	22,121
5	53,892	52,043	50,363	48,830	47,422	46,124	44,923	43,806	42,765	41,792	40,879	40,021
6	87,263	84,269	81,550	79,067	76,787	74,686	72,740	70,932	69,247	67,671	66,193	64,803

FIGURE 8.9 This table shows capacity of different-sized pipes at a pressure of 5.0 psi. *(Reprinted from the 2000 Uniform Plumbing Code with the permission of the International Association of Plumbing and Mechanical Officials)*

Medium Pressure Natural Gas Systems for Sizing Gas Piping Systems Carrying Gas of 0.60 Specific Gravity

Capacity of Pipes of Different Diameters and Lengths in Liters Per Second for Gas Pressure of 34.5 kPa with a Drop to 10.3 kPa

Pipe Size, mm	Length in Meters											
	15.2	30.4	45.6	60.8	76.0	91.2	106.4	121.6	136.8	152.0	167.2	182.4
15	11.0	7.6	6.1	5.2	4.6	4.2	3.8	3.6	3.4	3.2	3.0	2.9
20	23.0	15.8	12.7	10.9	9.6	8.7	8.0	7.5	7.0	6.6	6.3	6.0
25	46.3	29.8	23.9	20.5	18.1	16.4	15.1	14.1	13.0	12.5	11.8	11.3
32	89.0	61.2	49.1	42.0	37.3	33.8	31.1	28.9	27.1	25.6	24.3	23.2
40	133.3	91.6	73.6	63.0	55.8	50.6	46.5	43.3	40.6	38.4	36.4	34.8
50	256.8	176.5	141.7	121.3	107.5	97.4	89.6	83.4	78.2	73.9	70.2	66.9
65	409.2	281.3	225.9	193.3	171.3	155.2	142.8	132.9	124.7	117.8	111.8	106.7
80	723.5	497.2	399.3	341.7	302.9	274.4	252.5	234.9	220.4	208.2	197.7	188.6

Pipe Size, mm	197.6	212.8	228.0	243.2	258.4	273.6	288.0	304.0	334.4	364.8	395.2	425.6
15	2.7	2.6	2.5	2.5	2.4	2.3	2.2	2.2	2.1	2.0	1.9	1.8
20	5.7	5.5	5.3	5.1	5.0	4.8	4.7	4.6	4.3	4.1	3.9	3.8
25	10.8	10.4	10.0	9.7	9.4	9.1	8.8	8.6	8.1	7.8	7.4	7.1
32	22.2	21.3	20.6	19.9	19.2	18.6	18.1	17.6	16.7	15.9	15.3	14.7
40	33.3	32.0	30.8	39.7	28.8	27.9	27.1	26.4	25.0	23.9	22.9	22.0
50	64.1	61.6	59.3	57.3	55.4	53.8	52.2	50.8	48.2	46.0	44.1	42.3
65	102.2	98.2	94.6	91.3	88.4	85.7	83.2	80.9	76.9	73.3	70.2	67.5
80	180.6	173.5	167.2	161.4	156.2	151.5	147.1	143.1	135.9	129.6	124.1	119.3
100	368.4	353.9	341.0	329.3	318.6	308.9	300.0	291.8	277.2	264.4	253.2	243.3

Pipe Size, mm	456.0	486.4	516.8	547.2	577.6	608.0	638.4	668.8	699.2	729.6	760.0	790.4
15	1.7	1.7	1.6	1.6	1.5	1.5	1.5	1.4	1.4	1.4	1.3	1.3
20	3.7	3.5	3.4	3.3	3.2	3.1	3.0	3.0	2.9	2.8	2.8	2.7
25	6.9	6.6	6.4	6.2	6.1	5.9	5.7	5.6	5.5	5.3	5.2	5.1
32	14.1	13.6	13.2	12.8	12.4	12.1	11.8	11.5	11.2	11.0	10.7	10.5
40	21.2	20.4	19.8	19.2	18.6	18.1	17.6	17.2	16.8	16.4	16.1	15.7
50	40.8	39.4	38.1	36.9	35.9	34.9	34.0	33.1	32.4	31.6	30.9	30.3
65	65.0	62.8	60.7	58.9	57.2	55.6	54.2	52.8	51.6	50.4	49.3	48.3
80	114.9	111.0	107.4	104.1	101.1	98.3	95.8	93.4	91.2	89.1	87.2	85.3
100	234.3	226.3	219.0	212.3	206.2	200.6	195.3	190.5	186.0	181.7	177.8	174.0
125	424.0	409.4	396.2	384.1	373.1	362.9	353.4	344.6	336.4	328.8	321.6	314.8
150	686.5	662.9	641.6	622.0	604.1	587.6	572.2	558.0	544.8	532.4	520.7	509.8

FIGURE 8.10 This table shows the metric version of the previous table. *(Reprinted from the 2000 Uniform Plumbing Code with the permission of the International Association of Plumbing and Mechanical Officials)*

Maximum Capacity of Pipe in Thousands of BTU per Hour of Undiluted Liquified Petroleum Gases

(Based on a Pressure Drop of 0.5 Inch Water Column)
Low Pressure 11" Water Column

Length In Feet

Pipe Size, Inches	10	20	30	40	50	60	70	80	90	100	125	150	200
1/2	275	189	152	129	114	103	96	89	83	78	69	63	55
3/4	567	393	315	267	237	217	196	185	173	162	146	132	112
1	1071	732	590	504	448	409	378	346	322	307	275	252	213
1-1/4	2205	1496	1212	1039	913	834	771	724	677	630	567	511	440
1-1/2	3307	2299	1858	1559	1417	1275	1181	1086	1023	976	866	787	675
2	6221	4331	3465	2992	2646	2394	2205	2047	1921	1811	1606	1496	1260

FIGURE 8.11 The maximum capacity of pipes in thousands of Btu per hour for petroleum gas is shown here. *(Reprinted from the 2000 Uniform Plumbing Code with the permission of the International Association of Plumbing and Mechanical Officials)*

Maximum Capacity of Pipe in Thousands of Watts of Undiluted Liquified Petroleum Gases

(Based on a Pressure Drop of 12.7 mm Water Column)
Low Pressure 279 mm Water Column

Length in Millimeters

Pipe Size, mm	3048	6096	9144	12192	15240	18288	21336	24384	27432	30480	38100	45720	60960
15	80.6	55.4	44.5	37.8	33.4	30.2	28.1	26.1	24.3	22.9	20.2	18.5	16.1
20	166.1	115.2	92.3	78.2	69.4	63.6	57.4	54.2	50.7	47.5	42.8	38.7	32.8
25	313.8	214.5	172.9	147.7	131.3	119.8	110.8	101.4	94.4	90.0	80.6	73.8	62.4
32	646.1	438.3	355.1	304.4	267.5	244.4	225.9	212.1	198.4	184.6	166.1	149.7	128.9
40	969.0	673.6	544.4	456.8	415.2	373.6	346.0	318.2	299.7	286.0	253.7	230.6	197.8
50	1822.8	1269.0	1015.2	876.7	775.3	701.4	646.1	600.0	562.9	530.6	470.6	438.3	369.2

FIGURE 8.12   This table shows the metric version of the previous table. (*Reprinted from the 2000 Uniform Plumbing Code with the permission of the International Association of Plumbing and Mechanical Officials*)

For Undiluted Liquified Petroleum Gas Pressure of 10.0 psi with Maximum Pressure Drop of 3.0 psi

Maximum Delivery Capacity in Cubic Feet of Gas per Hour of IPS Pipe of Different Diameters and Lengths Carrying Undiluted Liquified Petroleum Gas of 1.52 Specific Gravity

Pipe Size, Inches	Length in Feet											
	50	100	150	200	250	300	350	400	450	500	550	600
1/2	1000	690	550	470	420	390	350	325	300	285	272	260
3/4	2070	1423	1142	978	867	785	722	672	631	596	566	540
1	3899	2680	2152	1842	1632	1479	1361	1266	1188	1122	1066	1017
1-1/4	8005	5502	4418	3782	3351	3037	2794	2599	2439	2303	2188	2067
1-1/2	11,994	8244	6620	5666	5022	4550	4186	3894	3654	3451	3278	3127
2	23,100	15,877	12,750	10,912	9671	8763	8062	7500	7037	6647	6313	6023
2-1/2	36,818	25,305	20,321	17,392	15,414	13,966	12,849	11,953	11,215	10,594	10,062	9599
3	65,088	44,734	35,923	30,746	27,249	24,690	22,714	21,131	19,827	18,728	17,787	16,969

	650	700	750	800	850	900	950	1000	1100	1200	1300	1400
1/2	250	240	230	222	215	208	202	196	188	180	171	164
3/4	517	496	478	462	447	433	421	409	389	371	355	341
1	973	935	901	870	842	816	793	771	732	699	669	643
1-1/4	1999	1920	1850	1786	1729	1676	1628	1583	1504	1434	1374	1320
1-1/2	2995	2877	2772	2676	2590	2511	2439	2372	2253	2149	2058	1977
2	5767	5541	5338	5155	4988	4836	4697	4568	4339	4139	3964	3808
2-1/2	9192	8831	8507	8215	7950	7708	7486	7281	6915	6597	6318	6069
3	16,250	15,611	15,040	14,523	14,055	13,627	13,234	12,872	12,225	11,663	11,169	10,730
4	33,145	31,842	30,676	29,623	28,667	27,795	26,993	26,255	24,935	23,789	22,780	21,885

	1500	1600	1700	1800	1900	2000	2100	2200	2300	2400	2500	2600
1/2	158	152	148	143	139	136	133	130	127	124	121	118
3/4	329	317	307	298	289	281	274	267	261	255	249	244
1	619	598	579	561	545	530	516	503	491	480	470	460
1-1/4	1271	1228	1188	1152	1119	1088	1060	1033	1009	986	964	944
1-1/2	1905	1839	1780	1726	1676	1630	1588	1548	1512	1477	1445	1415
2	3669	3543	3428	3324	3228	3140	3058	2982	2911	2845	2783	2724
2-1/2	5847	5646	5464	5298	5145	5004	4874	4753	4640	4534	4435	4342
3	10,337	9982	9680	9366	9096	8847	8616	8402	8203	8016	7841	7676
4	21,083	20,360	19,705	19,103	18,552	18,045	17,575	17,138	16,731	16,350	15,993	15,657
5	38,143	36,834	35,645	34,560	33,564	32,645	31,795	31,005	30,268	29,579	28,933	28,325
6	61,762	59,643	57,718	55,961	54,348	52,860	51,483	50,204	49,011	47,895	46,849	45,865

FIGURE 8.13 Pipe size and length for carrying undiluted liquefied petroleum gas of 1.52 specific gravity is examined. *(Reprinted from the 2000 Uniform Plumbing Code with the permission of the International Association of Plumbing and Mechanical Officials)*

For Undiluted Liquified Petroleum Gas Pressure of 68.9 kPa with Maximum Pressure Drop of 20.7 kPa

Maximum Delivery Capacity in Liters per Second of IPS Pipe of Different Diameters and Lengths Carrying Undiluted Liquified Petroleum Gas of 1.52 Gravity

Pipe Size,	Length in Meters											
mm	15.2	30.4	45.6	60.8	76.0	91.2	106.4	121.6	136.8	152.0	167.2	182.4
15	8.0	5.5	4.4	3.8	3.4	3.0	2.8	2.6	2.4	2.3	2.2	2.1
20	16.6	11.4	9.1	7.8	6.9	6.3	5.8	5.4	5.1	4.8	4.5	4.3
25	31.2	21.4	17.2	14.7	13.1	11.8	10.9	10.1	9.5	9.0	8.5	8.1
32	64.0	44.0	35.3	30.3	26.8	24.3	22.4	20.8	19.5	18.4	17.5	16.7
40	96.0	66.0	53.0	45.3	40.2	36.4	33.5	31.2	29.2	27.6	26.2	25.0
50	184.8	127.0	102.0	87.3	77.4	70.1	64.5	60.0	56.3	53.2	50.5	48.2
65	294.5	202.4	162.6	139.1	123.3	111.7	102.8	95.6	89.7	84.8	80.5	76.8
80	520.7	357.9	287.4	246.0	218.0	197.5	181.7	169.1	158.6	149.8	142.3	135.8
	197.6	212.8	228.0	243.2	258.4	273.6	288.8	304.0	334.4	364.8	395.2	425.6
15	2.0	1.9	1.8	1.8	1.7	1.7	1.6	1.6	1.5	1.4	1.4	1.3
20	4.1	4.0	3.8	3.7	3.6	3.5	3.4	3.3	3.1	3.0	2.8	2.7
25	7.8	7.5	7.2	7.0	6.7	6.5	6.3	6.2	5.9	5.6	5.4	5.1
32	16.0	15.4	14.8	14.3	13.8	13.4	13.0	12.7	12.0	11.5	11.0	10.6
40	24.0	23.0	22.2	21.4	20.7	20.1	19.5	19.0	18.0	17.2	16.5	15.8
50	46.1	44.3	42.7	41.2	39.9	38.7	37.6	36.5	34.7	33.1	31.7	30.5
65	73.5	70.7	68.1	65.7	63.6	61.7	60.0	58.2	55.3	52.8	50.5	48.6
80	130.0	124.9	120.3	116.2	112.4	109.0	105.9	103.0	97.8	93.3	89.4	85.8
100	265.2	254.7	245.4	237.0	229.3	222.4	215.9	210.0	199.5	190.3	182.2	175.1
	456.0	486.4	516.8	547.2	577.6	608.0	638.4	668.8	699.2	729.6	760.0	790.4
15	1.3	1.2	1.2	1.1	1.1	1.1	1.1	1.0	1.0	1.0	1.0	0.9
20	2.6	2.5	2.5	2.4	2.3	2.2	2.2	2.1	2.1	2.0	2.0	2.0
25	5.0	4.8	4.6	4.5	4.4	4.2	4.1	4.0	3.9	3.8	3.8	3.7
32	10.2	9.8	9.1	9.2	9.0	8.7	8.5	8.3	8.1	7.9	7.7	7.6
40	15.2	14.7	14.2	13.8	13.4	13.0	12.7	12.4	12.1	11.8	11.6	11.3
50	29.4	28.3	27.4	26.6	25.8	25.1	24.5	23.9	23.3	22.8	22.3	21.8
65	46.8	45.2	43.7	42.4	41.2	40.0	39.0	38.0	37.1	36.3	35.5	34.7
80	82.7	79.9	77.3	74.9	72.8	70.8	68.9	67.2	65.6	64.1	62.7	61.4
100	168.7	162.9	157.6	152.8	148.4	144.4	140.6	137.1	133.8	130.8	127.9	125.3
125	305.1	294.7	285.2	276.5	268.5	261.2	254.4	248.0	242.1	236.6	231.5	226.6
150	494.1	477.1	461.7	447.7	434.8	422.9	411.9	401.6	392.1	383.2	374.8	366.9

FIGURE 8.14 This is the metric version of the previous table. *(Reprinted from the 2000 Uniform Plumbing Code with the permission of the International Association of Plumbing and Mechanical Officials)*

Capacities of Listed Metal Appliance Connectors for Use with Gas Pressures Not Less Than an Eight Inch Water Column

Capacities for Various Lengths, in Thousands Btu/h
(Based on Pressure Drop of 0.2" Water Column Natural Gas of 1100 Btu/cu. ft.)

Semi-Rigid Connector O.D., Inches	Flexible Connector Nominal I.D., Inches	1 foot	1-1/2 feet	2 feet	2-1/2 feet	3 feet	4 feet	5 feet	6 feet
				All Gas Appliances			Ranges and Dryers Only		
3/8	1/4	40	33	29	27	25			
1/2	3/8	93	76	66	62	58			
5/8	1/2	189	155	134	125	116	101	90	80
	3/4	404	330	287	266	244			
	1	803	661	573	534	500			

Notes:
1. Flexible connector listings are based on the nominal internal diameter.
2. Semi-rigid connector listings are based on the outside diameter.
3. Gas connectors are certified by the testing agency as complete assemblies, including the fittings and valves. Capacities shown are based on the use of fittings and valves supplied with the connector.
4. Capacities for LPG are 1.6 times the natural gas capacities shown.

Example: Capacity of a 1/4" flexible connector one (1) foot long is 40,000 x 1.6 = 64,000 Btu/h

FIGURE 8.15 This table shows the capacities of various metal appliance connectors. *(Reprinted from the 2000 Uniform Plumbing Code with the permission of the International Association of Plumbing and Mechanical Officials)*

Capacities of Listed Metal Appliance Connectors for Use with Gas Pressures Not Less Than a 203 mm Water Column

Semi-Rigid Connector O.D., mm	Flexible Connector Nominal I.D., mm	Capacities for Various Lengths, in Thousands Watts (Based on Pressure Drop of 50 Pa mm Water Column Natural Gas of 11.4 W/L)							
		305 mm	457 mm	610 mm	762 mm	914 mm	1219 mm	1524 mm	1829 mm
		All Gas Appliances					Ranges and Dryers Only		
10	6.4	11.7	9.7	8.5	7.9	7.3			
15	9.5	27.3	22.3	19.3	18.2	17.0			
18	12.7	55.4	45.4	39.3	36.6	34.0	29.6	26.4	23.4
	19.1	118.4	96.7	84.1	77.9	71.5			
	25.4	235.3	193.7	167.9	156.5	146.5			

Note: See Table 12-9 (English units).

Example: Capacity of a 6.4 mm flexible connector 0.3 m long is 11720W x 1:6 = 18752W

FIGURE 8.16 This is the metric version of the previous table. *(Reprinted from the 2000 Uniform Plumbing Code with the permission of the International Association of Plumbing and Mechanical Officials)*

Copper Tube – Low Pressure
Maximum Delivery Capacity* of Cubic Feet of Gas Per
Hour of Copper Tube Carrying Natural Gas of 0.60**
Specific Gravity at Low Pressure (Less than 14 inches
Water Column) Based on Pressure Drop of 0.50 Inch
Water Column:

Length of Tube, feet	Outside Diameter of Tube, Inches						
	3/8 (10 mm)	1/2 (15 mm)	5/8 (18 mm)	3/4 (20 mm)	7/8 (22 mm)	1-1/8 (28 mm)	1-3/8 (34 mm)
10	24	50	101	176	250	535	963
20	17	34	69	121	172	368	662
30	13	27	56	97	138	295	531
40	11	23	48	83	118	253	455
50	10	21	42	74	105	224	403
60	9.1	19	38	67	95	203	365
70	8.4	17	35	62	84	197	336
80	7.8	16	33	57	81	174	313
90	7.3	15	31	54	76	163	293
100	6.9	14	29	51	72	154	277
125	6.1	13	26	45	64	136	245
150	5.6	11	23	41	58	124	222
175	5.1	11	21	38	53	114	205
200	4.8	10	20	35	50	106	190
250	4.2	8.7	18	31	44	94	169

*Includes 20% factor for fittings.
**For other pressure drops values see Table 12-12.

FIGURE 8.17   This illustrates the use of copper tube for carrying gas at low pressure. *(Reprinted from the 2000 Uniform Plumbing Code with the permission of the International Association of Plumbing and Mechanical Officials)*

Specific Gravity
Multipliers to be Used with Copper Tube when Specific
Gravity of Gas is other than 0.60.

Specific Gravity	Multiplier	Specific Gravity	Multiplier
.35	1.31	1.00	.78
.40	1.23	1.10	.74
.45	1.16	1.20	.71
.50	1.10	1.30	.68
.55	1.04	1.40	.66
.60	1.00	1.50	.63
.65	.96	1.60	.59
.70	.93	1.70	.58
.75	.90	1.80	.56
.80	.87	1.90	.56
.85	.84	2.00	.55
.90	.82	2.10	.54

Adjustment for a gas with an average specific gravity
(relative density) other than 0.60 is achieved by
multiplying the CFH values of Tables 12-11, 12-13, or
12-14 by the appropriate multiplier.

FIGURE 8.18   Certain figures need to be used when specific gravity of gas is other than 060. *(Reprinted from the 2000 Uniform Plumbing Code with the permission of the International Association of Plumbing and Mechanical Officials)*

Copper Tube – Medium Pressure
Maximum Delivery Capacity* of Cubic Feet of Gas
Per Hour of Copper Tube Carrying Natural Gas of
0.60** Specific Gravity at Medium Pressure (2.0 psig)
Based on Pressure Drop of 1.0 psig:

Length of Tube, feet	Outside Diameter of Tube, inches						
	3/8 (10 mm)	1/2 (15 mm)	5/8 (18 mm)	3/4 (20 mm)	7/8 (22 mm)	1-1/8 (28 mm)	1-3/8 (34 mm)
10	222	458	932	1629	2311	4937	8889
20	153	315	641	1120	1589	3393	6109
30	123	253	515	899	1276	2725	4906
40	105	216	440	770	1092	2332	4199
50	93	192	390	682	968	2067	3721
60	84	174	354	618	877	1873	3372
70	78	160	325	569	807	1723	3102
80	72	149	303	529	750	1603	2886
90	68	140	254	496	704	1504	2708
100	64	132	268	469	665	1421	2558
125	57	117	238	415	589	1259	2267
150	51	106	215	376	534	1141	2054
175	44	91	184	322	457	976	1758
200	39	80	163	286	405	865	1558
250							

*Includes 20% factor for fittings.
**For other pressure drops values see Table 12-12.

FIGURE 8.19 This illustrates the use of copper tube at medium pressure. *(Reprinted from the 2000 Uniform Plumbing Code with the permission of the International Association of Plumbing and Mechanical Officials)*

Length of Tube, feet	\multicolumn{7}{c}{Outside Diameter of Tube, Inches}						
	3/8 (10 mm)	1/2 (15 mm)	5/8 (18 mm)	3/4 (20 mm)	7/8 (22 mm)	1-1/8 (28 mm)	1-3/8 (34 mm)
10	462	954	1941	3392	4812	10279	18504
20	318	656	1334	2331	3307	7064	12718
30	255	527	1071	1872	2656	5673	10213
40	218	451	917	1602	2273	4855	8741
50	194	399	812	1420	2015	4303	8741
60	175	362	736	1287	1825	3899	7019
70	161	333	677	1184	1679	3587	6458
80	150	310	630	1101	1562	3337	6008
90	141	291	591	1033	1466	3131	5637
100	133	274	558	976	1385	2958	5324
125	118	243	495	865	1227	2621	4719
150	107	220	448	784	1112	2375	4276
175	98	203	413	721	1023	2185	3934
200	91	189	384	671	952	2033	3659
250	81	167	340	594	843	1802	3243

Copper Tube – High Pressure
Maximum Delivery Capacity* of Cubic Feet of Gas Per Hour of Copper Tube Carrying Natural Gas of 0.60** Specific Gravity at High Pressure (5.0 psig) Based on Pressure Drop of 3.50 psig:

*Includes 20% factor for fittings.
**For other pressure drops values see Table 12-12.

FIGURE 8.20 This shows the use of copper tube at high pressure. *(Reprinted from the 2000 Uniform Plumbing Code with the permission of the International Association of Plumbing and Mechanical Officials)*

Support of Piping

	\multicolumn{2}{c}{Size of Pipe}			
	Inches	mm	Feet	mm
	1/2	15	6	1829
	3/4 or 1	20 or 25	8	2438
Horizontal	1-1/4 or larger	32 or larger	10	3048
Vertical	1-1/4 or larger	32 or larger	Every floor level	

FIGURE 8.23 Different size pipes are needed for either horizontal or vertical use at each floor level. *(Reprinted from the 2000 Uniform Plumbing Code with the permission of the International Association of Plumbing and Mechanical Officials)*

Capacities of Listed Metal Appliance Connectors for Use with Gas Pressures Less Than an Eight Inch Water Column

Capacities for Various Lengths, in Thousands Btu/h
(Based on Pressure Drop of 0.2" Water Column Natural Gas of 1100 Btu/cu. ft.)

Semi-Rigid Connector O.D, Inches	Flexible Connector Nominal I.D., Inches	1 foot	1-1/2 feet	2 feet	2-1/2 feet	3 feet	4 feet	5 feet	6 feet
		All Gas Appliances					Ranges and Dryers Only		
3/8	1/4	28	23	20	19	17			
1/2	3/8	66	54	47	44	41			
5/8	1/2	134	110	95	88	82	72	63	57
–	3/4	285	233	202	188	174			
–	1	567	467	405	378	353			

Notes:
1. Flexible connector listings are based on the nominal internal diameter.
2. Semi-rigid connector listings are based on the outside diameter.
3. Gas connectors are certified by the testing agency as complete assemblies, including the fittings and valves. Capacities shown are based on the use of fittings and valves supplied with the connector.
4. Capacities for LPG are 1.6 times the natural gas capacities shown.

Example: Capacity of a 1/4" flexible connector 1 foot long is 28,000 x 1.6 = 44,800 Btu/h

FIGURE 8.21 The capacities of different metal appliance connector is listed. *(Reprinted from the 2000 Uniform Plumbing Code with the permission of the International Association of Plumbing and Mechanical Officials)*

Capacities of Listed Metal Appliance Connectors for Use with Gas Pressures Less Than a 203 mm Water Column

Semi-Rigid Connector O.D, mm	Flexible Connector Nominal I.D., mm	Capacities for Various Lengths, in Thousands Watts (Based on Pressure Drop of 50 Pa Water Column Natural Gas of 11.4 W/L)							
		305 mm	457 mm	610 mm	762 mm	914 mm	1219 mm	1524 mm	1829 mm
		All Gas Appliances					Ranges and Dryers Only		
10	6.4	8.2	6.7	5.9	5.6	5			
15	9.5	19.3	15.8	13.8	12.9	12			
18	12.7	39.3	32.2	27.8	25.8	24	21.1	18.5	16.7
	19.1	83.5	68.3	59.2	55.1	51			
	25.9	166.1	136.8	118.7	110.8	103.4			

Note: See Table 12-10 (English units).

Example: Capacity of a 6.4 mm flexible connector 305 mm long is 8204 x 1.6 = 13126.4W

FIGURE 8.22 This is the metric version of the previous table. *(Reprinted from the 2000 Uniform Plumbing Code with the permission of the International Association of Plumbing and Mechanical Officials)*

Minimum Demand of Typical Gas Appliances in Btus (Watts) Per Hour

Appliance	Demand in Btu/h	Watts/h
Barbecue (residential)	50,000	14,650
Bunsen Burner	3,000	879
Domestic Clothesdryer	35,000	10,255
Domestic Gas Range	65,000	19,045
Domestic Recessed Oven Section	25,000	7,325
Domestic Recessed Top Burner Section	40,000	11,720
Fireplace Log Lighter (commercial)	50,000	14,650
Fireplace Log Lighter (residential)	25,000	7,325
Gas Engines (per horsepower)	10,000	2,930
Gas Refrigerator	3,000	879
Mobile Homes (see Appendix E)	—	—
Steam Boilers (per horsepower)	50,000	14,650
Storage Water Heater up to 30 gallon (114 L) tank	30,000	8,790
Storage Water Heater 40 (151 L) to 50 gallon (189 L) tank	50,000	14,650

FIGURE 8.24 Appliances place demands in Btu per hour. Here are some of the numbers you can work with. (*Reprinted from the 2000 Uniform Plumbing Code with the permission of the International Association of Plumbing and Mechanical Officials*)

RADIANT HEAT PIPING

The components that go into a hydronic heating system are both numerous and important. There are many pieces of equipment needed to make a safe, functional hydronic heating system. People generally think of a boiler and some type of heat emitter, such as baseboard heating units. Rarely is much thought given to circulating pumps, expansion tanks, zone valves, relief valves, and other essential elements of a safe system. The components of a hydronic heating system that may seem inconsequential are not. They are key elements of the system. Failing to install an expansion tank or relief valve could prove disastrous. Without the use of a circulating pump, a modern hydronic system will not function very well. Zone valves are an inexpensive alternative to multiple circulating pumps when creating multiple heating zones with a system. A number of factors come into play when designing and installing a hydronic heating system. Learning what is needed, where it is needed, and how to install it is essential if you wish to install effective hydronic heating systems.

MODERN PIPING MATERIALS

Modern piping materials in hydronic heating systems most often include copper tubing and cross-linked polyethylene tubing (PEX). Another type of tubing sometimes used is polybutylene (PB) tubing. Copper tubing is the most popular type of piping used for general convection heating, such as systems utilizing baseboard heating elements, kick-space heaters, and space heaters. PEX tubing is most often used for radiant floor heating systems. PB tubing was being used before PEX tubing, and it is still in use, but PEX is replacing it quickly as a prime choice.

The copper tubing used most in heating systems is type-M copper in a hard-drawn (rigid) form. Type-L copper is sometimes used in tight spots when a flexible, rolled copper tubing is more feasible. Distribution tubing typically has a diameter of ¾ inch. Since copper expands and contracts with temperature variations, the tubing must be supported properly to maintain a quiet heating system.

PEX tubing is a polymer (plastic) material. It is extremely flexible, sold in long coils, and suitable for many hydronic applications. One of the most effective applications for PEX tubing is found in radiant floor heating. Standard PEX tubing can handle water with a temperature of 180 degrees at 100 psi. If the pressure is reduced to 80 psi, the tubing can handle water temperatures up to 200 degrees.

Both copper and PEX tubing have their places in heating systems. Matching the proper tubing to the job is important. A general rule of thumb is to use copper tubing for general heating applications and to use PEX for radiant floor heating. There are exceptions to this of course, but in general, the formula works.

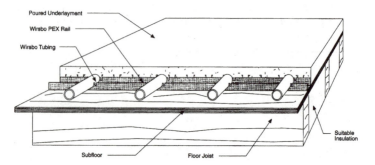

FIGURE 9.1 Thin-slab radiant heating system. *(Courtesy of Wirsbo)*

FITTINGS

The fittings used in heating systems are often the same as those used in plumbing jobs. There are, however, some special fittings that are used most frequently with heating systems. Typical, generic fittings include couplings, slip couplings, reducing couplings, 45 degree elbows, 90 degree elbows, male adapters, female adapters, unions, and tees of both full size and reducing sizes. All of these are basic plumbing fittings. Now, let's look at the fittings that are primarily used with heating systems.

Baseboard tees, also called baseboard ells, are fittings that are shaped like a standard 90 degree elbow. However, the fitting has a threaded fitting in the bend of the elbow. The threads are there to accept the installation of an air vent/purger. These fittings are often used when copper tubing rises vertically through a floor and is turning on a 90-degree angle into a section of baseboard heat. The fittings are usually made of wrought copper or cast brass.

Diverter tees create a flow through a branch piping path that passes through one or more heat emitters before reconnecting to a primary piping circuit. The use of diverter tees can involve pushing or pulling water through a piping path. It can be difficult to tell a diverter tee from a regular tee when relying only on outward appearance. However a peek inside will reveal the diversion section of the tee. When diverter tees are used, they must be installed in the proper location and direction. Diverter tees are equipped with arrows to indicate proper positioning for water flow. There are some installers in the trade who call the diverter tees venturi tees. Most commonly, they are known by the registered trademark name of MonoFlo tees, which is a product and trademark of the Bell & Gossett Company.

Dielectric Unions are often used in plumbing applications. They are installed often on electric water heaters. These same unions are sometimes used with heating systems. A dielectric union mates together with two dissimilar metals. This breaks the continuity of a conductive reaction. In turn, it avoids galvanic corrosion. The unions are used to fit copper materials to steel materials.

HEATING VALVES

Valves used in heating systems can be of the same type used in plumbing systems, but this is not always the case. Selecting a valve for a heating system may seem like a simple task, but it can be more important than you might think. Valves are typically used for either component isolation or flow regulation. Gate valves and globe valves are two common types of valves used in heating systems. These same valves are also used in plumbing systems. Understanding which valves to use, why they should be used, and when they should be used will be of value to you as you install hydronic heating systems. So, let's explore the different types of valves that may be of interest to you in your heating jobs.

Gate Valves

Gate valves are used frequently in both plumbing and heating. These valves are intended for use as isolation valves. They are not meant to be used as flow regulators. This means that gate valves should either be fully open or completely closed. When open, a gate valve does not affect flow velocity very much. Don't install gate valves where flow regulation is required. If the gate is partially down in a gate valve, to regulate flow, there is a strong likelihood that vibration or chatter will be the result.

Globe Valves

Globe valves can be used to regulate flow. In fact, that's what they are intended to do. There is a right and a wrong way to install a globe valve. Make sure that any globe valve you install is positioned so that water enters the lower body chamber. If water comes in from the other end, noise in the system is a strong possibility. Globe valves should not be used to isolate equipment. The design of a globe valve does not make it ideal for isolation. While globe valves can be used for isolation, they shouldn't be, since they are not the most efficient choice.

Ball Valves

Ball valves are probably one of the most used valves in the heating industry. These valves can be used for isolating components or regulating flow. The positive closing action of a ball valve makes it a fine choice as an isolation valve. While ball valves can be used for flow regulation, it is not considered wise to use ball valves to control flow when the flow must be reduced by more than 25 percent. There is no hard and fast rule on this. It is a recommendation to maintain valve condition.

Check Valves

Check valves are needed to insure that fluids do not flow in an unwanted direction. There are two common types of check valves used in heating systems. One is a swing check and the other is a spring-loaded check valve. When a swing check valve is installed, it must be installed on a horizontal line with its bonnet pointing straight up. The operation of a swing check is simple. Fluid flowing through the valve in the proper direction holds the flap of a swing check open. If, for any reason, a backflow situation occurs, the flap of the check valve closes, preventing the backflow.

Spring-loaded check valves are not so sensitive to orientation as swing checks are. This is one reason why installers like spring-loaded valves. Due to the spring action, these valves can be use in any orientation. While it is not important that a spring-check be installed in a straight-up position, it is essential that the valve be installed with the right direction of flow. An arrow on the valve makes it easy to know which direction to install the valve in.

Pressure-Reducing Valves

Pressure-reducing valves are used to lower the pressure of water in a system before it enters a boiler or water distribution system. In the case of heating, the pressure-reducing valve, also known as a boiler feed valve, is used to lower pressure from the water distribution system prior to it entering a boiler. As with check valves, pressure-reducing valves must be installed in the proper direction. The valves have arrows to point them in the direction of flow. There is usually a lever on the top of a pressure-reducing valve that can be lifted manually to speed up the filling process for a boiler during a set-up and start-up procedure.

Pressure-Relief Valves

Pressure-relief valves are important safety valves. When temperature or pressure reaches a risky level, these valve will open to release water that might otherwise cause damage, or even an explosion. These valves are required by code on all hydronic heating systems.

It's common for new boilers to be shipped with a relief valve already installed. A typical rating for a relief valve in a small boiler is 30 psi. Never install a heating system without a relief valve. Relief valves are equipped with threads to accept a discharge tube. This tube is very important. If a relief valve blows off without a proper safety tube installed, people could be seriously injured from hot water or steam. I see far too many relief valves that are not equipped with discharge tubes. The tube should run from the valve to a point about 6 inches above a floor drain. If a drain is not available, at least extend the pipe to within approximately 6 inches above the finished floor level.

Backflow Prevention

Backflow prevention is an important part of a hydronic heating system. It simply would not do to have boiler water mixing with potable water. Local codes require backflow prevention, and there are various types of devices available to achieve code requirements. Most systems utilize in-line devices. These backflow preventers must be installed in the proper direction. They have arrows on the valve bodies to indicate the direction of flow. It is common for this type of device to have a threaded opening about midway on the valve that will act as a vent. Again, a safety tube should be installed on the threaded vent opening to prevent spraying or splashing if the valve discharges.

Flow Checks

Flow-check valves are another type of valve used in heating systems. These valves have weighted internal plugs that are heavy enough to stop thermosiphoning or gravity flow when a system's circulating pumps are not running. Some flow checks have two ports, while others have three.

Valves with two ports are intended for use in horizontal piping. When three ports exist, one can be plugged and the valve can be used in a vertical application. A small lever on top of the valve allows the valve to be opened manually, in the event of a circulator failure. The lever, however, cannot be used to control flow rate. Again, the direction of flow is important when installing a flow check. An arrow on the body of the valve will indicate the direction of flow as it should be used with the valve.

Mixing Valves

Mixing valves are used to mix cold water with hot water to create a regulated temperature in water being delivered from the mixing valve. This might be the case in a hydronic system that uses both baseboard heat emitter and radiant floor heating. The temperature of water for the floor heating would need to be lower than that of the water used for the baseboard system. This is possible with the use of a mixing valve. Using knobs or levers on the outside body of a mixing valve is all that is required to regulate the temperature of water being delivered from the valve.

Zone Valves

Zone valves can be used in place of additional circulators when a hydronic heating system is being zoned off into different zones. Circulating pumps cost more than zone valves, so the zone valves are often used in place of additional circulators. There are old school installers who don't like zone valves. A preference between circulators and zone valves is a personal matter. Most installers are comfortable using zone valves, and many contractors use them extensively.

There are two types of zone valves. One type uses a small electric motor combined with gears to produce a rotary motion of a valve shaft. The other type uses heat motors to produce a linear push-pull motion. Both types consist of a valve body and an actuator. Either type of valve must be operated either fully open or completely closed.

Zone valves for residential systems are located on the supply pipe of each zone circuit. They are generally positioned near the heat source. In most cases, the zone valves are equipped with transformers that allow them to be wired with regular thermostat wire. There are, however, some zone valves that are designed to work with full, 110 volt power.

Other Types Of Valves

Other types of valves are sometimes used in hydronic heating systems. For example, a differential pressure bypass valve might be used in a system where there are numerous individual zones. When a large circulating pump is used for the entire system, pressure can build up if several of the zones are shut down. This is not good, due to high flow rates and possible noise in the zones that are running. The bypass valve eliminates the problem.

Metered balancing valves are another type of valve that may be encountered in a hydronic heating system. These valves may be used in multi-zone systems where more than one parallel piping path exists. The flow rates in the pipes must be balanced to produce desirable heating conditions, and metered balancing valves make this possible.

Another type of valve that is sometimes used is the lockshield-balancing valve. This is a valve that allows a system to be isolated, balanced, or even drained at individual heat emitters. These valves can be purchased in a straight, in-line fashion, or in an angle version. The installation of lockshield-balancing valves are not common in typical residential applications, but they may be installed in such systems.

CIRCULATING PUMPS

Circulating pumps are to a heating system what the heart is to the human body. The pumps move fluid through the pipes of a heating system. Closed-loop, fluid-filled, hydronic heating systems may be equipped with a single circulating pump or with many. When zone valves are use, it is common for only a single pump to be installed. Most circulating pumps are centrifugal pumps. The design and types of circulating pumps vary. Many of them are of a three-piece design while others are in-line designs.

Wet rotor circulators are quite common in small heating systems. This type of pump has a motor, a shaft, and an impeller fitted into a single assembly. The assembly is housed in a chamber that is filled with system fluid, and the motor of the pump is cooled and lubricated by the system fluid. There are no fans or oiling caps. Maintenance of a wet rotor circulator is minimal. Due to their quiet operation and their worry-free maintenance, wet rotor circulators rule the roost when it comes to residential and light commercial heating systems.

Three-piece circulators are also common in residential and light commercial applications. One advantage to this type of pump is that the motor is not housed within the system fluid. If a problem exists, the motor can be worked on without opening the wet system. The disadvantage of a three-piece circulator is that it must be oiled periodically and more noise is present during operation, since the motor is mounted externally.

Pump Placement

Deciding on where to place circulating pumps is not a job that should be taken lightly. Proper placement has much to do with the quality of service derived from a pump. A rule of thumb to follow is to keep all circulators located in a manner so that their inlet is close to the connection point of the system's expansion tank. The reason for this is that the expansion tank is responsible for controlling the pressure of a system's fluid. By keeping circulators near the part of the system that is considered to be the point-of-no-pressure-change, which is the location of the expansion tank, the circulators are always working with a constant pressure. Therefore, the circulators can give better, more uniform service.

How a circulating pump is mounted in a system is also important. It is not wise to hang a circulator on flimsy piping. Many installers build their headers, the section of piping where circulators are mounted, with steel pipe and then switch to copper tubing as they distribute water to heat emitters. This is a good idea. The circulators can, however, be mounted in copper pipe or tubing lines, but the pump should be well supported with some type of hanger or bracket.

Most circulators used in residential heating systems are designed to be installed with their shafts in horizontal positions. In doing this, pressure from the thrust load on bushings, due to the weight of the rotor and impeller, is reduced. Like so many other in-line compo-

fastfacts

➤ *Keep circulators located so that their inlet is close to the connection point of the system's expansion tank*

➤ *Most circulators used for residential jobs are meant to be installed so that the shaft is in a horizontal position*

➤ *Circulators must be installed with the proper orientation to water flow*

➤ *Install valves on each side of circulators to facilitate the easy replacement of circulators when needed*

nents of a heating system, circulators must be installed with the proper orientation to water flow. There are arrows on the pump housings to indicate the proper direction of flow for installation.

Circulators are mounted to a system with the use of flanges. The flanges are bolted to the circulator, so that the pump may be removed for repair or replacement. Smart installers place valves on either side of circulators to make the replacement procedure easier if it is ever required. By having isolation valves both above and below a pump, it is a simple matter to remove the pump without draining the entire system.

EXPANSION TANKS

Any hydronic heating system must be equipped with an expansion tank. When water is heated in a heating system, the liquid expands. Since thermal expansion of this type is unavoidable, it must be given a means to occur without damage to the heating system. This is done with an expansion tank.

The air cushion that is provided in an expansion tank allows water to expand and contract naturally, without fear of damage to the heating system or people in its vicinity. Without an expansion tank, a hydronic heating system could rupture or explode, causing severe damage to both property and people.

The concept behind an expansion tank is easy to understand. The tank has a specified amount of air in it. When water is forced into the tank, the air is compressed. The volume of water in the tank increases, but the air cushion creates a buffer for the expanding water. For example, if the water temperature in the expansion tank is 70 degrees, the tank might be half full of water. The same tank holding water at a temperature of 160 degrees might have only one-fourth of its space not filled with water. These are not scientific numbers, just examples. Basically, the cooler the water is, the less water there will be in the tank.

Old style expansion tanks were basically just metal tanks with an air valve on them where air could be injected. A common problem with this type of tank was related to keeping air in the tank. As air escaped, the tank filled with more water than it should have. This condition is known as waterlogging. It used to be a common problem in both heating and well systems.

The problem of waterlogging has been all but eliminated with new technology in the form of diaphragm-type tanks. These tanks have a flexible diaphragm built into them that regulates air charges and greatly reduces, or eliminates, waterlogging.

How do diaphragm tanks work? They are fitted with a synthetic diaphragm that separates the captive air in the tank from water in the tank. By doing this, air loss is greatly diminished. These tanks are the rule, rather than the exception in modern heating and well systems. It is important that the diaphragm material used in an expansion tank be compatible with fluids used in the heating system.

All diaphragm materials are safe to use with water, but butyl rubber, which is one type of diaphragm material, is not compatible with glycol-based antifreezes, which may be present in some heating systems. If the system you are installing a tank for will contain glycol-based antifreezes, a tank with an EPDM diaphragm material is a better choice. Hydrin diaphragm materials are most commonly used with solar-powered, closed-loop systems.

Check the pressure and temperature ratings assigned to any tank you are planning to use with a heating system. A typical rating for residential use might be 60 psi and 240 degrees. These ratings must be matched to the safety temperature-and-pressure-relief (T&P) valves used with a system. If a typical T&P valve for a residential water heater, rated for 150 psi, were used with an expansion tank system that is rated for 60 psi, the tank could rupture before the T&P valve discharged.

Selection And Installation

The selection and installation of expansion tanks can be confusing. There are tanks available in many shapes and sizes. Most residential heating systems will work fine with tanks having capacities of 10 gallons or less. Sometimes the tanks are mounted vertically. Others are mounted horizontally. Most residential systems have the expansion tank suspended either from piping or ceiling joists. Large expansion tanks are usually floor mounted. In any case, the air-inlet valve for any expansion tank should be readily accessible. This is required in case air must be added to the tank.

It's a good idea to install a pressure gauge near the inlet of an expansion tank. The gauge makes it possible to monitor the static fluid pressure in the system. In all cases, regardless of tank style, the expansion tank must be supported properly to avoid operation problems.

CONTROLS FOR HEATING SYSTEMS

Controls for heating systems are integral parts of a functioning system. Every hydronic heating system relies on controls to work properly. There are four basic categories for the controls to fall into. There are controls that are merely used to turn a system on or off. Other types of controls are staged controls, modulating controls, and outdoor reset controls.

The choice and installation of controls is, to a large extent, a matter of the heating system being fitted with the controls. The best advice is to refer to manufacturers' recommendations and follow them when selecting and installing controls for heating systems. There are far too many facets of controls to cover completely in this chapter. However, we will overview some of the controls, so let's do that now.

Thermostats

Thermostats, like those on the walls of homes where heating systems are installed, are examples of on-off controls. Devices used to open or close an electrical contact are the most common type of controls used in heating systems.

Burner relays and setpoint controls are also examples of on/off controls. These types of devices simply allow a system to cut on or off. The devices do not control or regulate the heat output of a system. This confuses some people. Many people think that a thermostat regulates heat. This is not the case. A thermostat that is set for a high temperature allows a heat source to run longer, but not to burn hotter.

Controlling In Stages

One way to match the output of a heat source with a building's need for heat is to control the heat output in stages. This is known as staged control. It means that the controls used for this type of control cut on and off in stages. The stages can range from zero to maximum heat output. The stages come on and go off in sequence. The use of staged sequences makes it easier to balance heat output with heat need. This type of control is rarely needed in typical residential applications.

Modulating Control

The ultimate in matching output with heat need comes from modulating control. The advantage of modulating control is that heat output is continuously variable over a range from zero to full output. Modulating controls make this possible. Most modulating controls do their job by controlling the temperature of water passing into and through heat emitters. There are several different ways to accomplish modulating control. The most common involves the use of mixing valves or variable speed pumps. But, electrical elements and even thermostatic radiator valves can produce the effects desired with modulating control.

Outdoor Temperatures

Outdoor temperatures affect the effectiveness of a heating system. When a standard hydronic system is designed, it is set up to provide a certain room temperature based on a certain outdoor temperature and the temperature of water in the heating system. An outdoor reset control is ideal for balancing outdoor temperatures with the water temperature in hydronic heating systems. With this type of balancing, a building enjoys near perfect heating control, even when outside temperatures are uncooperative. When outdoor temperatures plummet, the temperature of water in the heating system is raised through the use of an outdoor rest control. In reverse, if outdoor temperatures warm up considerably, the temperature of water in the heating system is lowered. This balancing act combines to make a more efficient and more comfortable heating system.

There all sorts of controls used in heating systems. Switches, relays, time delay devices, high limit switches, triple action controls, multi-zone relays, resets, and so forth. The safest bet when dealing with controls is to follow the recommendations of manufacturers. The list of potential controls is a long and complicated one. We simply don't have the space here to delve into all of them.

HOT WATER RADIANT HEAT

Hot water radiant heat is the heat of choice in regions where winter temperatures are brutal. This type of heat has proved itself over and over again under such circumstances. I live in Maine, where temperatures often drop well below zero on any given day or night during the winter months. Seeing the outside thermometer dip to 20 degrees below zero, not counting windchill factors, is not uncommon. This type of cold requires a strong heating system if a home or business is to remain at comfortable indoor temperatures. Maine, of course, is not the only region where extremely cold temperatures put pressure on heating systems.

Having been a plumbing and heating contractor for over twenty years, I've seen a number of heating systems in a variety of situations. Most heating professionals agree that hot-water baseboard heat is one of the best heating systems available for cold climates. Radiant floor heating is also gaining ground rapidly as a desirable heating system. There are, to be sure, other types of heating systems that will maintain comfortable indoor temperatures. However, the efficiency and operating costs of some systems leaves much to be desired when compared to radiant or hot water heat.

Radiant floor heating systems are rapidly growing in my region, as they are in other areas. This type of heating system heats through flooring, so floors stay warm and heat is evenly distributed when the systems are installed properly. Hot water baseboard heating units are still the most popular within Maine, and probably in many other cold states. Combining the two types of heat emitters makes a lot of sense. In this chapter, we are going to take a look at both types of systems and what the requirements are for each of them. Let's start with baseboard units and work our way into radiant floor heating.

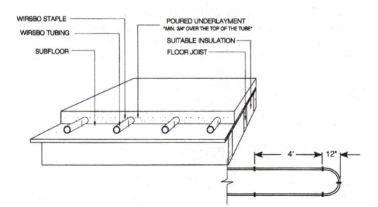

FIGURE 9.2 Radiant heating system on a suspended wood floor. *(Courtesy of Wirsbo)*

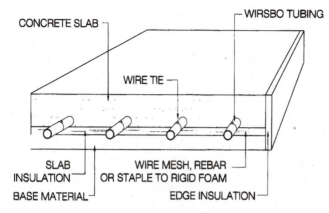

FIGURE 9.3 Attachment method for heat tubing when wire ties are used for radiant floor heating.

FINNED-TUBE BASEBOARD CONVECTORS

Finned-tube baseboard convectors are probably the most commonly used form of heat emitter for hot water heat. This is true of both residential and commercial applications. This type of heat consists of a housing and an element. The element is made of copper tubing that is surrounded by aluminum fins. The length of an element usually ranges from 2 feet to 10 feet and is often sold in increments of one foot. For example, you can buy a 5 foot section of element or a 6 foot section. The elements can be cut with regular tubing cutters for customized fits. The copper tubing used in elements usually has a diameter of ¾ inch, but there are 2 versions available. Commercial elements exist in larger sizes. It is also possible to buy commercial elements where the fins surround steel pipe. But, due to size and cost, commercial elements are not normally used in residential installations.

Heating elements are housed in sheet-metal enclosures. These enclosures consist of a back, which is screwed to a wall. The back has movable brackets which hold the heating element. There is also a movable damper built into the back section, so that the heating housing can be closed to reduce heat output. A front cover snaps into place on special retainer brackets that are a part of the back section. In addition to the front and back covers, there are trim pieces which are used to join sections of baseboard and to terminate runs.

The pieces used to join sections of baseboard are called splicers. The termination pieces are called end caps. There are also pieces called inside corners and outside corners which are used when the baseboard sections change direction.

Finned-tube heating elements do not rely on a fan for heat distribution. Instead, the convectors operate on the basis that warm air rises. Cool air is pulled in from the bottom of the heating element housing, comes into contact with the finned tube, is warmed, and then rises. It is a simple, yet very effective process that produces good heat. During peak operation, the damper on a heater housing should be open. If the damper is closed, the amount of heat output is reduced by up to 50 percent. Having an upward draft, the warm air from finned convectors negates the effects of downward drafts from exterior walls and windows. This makes the heat feel quite comfortable.

Due to the nature of convection heat, there can be a problem known as stratification. This is basically a situation where warm air rises to a ceiling while cool air lingers lower in a room. When the water temperature in a heating system of this type is very hot, the rising of the air is more extreme. Homes with vaulted ceilings can suffer greatly from stratification. One way to over come this is the installation of ceiling fans that will push the warm air back down into the living space.

It is most common for baseboard heating units to be placed on outside walls and under windows. Since baseboard heating units are attached to walls, the placement of furniture can be a concern. There should be at least 6 inches of open space between heating units and any furniture that may block the heat emission. Knowing how many linear feet of heating elements will be needed is a major part of a heat design. Most wholesalers who sell heating systems to contractors will perform heating loads for houses and provide a drawing of where to place heating units and dictate how large the units should be.

INSTALLATION

The installation of common baseboard heating systems is not difficult. Once a plan has been determined for the size and placement of baseboard units, an installer can set to work. Baseboard enclosures are often installed prior to finish floor coverings. This means that an allowance must be made for the planned flooring installation. The back sections of baseboard housing should be nailed, or

fastfacts

➤ *Finned-tube heating elements do not rely on a fan for heat distribution.*

➤ *If the damper is closed on a fined-tube heating unit, the amount of heat output can be reduced by up to 50 percent.*

➤ *Baseboard heating units are usually mounted under windows and along exterior walls.*

➤ *There should be at least 6 inches of open space between heating units and any object that may block heat output.*

preferably screwed, to wall studs. When this is done, an allowance must be made for the bottom of the enclosure. As a rule of thumb, many installers mount the baseboard backs so that the bottom edge is about 1 inch above the subflooring. This allows what is usually adequate room for carpet and pad to be installed. It's best, however, to consult with the builder or flooring contractor to make sure that any baseboard installed will not interfere with forthcoming flooring.

Holes must be drilled to get distribution piping to heating elements. The holes should be oversized to prevent squeaking in the heating pipes as they expand and contract. Many contractors feel that the holes should be a full 2 inches larger than the pipe passing through the hole. Some contractors use flexible sleeves to hold pipes firmly in place.

In most cases, distribution piping will start at a boiler and run in a continuous loop. The piping will enter a heating element, run through it, exit it, and then travel on to the next heating element. In the end, the loop winds up back at the boiler. When multiple zones are used, such as one zone for general living space and one zone for sleeping areas, there will be multiple loops. There are other ways to pipe a system, but the continuous loop is the most popular and it's very effective.

Plumbers who are not used to working with baseboard heating elements often find themselves fixing leaks in their solder joints. Unlike plumbing joints, where both sections of pipe being joined are the same thickness, the thin-wall copper used with heating elements gets hot much faster than the type-M copper tubing usually used to supply water to the element. Since the supply pipe and the heating element will reach soldering temperature at different times, a successful joint requires a little experience. A plumber or heating mechanic who bases the joint on a typical plumbing joint will find leak after leak. The distribution pipe takes longer to bring up to heat for soldering than the heating element does.

ESTABLISHING HEAT ZONES

Establishing heat zones is as easy as determining what parts of a home or building should be controlled by separate thermostats. In average houses, there may be only two zones. One zone will serve general living space, while a second zone heats sleeping areas. It's not uncommon to find small homes with a single heat zone. Two-story homes may have a heat zone for each level of living area. Larger homes can have many heat zones. In my home, I installed one zone for bedrooms, one zone for general living space on the ground floor, one zone for offices upstairs, and a fourth zone for the heated garage. If I had a basement, it probably would have had a separate heating zone from the rest of the house.

Adding zones to a home make the heating system more expensive. Either additional circulating pumps or zone valves will be needed. More copper tubing is required to feed the various zones. Yet, the money saved in operating cost by heating zones of a home as they are being used can help to recover installation costs. The critical thing about zones is to establish them during the planning stages of a job and lay them out accordingly. How a home or building is zoned off for heating is largely the choice of the occupant.

FAN-ASSISTED HEATERS

Fan-assisted heaters, or fan-coil heaters as they are also know, see frequent use in kitchens when hot-water heat is installed. Since kitchens have base cabinets in them, it can be difficult to find wall space to mount baseboard units. The solution to this problem is a small, fan-coil heater that is best known as a kick-space heater. This little heater installs under a base cabinet and requires only about 4 inches of height. An average kick-space heater will be about 1 foot wide, but it will deliver the heat output of a 10 foot long piece of radiant baseboard heating element. Wider kick-space units are available for greater heat output when space allows.

How does such a small package deliver so much heat? The heaters have a built-in coil of ½ inch copper tubing. Air is drawn in through the top portion of the heater grill by means of a transverse blower mounted in the back of the unit. This blower reverses air flow and sends it out through the horizontal coil in the lower part of the heater. It is this lower part that contains the heating coil. Since air is moved over the coil via a blower, more air flow is generated, and thus, more heat is created. A low-limit aquastat control prevents the blower from operating until the water temperature in the coil reaches a desired temperature. Normally, the temperature setting is pre-set by the heater manufacturer and is not adjustable. Most settings range from 110 degrees to 140 degrees. In cases where low-temperature water is used for a system, the low-limit control can be bypassed.

Wall-Mounted Fan-Coil Heaters

Wall-mounted fan-coil heaters are available for use in both residential and commercial applications. There are types that are mounted to walls and types that are recessed into walls. The recessed types take up less floor space and often provide a more attractive appearance. When a recessed unit is used, there must be a cavity built into the wall framing that will accept the size of the unit. Both types usually have copper coils that are connected with ½ inch copper tubing. It's possible for a single wall-mounted fan-coil heater to heat an entire room, but this does not come without some potential problems.

Since the wall heaters depend on a blower to push air into a room, two potential problems exist from the design. First, the heating of the room will not be as even as it would be with baseboard convectors. This is obvious, since all of the heat is coming from a single source and location. Secondly, furniture cannot be placed in front of the heating unit.

Wall-mount heaters are available in different sizes, both in width and height. Even so, finding a location where the heater can operate without blockage may be a problem, especially in a residential home. Since the heaters use blowers, there is some operational noise which must be expected. And, dust circulation may be a problem. Good heaters have dust filters, but dust still escapes in some quantity.

Wall-mount fan-coil heaters are much more effective than finned-tube convectors when it comes to producing and distributing heat.

Due to the design of fan-coil heaters, they can produce more heat with less temperature in the heating coils. This is a big advantage when the heat source is a hydronic heat pump or an active solar system. To use finned-baseboard heat with the heat pump or solar set-up would be risky. There is a great chance that either heat source would not produce water hot enough to heat the finned-tube heating elements sufficiently. Finned-tube baseboard units, used with boilers, are much more common that wall-mounted fan-coil units, but the fan-coil units definitely have their place in the heating world.

HEATED GARAGES

Heated garages are a grand convenience for people who live in extremely cold climates. There are many ways to heat a garage. Finned-baseboard can be use, as can a wall-mount fan-coil unit. However, an overhead fan-coil unit is often the most practical. Just one of these out of the way heaters produces plenty of heat for even a large garage. These same heaters are frequently used in commercial installations. These heaters are sometimes called unit heaters. Louvers on the face of the heaters allow air to be blown in an angled horizontal direction. Overhead unit heaters can be hung from ceiling supports with brackets or threaded rod. The units require very little space, minimal piping, and produce substantial heat.

PANEL RADIATORS

Panel radiators are a type of hydronic heat that many Americans may never have seen. These heating units have been used for years in Europe, but they are fairly new to the scene in the United States. When horizontal wall space is at a premium, panel radiators are a good choice to consider. Unlike fan-coil units, panel radiators don't depend on blowers for operation. The heaters are both attractive, unobtrusive, and very effective.

Installation of panel radiators is not difficult. In most cases, the radiators are mounted on walls and extend only about 2 inches into living space. The units are generally mounted so that the bottom of the heater is between 6 to 8 inches above finished floor level. Wall brackets, which must be secured to wood members or securely into solid masonry, hold the panels in place. When the installation of panel heaters is planned in conjunction with new construction, wood backing blocks can be installed horizontally between wall studs to provide a strong surface for attachment.

Water connections to the panels may be made with either ½ inch or ¾ inch piping. It is common to install a manual or thermostatic radiator valve in the upgoing riser, directly to the inlet connection of the radiator, to allow individual control of the heating unit. Valves and unions can be installed on risers to allow the removal of radiator panels when the time comes to paint the room where the heater is located.

Radiator panels can be purchased as either vertical units or in a horizontal design. This allow maximum flexibility in mounting locations. Since there are numerous size combinations available, it's possible to install panel radiators in almost any location. Panel radiators are subject to rust and should be used only in closed-loop systems.

RADIANT BASEBOARD HEAT

Radiant baseboard heat is another relatively new addition to the U.S. market. The baseboard is only about 1 inch wide and about 5 inches high. PEX tubing is most often used in the baseboard, but copper tubing is sometimes used. There are no fin-tube elements in radiant baseboard. The heat output from radiant baseboard is about 85 percent thermal radiation and about 15 percent convection. Radiant baseboard heats in such a way that stratification is hardly a problem. And, the baseboard units are sold in lengths from 1 to 10 feet, in 6 inch increments.

Due to their thin line design, radiant baseboard heat is intended to act as a replacement for traditional baseboard. Essentially, the heating units are installed in a full perimeter around a room. When baseboard units are connected to each other, the connections are made with brass compression fittings. This is an advantage as there is no soldering required. It's common to run the supply and return for each room from a central manifold. Radiant baseboard heat does not produce as much heat on a per-foot basis as fin-tube baseboard, but the fact that radiant baseboard is installed completely around the room, this problem turns out not to be a problem at all.

There are plenty of choices when it comes to using hot-water for a heating system. While fin-tube baseboard is currently the most popular way to heat homes with hot water, there are certainly other options available, and some of them are gain popularity quickly. People are looking into the use of heat pump and solar-powered systems. Radiant baseboard and radiator panels are on the move. Another major mover in the market is radiant floor heating.

RADIANT FLOOR HEATING SYSTEMS

Radiant floor heating systems have been around for quite awhile. If you want to go back to the origins of heating buildings by having heated floors, you can trace the process all the way back to the Roman Empire. The Romans heated their floors by directing exhaust gases from wood fires to open space under raised floors. Today's radiant floor systems depend on hot water, rather than exhaust gases, to heat floors. For many years, the concept of heating homes and buildings by putting heat in the floors of the structures was largely ignored. This fact is changing. With modern technology, the cost of installing radiant floor heat is lower. Additionally, the problems that were often associated with threaded pipe connections and steel pipe, as well as the pin-hole leaks sometimes experienced with copper tubing, have been greatly reduced with the introduction of PEX tubing.

The benefits of radiant floor heat are numerous. Such systems give unsurpassed thermal comfort, produce no noise, can operate with low-temperature water, and often uses less energy to function. Another advantage of radiant floor heating is the fact that the heating pipes are out of sight, out of mind, and don't interfere with furniture placement. Dust can be a problem with fin-tube baseboard heat, and it can be a big problem with some fan-assisted heating

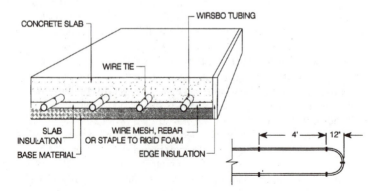

FIGURE 9.4 Tubing installed deep within a slab. *(Courtesy of Wirsbo)*

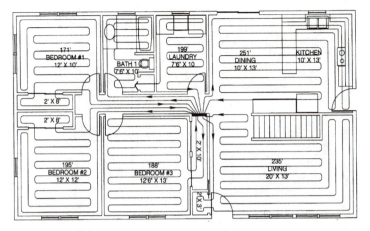

FIGURE 9.5 Typical heat design. *(Courtesy of Wirsbo)*

units. This is not a problem with radiant floor heating. Stratification is another problem that may occur with baseboard heating systems, but it is not a problem with radiant floor heating. The key to having an enjoyable experience with radiant floor systems is to have the system designed and installed properly.

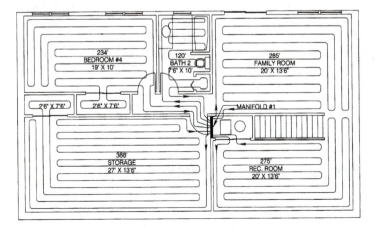

FIGURE 9.6 Basement heating design. *(Courtesy of Wirsbo)*

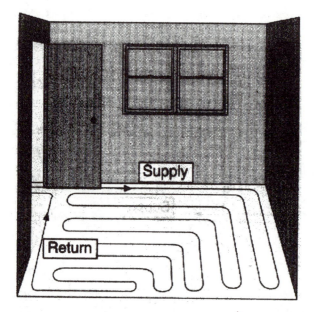

FIGURE 9.7 Supply and return heating layout. *(Courtesy of Wirsbo)*

Radiant floor heating is not new, but the methods and materials used to install the systems are new. When this type of heating system first gained acceptance in the U.S., it was most often installed in concrete slabs. Concrete is a good place to embed radiant heat pipes. The concrete gathers the heat and stores it. As the desire for in-floor heating increased, so did new installation methods. The three types of installations used today are: slab-on-grade systems, thin-slab systems, and dry systems. For a long time, slab-on-grade systems were the most often used type of radiant floor heating. This is still a popular type of heating installation, but other ways are growing in popularity, as people want more of the advantages possible with radiant floor heating.

PUTTING HEAT IN A CONCRETE SLAB

Putting heat in a concrete slab is not a huge undertaking. When you or a customer are paying to have a concrete slab installed, it makes a lot of sense to consider putting radiant heat in the slab. Since there

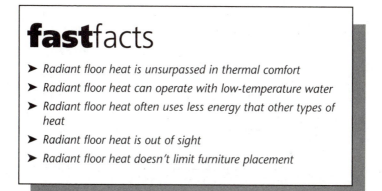

- ➤ *Radiant floor heat is unsurpassed in thermal comfort*
- ➤ *Radiant floor heat can operate with low-temperature water*
- ➤ *Radiant floor heat often uses less energy that other types of heat*
- ➤ *Radiant floor heat is out of sight*
- ➤ *Radiant floor heat doesn't limit furniture placement*

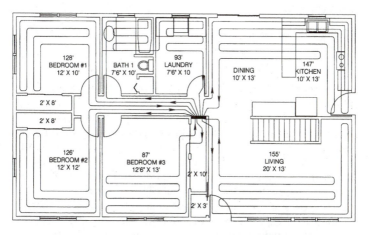

FIGURE 9.8 Underfloor heating doesn't compromise furniture placement. *(Courtesy of Wirsbo)*

will be a slab regardless of whether there is heat or not, the only real cost of adding the heat is some inexpensive PEX tubing and some labor cost in having it installed. This, of course, is assuming that some form of hot-water heat will be used for the remainder of the heating system. If the slab will hold the entire living space of a home, the situation is even better. This means that there will be no need for any other type of above-floor heating units.

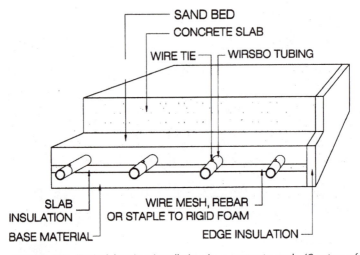

FIGURE 9.9 Typical heating installation in a concrete pad. *(Courtesy of Wirsbo)*

A very important part of installing radiant heat in a concrete slab is the protection of the tubing used to transport hot water through the concrete. It is normal to install PEX tubing on the reinforcement wire that is generally placed in the slab area prior to pouring concrete. The tubing is looped throughout the slab area and can be attached to reinforcing wire with wire ties that are specifically designed for the purpose. Another method, and one that many contractors feel is better, is to use special clips to support the PEX tubing. The clips are available as individual clips or as bars that are notched to accept the tubing. Regardless of your preference for securing the tubing, refer to the instructions provided by the tubing manufacturer. Failure to secure the pipe in compliance with manufacturer's recommendations can result in a voided warranty.

The spacing between loops of radiant piping can vary a great deal. Loops that are placed close together will boost heat output. Loops may be placed as far as two feet apart in some jobs and much closer in different jobs. Most installers attempt to keep the loops about one foot apart. This makes bending PB or PEX tubing much easier, without as much fear of kinking the tubing. The distance between loops is determined when a heat load is computed and a system is designed.

Just as the distance between loops varies, so does the depth as which radiant tubing is installed. It's generally considered best to position tubing so that it is concealed approximately mid way in a slab. However, tubing is often set lower in concrete. This can be due to many factors, such as reinforcing wire sinking during the pouring of the concrete. When tubing is placed deep within a concrete slab, the heating efficiency is reduced. This means that hotter water is needed to produce the same amount of heat that cooler water would produce if the tubing was closer to the surface of the slab. If a system has been shut down for awhile, it will take longer for tubing that is deep within a slab to warm the surface of the concrete.

In order to maximize heating efficiency, rigid foam insulation should be installed between the earth under the slab and the tubing installed for heating. Older homes where radiant floor heating was installed often did not have the benefit of such insulation. As a result, the heating systems were forced to work much harder to maintain a comfortable heating temperature. Without the insulation, much of the heat produced by the tubing is lost to the earth. In time, the ground reaches a level temperature that allows the tubing to heat the slab fairly well, but this lost heat can be greatly

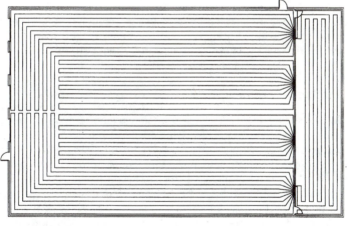

1. Building square footage:	9375
2. Tubing installed on center:	18"
3. Feet of tubing installed:	6075
4. Number of loops:	26
5. Number of manifold locations:	4
6. Number of zones:	1

FIGURE 9.10 Installation diagram. *(Courtesy of Wirsbo)*

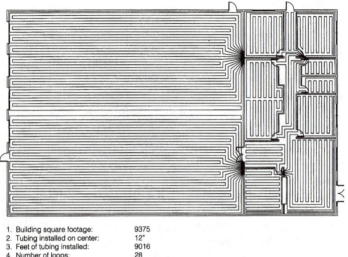

1. Building square footage: 9375
2. Tubing installed on center: 12"
3. Feet of tubing installed: 9016
4. Number of loops: 28
5. Number of manifold locations: 4
6. Number of zones: 3

FIGURE 9.11   Heat tubing layout with manifold locations noted. *(Courtesy of Wirsbo)*

reduced when insulation is installed below the tubing. A foam insulation board with a thickness of just 1 inch will make a huge difference in how well a radiant floor heating system performs.

Site Preparation

Before heating tubing is installed in anticipation of concrete being poured, there are some site preparations to be attended to. It is common for plumbing and electrical work to be covered with concrete. When this is the case, make sure that all of the plumbing and electrical work is completed before installing heating tubing. The earth in the slab area should be fully prepared for concrete prior to heating tubing installation. Before tubing is placed for heating, all aspects of the slab preparation should be complete. Check to make sure that foam insulation is in place, that reinforcing wire is installed, and that there are no rocks or dirt clumps laying on top of the insulation board. Sweep the insulation, if necessary, to provide a clean surface for your tubing. Rocks or other sharp objects could cut or crimp tubing when concrete is poured.

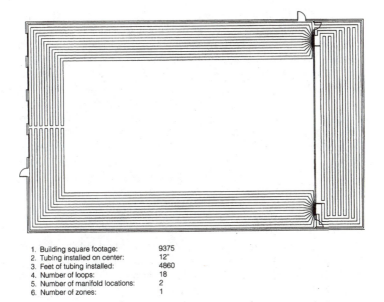

1. Building square footage: 9375
2. Tubing installed on center: 12"
3. Feet of tubing installed: 4860
4. Number of loops: 18
5. Number of manifold locations: 2
6. Number of zones: 1

FIGURE 9.12 Perimeter heating system for radiant floor heat. *(Courtesy of Wirsbo)*

Manifold Risers

Manifold risers are the sections of tubing that are turned up to be exposed above a slab when concrete is poured. These are the feed and return pipes for the heating system. Assuming that you are working with a detailed heating design, and you should be, the locations for manifold risers will be shown on the layout drawing. Precise placement of the manifold risers is usually critical. These pipes are typically designed to turn up in wall cavities. Since installers are working prior to full construction, the walls that the risers are to turn up in are not yet in place. This requires the installer to read plans perfectly and to create mock walls. The mock wall locations can be created with string and stakes. Some installers use what are called block guides. These guides are usually the same width as the walls to be installed. Not everyone likes the idea of using wood below grade. This is due to the potential risk of termite infestation. For this reason, a lot of installers pull strings to indicate wall locations and then stake the risers with non-wood stakes.

Plastic bend supports are used to turn risers up through the concrete. The supports act as sleeves to protect tubing that is to penetrate concrete. Another purpose of the supports is to guide the tubing through a tight bend to an upright position without kinking the tubing. If stakes are used to hold risers in place, the supports are generally attached to the stakes with several wraps of duct tape. It is essential that the risers not move during the pouring and finishing of the concrete. If the tubing is moved accidentally, it may wind up sticking up through the finished floor in a place that is not acceptable. The result of this could be using a jackhammer and losing a good deal of money to correct the problem. To avoid this, make sure that all risers are secured in a way that is sure to produce desired results.

Follow The Design

When you begin to layout tubing in a groundworks (the pre-pour slab area), you should always follow the heating diagram provided to you by your design engineer. After finding locations for manifold risers, it's time to run the tubing system in the groundworks. All tubing should be installed without joints or couplings. When it is absolutely necessary to install a joint below a slab, there are approved fittings for the job, but they should only be used in extreme circumstances. Use full lengths of tubing whenever you can. When joints don't exist, they can't leak, and this is a big advantage, especially in underfloor systems.

A typical slab will contain a lot of heating tubing installed in it. PEX is the tubing most often used for heating, but PB tubing is also used, and was used more than PEX for a long time in the U.S. Both types of tubing are easy to work with, and neither of them crimp easily. But, crimping or kinking is a possibility that must be avoided. Since there is a large volume of tubing needed for most jobs, you need some way to manage the tubing. A good way of doing this is using an uncoiler. The uncoiler is basically a rotating spool that allows tubing to come off a large roll with little likelihood of kinking. It's similar to the uncoilers often used by electricians for their large rolls of electrical wire.

All tubing installed should be installed in compliance with the manufacturer's recommendations. In typical, straight runs, the tubing should be secured in increments of 24 to 30 inches. It is common practice to install three fasteners on return bends. But, it is important to read and follow the fastening instructions provided by manufacturers. If you are securing tubing to reinforcing wire, make sure that the wire where it comes into contact with tubing is not sharp.

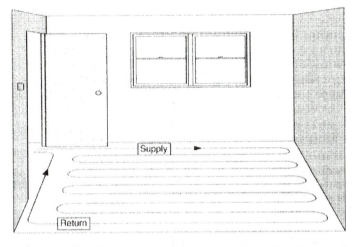

FIGURE 9.13 Single-wall serpentine design. *(Courtesy of Wirsbo)*

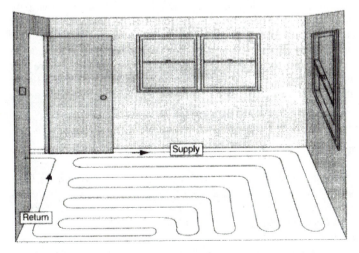

FIGURE 9.14 Double-wall serpentine design. *(Courtesy of Wirsbo)*

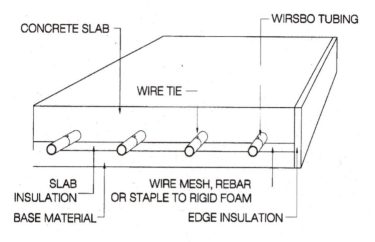

FIGURE 9.15 Foam insulation used with a wet-system design. *(Courtesy of Wirsbo)*

All tubing should be protected from any anticipated damage. If you have tubing in an area where damage may occur, such as where it is passing through an area that will become a foundation wall or is in an area where expansion joints or sawn control joints might exist, sleeve the tubing. Any sleeve used should be at least two pipe sizes larger than the tubing it is protecting. The pipe used by plumbers for drains and vents work well for sleeving PEX or PB tubing. This could be polyvinyl chloride (PVC) or acrylonitrile butadiene styrene (ABS) plastic pipe. If tubing is already installed before you realize the need for a sleeve, you can cut a section out of the sleeve material and place it over the tubing. The main thing is to make sure that all tubing is protected from potential damage.

Testing

Testing a new installation before concrete is poured is important. Even when there are no joints in a system, the tubing should still be tested with air pressure to make sure that there are no defects in the system. Procedures for testing are easy, don't take long to perform, and assure a much better job. Once you create a test rig, or rigs, the testing process is fast. You can test each line individually, or you can tie the tubing together with temporary connections to test all of it at once.

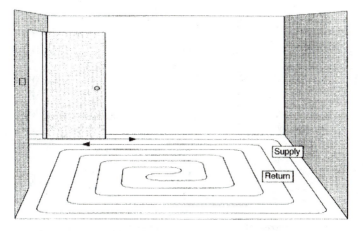

FIGURE 9.16   Example of counterflow design. *(Courtesy of Wirsbo)*

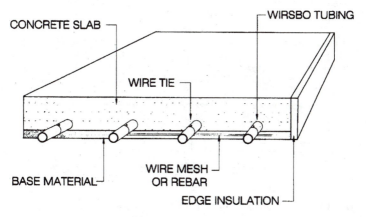

FIGURE 9.17   Foam edge insulation used with a wet system. *(Courtesy of Wirsbo)*

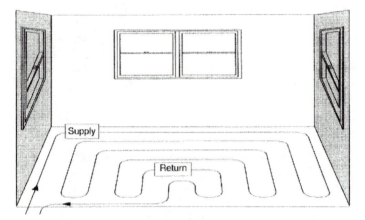

FIGURE 9.18 Triple-wall serpentine design. *(Courtesy of Wirsbo)*

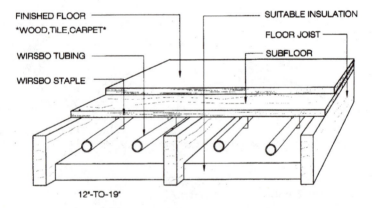

FIGURE 9.19 Dry heating system installed between floor joists. *(Courtesy of Wirsbo)*

What amount of air pressure should you use when testing a system? A standard test pressure is 50 psi. A system should be able to maintain this pressure for 24 hours. It's not uncommon for the tubing to expand during a test period. If this happens, the air pressure on an air gauge will drop. It is also possible that leaks may be present in the temporary connections used to join a system for testing. If you suspect a leak at a test joint, use some soapy water to paint the connection points. When a leak is present, bubbles will appear in the soapy water. Don't omit the test stage. Even without underfloor joints, tubing can have holes in it. This could be a factory defect, a hole that was incurred during storage and transportation, or a hole that was made during installation. Check all the pipes. It's much easier to make replacements before concrete is poured.

THIN-SLAB INSTALLATIONS

Thin-slab installations are pretty much what they sound like. Radiant heating systems installed above a large slab can be done with a thin-slab approach. This means installing the heating tubing on a wooden floor and then pouring a thin slab of concrete over the heating tubing. This procedure is common and effective. A thin-slab system is not the only way to use radiant floor heating above a slab, but it is a fine choice. The thickness of concrete used in thin-slab systems doesn't usually exceed 1½ inches.

Piping procedures for a thin-slab system are about the same as those used for larger concrete slabs. One difference is how the tubing is attached. In a slab-on-grade system, the tubing is usually attached to reinforcing wire or with special clamps. When installing a thin-slab system, the tubing is secured to the wood subflooring. The floor joist cavities below the subflooring should be filled with insulation. Batt, glass fiber insulation is the typical choice for insulation.

Manifold risers are needed with a thin-slab system, just as they are with a large slab. The risers usually turn up either in walls or inside closets. When the risers are in a wall, there should be an access door provided for service and repair. All tubing must be attached securely to the subflooring. Since thin slabs don't have much concrete cover for tubing to be protected and covered, the tubing must be held tightly to the floor. Any loose fitting tubing can rise up and protrude above the finished concrete covering. General principles call for tubing to be secured at intervals of no more than 30 inches on straight runs. It is a good idea to keep the fasteners closer together.

Construction Considerations

Special construction considerations have to be addressed when a thin-slab system is used. Weight is one major concern. The weight added when concrete is poured over subflooring can be quite substantial. This has to be considered during the planning stage of construction. Larger floor joists or other means of additional support are needed to accept the added weight of concrete. While weight is a major consideration with thin-slab systems, it is not the only thing to think about.

Adding about 1½ inches of concrete to a subfloor raises the level of a finished floor. This is not a big problem when it is planned for in advance, but it can mean trouble if allowances are not factored in for the increased height. For example, door openings and thresholds will be off if the added height is not allowed for. Plumbing fixtures and base cabinets will be affected by the increased height. As long as adjustments are made during rough construction, the finished product should turn out fine.

Concrete or Gypsum?

There are two types of cover to place over heating tubing in a thin-slab system. In my region, lightweight concrete is the leading cover for thin-slab systems. But, gypsum-based underlayment is another popular material for covering heating tubing. Lightweight concrete in a typical mix will add up to about 14 pounds per square foot to the dead-load weight rating of a floor. Believe it or not, this is actually a little less weight than what gypusm-based material will add. Either type of cover material will work fine, it's a matter of regional preference as to which material is most likely to be used.

When gypsum-based materials are used, a sealant is generally sprayed on the subflooring. By spraying a sealant and bonding agent on the subflooring, the floor surface is strengthened and made more resistant to moisture problems. The spraying process is done after all tubing is installed. Once the floor is prepared, the gypsum material is mixed in a concrete-type of mixer, and pumped to the location of the heating system through a hose. The material is loose enough in consistency to flow under and around the tubing. Material is pumped in until the level of it is equal to the tubing's diameter. This is called the first coat or the first lift. More material will be needed, after the first coat is dry.

The first layer of cover material should dry, depending upon site conditions, within a few hours. The material can be ready in as little as two hours, and you can apply the second lift or coat. When the second level is distributed, it should cover the tubing by at least three quarters of an inch. After a few hours, the second coat should be dry enough to walk on, but this does not mean that it is dry enough to install a finished floor covering. Depending upon site conditions, meaning air temperature, humidity, and so forth, it can take up to a full work week for the material to dry adequately for finish flooring.

There are pros and cons to gypsum-based cover material. A big advantage over concrete is the ease of application offered with gypsum-based material. Another advantage of the gypsum material is that it doesn't shrink or crack as badly as concrete might. The strength of a gypsum floor material is strong enough to support foot traffic and light equipment, but the material is subject to cuts and gouges, which must be avoided during construction. Any type of consistent water leak can ruin the gypsum material. Even a minor leak that goes undetected for a period of time can turn the material to mush. It is common for a finished gypsum-based cover material to be sealed. Another disadvantage of the gypsum material is that it does not have the thermal conductivity that concrete does. This means that the water temperature in heating tubing covered with gypsum material is likely to be higher than it would need to be in a concrete floor.

Lightweight Concrete

Lightweight concrete is a common cover material for thin-slab systems. The concrete can be delivered to the point of installation in buckets or wheelbarrows. It can also be pumped to the distribution point with a hose and grout pump. Concrete is not affected greatly by moisture, as a gypsum-based produce would be. But, concrete does have a tendency to crack, and this can be a problem for finished floor coverings. Most contractors install control joints in the material to reduce, or to at least control the effects of cracking. Control joints are often placed under door openings.

The finished floor covering on a floor containing radiant heat must be taken into consideration. Some types of floor coverings allow heat to rise better than others. For example, a tile floor will allow much more heat to escape from a radiant system than a padded, carpeted floor covering. To control this, insulation can be placed under the subflooring of a radiant system. When the R-value of the insulation under the floor is greater than the R-value of the finished floor covering, heat will rise through the floor, rather than escaping below it.

Dry Systems

When radiant heat is installed beneath a floor and is not covered with concrete or gypsum-based material, the installation is known as a dry system. The reason for this is simple. The fact that no material is poured over the tubing makes the system dry. These systems are great in that they don't add any appreciable weight to a flooring system. Since weight is not added in substantial amounts, a dry system can be used in remodeling without the need for additional floor support. However, a dry system does need some help in producing heat in desirable quantities.

Top View of Joist Installation

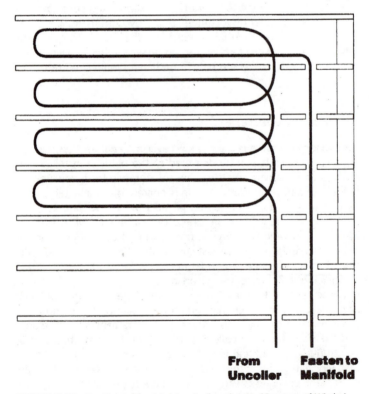

FIGURE 9.20   Routing of heat tubing in floor joists. *(Courtesy of Wirsbo)*

Since dry systems do not have concrete or gypsum to conduct heat, it's common to install heat transfer plates for lateral heat conduction. The plates may be installed above or below subflooring, depending upon the design. Heat plates are made of aluminum and work very well. The installation of below-floor systems is less expensive than above-floor systems. There is much less labor involved in below-floor systems, and this is one reason why they cost is less.

Above-Floor Systems

Above-floor systems are installed above subflooring and below finished flooring. Since space is needed for tubing, a sleeper system is required for above-floor systems. The sleeper system is a series of wood strips that provide cavities for the tubing to run through, and give flooring installers a nailing surface for an additional layer of subflooring. Heat transfer plates are installed for straight runs of tubing. The plates are placed between the sleeper members and stapled down on one side. Once the plates are in place, the tubing can be installed in them. Since sleepers and a second subflooring are installed in an above-floor system, the floor height is raised by an inch or more. This can be a problem for doors, plumbing fixtures, and base cabinets. Just as with a thin-slab system, these problems can be avoided with proper planning and adjustments.

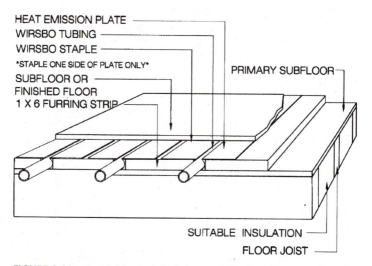

FIGURE 9.21   Heat tubing installed above subflooring. *(Courtesy of Wirsbo)*

Below-Floor Systems

Below-floor systems require less time and labor to install. They also require less material, since sleepers and a second subflooring are not needed. Whether remodeling or building, below-floor systems make a lot of sense. Heat tubing is usually placed in heat transfer plates that are stapled to the bottom of subflooring. When tubing penetrates floor joists, the holes should be drilled near the middle of the joist, rather than the top or bottom edge. This helps to maintain structural integrity. Also, the holes should be somewhat larger than the tubing diameter, to avoid squeaking as the tubing expands.

When heat tubing is attached to the underside of subflooring, there is a risk of damage from the work of others. A plumber or electrician who is not aware of the heat tubing may drill through it accidently. It's not practical to protect all tubing with nail plates, but it is wise to make all workers aware of the tubing under the subfloor. If a section of tubing is damaged, repair couplings can be used to correct the problem. However, whenever possible, avoid joints and connections in heating systems. It's best to maintain full lengths of tubing, rather than coupled sections.

Radiant floor heating systems are gaining popularity quickly. They can be quite cost-effective to install and operate. The comfort offered from a radiant system is, in many ways, unmatched. If you are not familiar with this type of heating system, you owe it to yourself and your customers to learn more about it. Radiant heating systems in floors are well worth consideration.

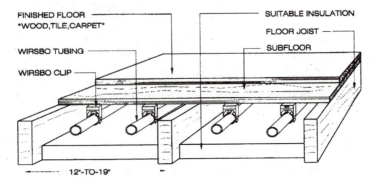

FIGURE 9.22 Installation below subflooring using clips to support heat tubing. *(Courtesy of Wirsbo)*

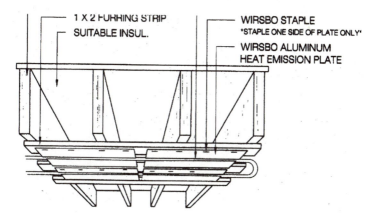

FIGURE 9.23   Heat transfer plates can support heat tubing. *(Courtesy of Wirsbo)*

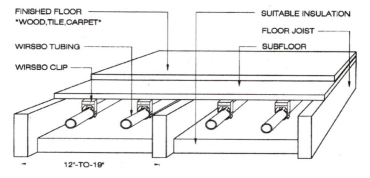

FIGURE 9.24   A sandwich method of using heat-transfer plates. *(Courtesy of Wirsbo)*

Chapter

10

SEPTIC SYSTEMS

Septic systems are common in rural housing locations. Many people who live outside the parameters of municipal sewers depend on septic systems to solve their sewage disposal problems. Plumbers who work in areas where private waste disposal systems are common often come into contact with problems associated with septic systems. Ironically, plumbers are rarely the right people to call for septic problems, but they are often the first group of professionals homeowners think of when experiencing septic trouble.

One reason that plumbers are called so frequently for septic problems is that the trouble appears to be a stopped-up drain. When a septic system is filled beyond capacity, back-ups occur in houses. Most homeowners call plumbers when this happens. Smart plumbers check the septic systems first and find out if they are at fault.

Back-ups in homes is not the only reason plumbers need to know a little something about septic systems. Customers frequently have questions about their plumbing systems that can be influenced by a septic system. For example, is it all right to install a garbage disposer in a home that is served by a septic system. Some people think it is, and others believe it isn't. The answer to this question may not be left up to a plumber's personal opinion. Many local plumbing codes prohibit the installation of food grinders in homes where a septic system will receive the discharge.

Considering all of the questions and concerns that customers might come to their plumbers with, I feel it is wise for plumbers to develop a general knowledge of septic system. This chapter will help you achieve this goal. With that said, let me show you what is involved with septic systems.

SIMPLE SYSTEMS

Simple septic systems consist of a tank, some pipe, and gravel. These systems are common, but they don't work well in all types of ground. Since most plumbers are not septic installers, I will not bore you will all of the sticky details for putting a pipe-and-gravel system into operation. However, I would like to give you a general overview of the system, so that you can talk intelligently with your customers.

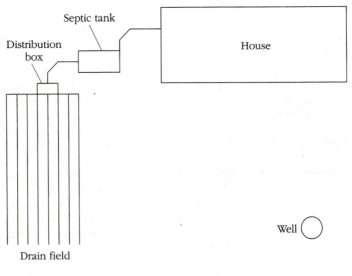

FIGURE 10.1 A typical site plan.

THE COMPONENTS

Let's talk about the basic components of a pipe-and-gravel septic system. Starting near the foundation of a building, there is a sewer. The sewer pipe should be made of solid pipe, not perforated pipe. I know this seems obvious, but I did find a house a few years ago where the person who installed the sewer used perforated drain-field pipe. It was quite a mess. Most jobs today involve the use of schedule 40 plastic pipe for the sewer. Cast-iron pipe can be used, but plastic is the most common and is certainly acceptable.

The sewer pipe runs to the septic tank. There are many types of materials that septic tanks can be made of, but most of tanks are constructed of concrete. It is possible to build a septic tank on site, but every contractor I've ever known has bought pre-cast tanks. An average size tank holds about 1,000 gallons. The connection between the sewer and the septic tank should be watertight.

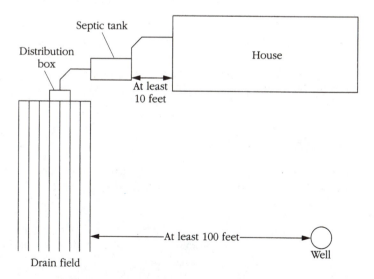

FIGURE 10.2 Recommended minimum distances between wells and septic systems and septic tanks and homes.

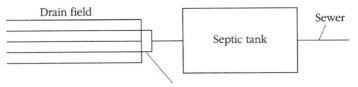

FIGURE 10.3 Common septic layout.

The discharge pipe from the septic tank should be made of solid pipe, just like the sewer pipe. This pipe runs from the septic tank to a distribution box, which is also normally made of concrete. Once the discharge pipe reaches the distribution box, the type of materials used changes.

The drain field is constructed according to an approved septic design. In basic terms, the excavated area for the septic bed is lined with crushed stone. Perforated plastic pipe is installed in rows. The distance between the drain pipes and the number of drain pipes is controlled by the septic design. All of the drain-field pipes connect to the distribution box. The septic field is then covered with material specified in the septic design.

As you can see, the list of materials is not a long one. Some schedule 40 plastic pipe, a septic tank, a distribution box, crushed stone, and some perforated plastic pipe are the main ingredients. This is the primary reason why the cost of a pipe-and-gravel system is so low when compared to other types of systems.

Types Of Tanks

There are many types of septic tanks in use today. Pre-cast concrete tanks are, by far, the most common. However, they are not the only type of septic tank available. For this reason, let's discuss some of the material options that are available.

Pre-cast concrete is the most popular type of septic tank. When this type of tank is installed properly and is not abused, it can last almost indefinitely. However, heavy vehicular traffic running over the tank can damage it, so this should be avoided.

Metal septic tanks were once prolific. There are still a great number of them in use, but new installations rarely involve a metal tank. The reason is simple, metal tends to rust out, and that's not good for a septic tank. Some metal tanks are said to have given twenty years of good service. This may be true, but there are no guarantees that a metal tank will last even ten years. In all my years of being a contractor, I've never seen a metal septic tank installed. I've dug up old ones, but I've never seen a new one go in the ground.

I don't have any personal experience with fiberglass septic tanks, but I can see some advantages to them. Their light weight is one nice benefit for anyone working to install the tank. Durability is another strong point in the favor of fiberglass tanks. However, I'm not sure how the tanks perform under the stress of being buried. I assume that their performance is good, but again, I have no first hand experience with them.

Wood seems like a strange material to use for the construction of a septic tank, but I've read where it is used. The wood of choice, as I understand it, is redwood. I guess if you can make hot tubs and spas out of it, you can make a septic tank out of it. However, I don't think I would be anxious to warranty a septic tank made of wood.

Brick and block have also been used to form septic tanks. When these methods are employed, some type of purging and water-proofing must be done on the interior of the vessel. Personally, I would not feel very comfortable with this type of setup. This is, again, material that I have never worked with in the creation of a septic tank, so I can't give you much in the way of case histories.

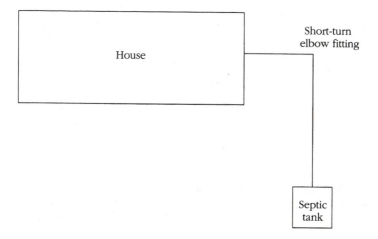

FIGURE 10.4   Avoid using short-turn fittings between house and septic system.

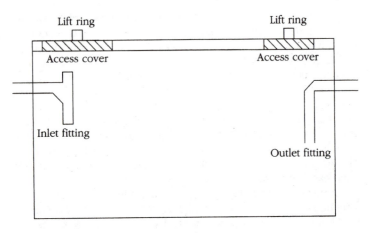

FIGURE 10.5 Side view of a septic tank.

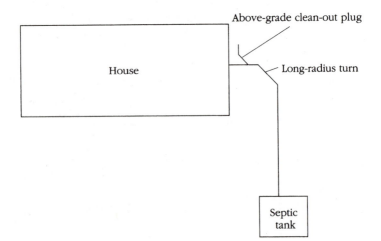

FIGURE 10.6 Outside cleanout installed in sewer pipe and sweep-type fittings used to avoid pipe stoppages.

CHAMBER SYSTEMS

Chamber septic systems are used most often when the perk rate on ground is low. Soil with a rapid absorption rate can support a standard, pipe-and-gravel septic system. Clay and other types of soil may not. When bedrock is close to the ground surface chambers are often used.

What is a chamber system? A chamber system is installed very much like a pipe-and-gravel system, except for the use of chambers. The chambers might be made of concrete or plastic. Concrete chambers are naturally more expensive to install. Plastic chambers are shipped in halves and put together in the field. Since plastic is a very durable material, and it's relatively cheap, plastic chambers are more popular than concrete chambers.

When a chamber system is called for, there are typically many chambers involved. These chambers are installed in the leach field, between sections of pipe. As effluent is released from a septic tank, it is sent into the chambers. The chambers collect and hold the effluent for a period of time. Gradually, the liquid is released into the leach field and absorbed by the earth. The primary role of the chambers is to retard the distribution rate of the effluent.

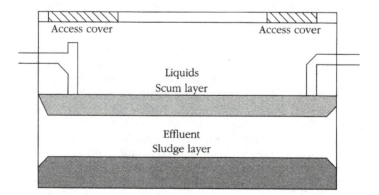

FIGURE 10.7 Example of the various levels of materials in a septic system.

Distribution box

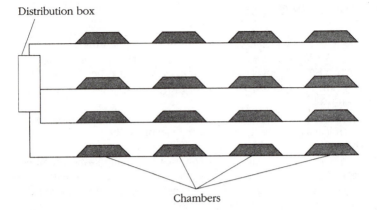

Chambers

FIGURE 10.8 Example of a chamber-type septic field.

Building a chamber system allows you to take advantage of land that would not be buildable with a standard pipe-and-gravel system. Based on this, chamber systems are good. However, when you look at the price tag of a chamber system, you may need a few moments to catch your breath. I've seen a number of quotes for these systems that pushed the $10,000 mark. Ten grand is a lot of money for a septic system. But, if you don't have any choice, what are you going to do?

A chamber system is simple enough in its design. Liquid leaves a septic tank and enters the first chamber. As more liquid is released from the septic tank, it is transferred into additional chambers that are farther downstream. This process continues with the chambers releasing a pre-determined amount of liquid into the soil as time goes on. The process allows more time for bacterial action to attack raw sewage, and it controls the flow of liquid into the ground.

If a perforated-pipe system was used in ground where a chamber system is recommended, the result could be a flooded leach field. This might create health risks. It would most likely produce unpleasant odors, and it might even shorten the life of the septic field.

Chambers are installed between sections of pipe within the drain field. The chambers are then covered with soil. The finished system is not visible above ground. All of the action takes place below grade. The only real downside to a chamber system is the cost.

TRENCH SYSTEMS

Trench systems are the least expensive version of special septic systems. They are comparable in many ways to a standard pipe-and-gravel bed system. The main difference between a trench system and a bed system is that the drain lines in a trench system are separated by a physical barrier. Bed systems consist of drain pipes situated in a rock bed. All of the pipes are in one large bed. Trench fields depend on separation to work properly. To expand on this, let me give you some technical information.

A typical trench system is set into trenches that are between one to five feet deep. The width of the trench tends to run from one to three feet. Perforated pipe is placed in these trenches on a six-inch bed of crushed stone. A second layer of stone is placed on top of the drain pipe. This rock is covered with a barrier of some type to protect it from the backfilling process. The type of barrier used will be specified in a septic design.

When a trench system is used, both the sides of the trench and the bottom of the excavation are an outlet for liquid. Only one pipe is placed in each trench. These two factors are what separate a trench system from a standard bed system. Bed systems have all of the drain pipes in one large excavation. In a bed system, the bottom of the bed is the only significant infiltrative surface. Since trench systems use both the bottoms and sides of trenches as infiltrative surfaces, more absorption is potentially possible.

Neither bed or trench systems should be used in soils where the percolation rate is either very fast or slow. For example, if the soil will accept 1 inch of liquid per minute, it is too fast for a standard absorption system. This can be overcome by lining the infiltrative surface with a thick layer (about two feet or more) of sandy loam soil. Conversely, land that drains at a rate of one inch an hour is too slow for a bed or trench system. This is a situation where a chamber system might be recommended as an alternative.

Because of their design, trench systems require more land area than bed systems do. This can be a problem on small building lots. It can also add to the expense of clearing land for a septic filed. However, trench systems are normally considered to be better than bed systems. There are many reasons for this.

Trench systems are said to offer up to five times more side area for infiltration to take place. This is based on a trench system with a bottom area identical to a bed system. The difference is in the depth and separation of the trenches. Experts like trench systems because digging equipment can straddle the trench locations during excavation.

This reduces damage to the bottom soil and improves performance. In a bed system, equipment must operate within the bed, compacting soil and reducing efficiency.

If you are faced with hilly land to work with, a trench system is ideal. The trenches can be dug to follow the contour of the land. This gives you maximum utilization of the sloping ground. Infiltrative surfaces are maintained while excessive excavation is eliminated.

The advantages of a trench system are numerous. For example, trenches can be run between trees. This reduces clearing costs and allows trees to remain for shade and aesthetic purposes. However, roots may still be a consideration. Most people agree that a trench system performs better than a bed system. When you combine performance with the many other advantages of a trench system, you may want to consider trenching your next septic system. It costs more to dig individual trenches than it does to create a group bed, but the benefits may outweigh the costs.

MOUND SYSTEMS

Mound systems, as you might suspect, are septic systems which are constructed in mounds that rise above the natural topography. This is done to compensate for high water tables and soils with slow absorption rates. Due to the amount of fill material to create a mound, the cost is naturally higher than it would be for a bed system.

Coarse gravel is normally used to build a septic mound. The stone is piled on top of the existing ground. However, top soil is removed before the stone is installed. When a mound is built, it con-

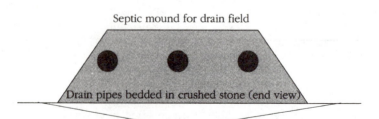

Septic mound for drain field

Drain pipes bedded in crushed stone (end view)

Normal grade level

FIGURE 10.9 Cut-away of a mount-type septic system.

tains suitable fill material, an absorption area, a distribution network, a cap, and top soil. Due to the raised height, a mound system depends on either pumping or siphoning action to work properly. Essentially, effluent is either pumped or siphoned into the distribution network.

As the effluent is passing through the coarse gravel and infiltrating the fill material, treatment of the wastewater occurs. This continues as the liquid passes through the unsaturated zone of the natural soil.

The purpose of the cap is to retard frost action, deflect precipitation, and to retain moisture that will stimulate the growth of ground cover. Without adequate ground cover, erosion can be a problem. There are a multitude of choices available as acceptable ground covers. Grass is the most common choice.

Mounds should be used only in areas that drain well. The topography can be level or slightly sloping. The amount of slope allowable depends on the perk rate. For example, soil that perks at a rate of one inch every 60 minutes or less, should not have a slope of more than 6 percent if a mound system is to be installed. If the soil absorbs water from a perk test faster than 1 inch in 1 hour, the slope could be increased to 12 percent. These numbers are only examples. A professional who designs a mound system will set the true criteria for slope values.

Ideally, about 2 feet of unsaturated soil should exist between the original soil surface and the seasonally saturated top soil. There should be 3 to 5 feet of depth to the impermeable barrier. An overall range of perk rate could go as high as 1 inch in 2 hours, but this, of course, is subject to local approval. Perk tests for this type of system are best when done at a depth of about 20 inches. However, they can be performed at shallow depths of only 12 inches. Again, you must consult and follow local requirements.

HOW DOES A SEPTIC SYSTEM WORK?

How does a septic system work? A standard septic system works on a very simple principal. Sewage from a home enters the septic tank through the sewer. Where the sewer is connected to the septic tank, there is a baffle on the inside of the tank. This baffle is usually a sanitary tee. The sewer enters the center of the tee and drops down through the bottom of it. The top hub of the tee is left open.

The bottom of the tee is normally fitted with a short piece of pipe. The pipe drops out of the tee and extends into the tank liquids. This pipe should never extend lower than the outlet pipe at the other end of the septic tank. The inlet drop is usually no more than 12 inches long. The outlet pipe for the tank also has a baffle, normally an elbow fitting. The drop from this baffle is frequently about 16 inches in length.

When sewage enters a septic tank, the solids sink to the bottom of the tank and the liquids float within the confines of the container. As the tank collects waste, several processes begin to take place.

Solid waste that sinks to the bottom becomes what is known as sludge. Liquids, or effluent as it is called, is suspended between the lower layer of sludge and an upper layer of scum. The scum layer consists of solids and gases floating on the effluent. All three of these layers are needed for the waste disposal system to function properly.

Anaerobic bacteria works inside the septic tank to break down the solids. This type of bacteria is capable of working in confinements void of oxygen. As the solids break up, they form the sludge layer. As the effluent level rises in the tank, it eventually flows out of the tank, through the outlet pipe. The effluent drains down a solid pipe to the distribution box. After entering the distribution box the liquid is routed into different slotted pipes that run through the leach field.

As the effluent mixes with air, aerobic bacteria begin to work on the waste. This bacteria attacks the effluent and eventually renders it harmless. Aerobic bacteria needs oxygen to do its job. Drain fields should be constructed of porous soil or crushed stone to ensure the proper breakdown of the effluent.

As the effluent works its way through the drain field, it becomes odorless and harmless. By the time the effluent passes through the earth and becomes ground water, it should be safe to drink.

SEPTIC TANK MAINTENANCE

Septic tank maintenance is not a time consuming process. Most septic systems require no attention for years at a time. However, when the scum and sludge layers combined have a depth of eighteen inches, the tank should be cleaned out.

Trucks equipped with suction hoses are normally used to clean septic tanks. The contents removed from septic tanks can be infested with germs. The disease risk of exposure to sludge requires that the sludge be handled carefully and properly.

WHAT HAPPENS IF THE DRAIN FIELD DOESN'T WORK?

What happens if the drain field doesn't work? When a septic field fails to do its job, a health hazard exists. This situation demands immediate attention. The main reason for a field to fail in its operation is clogging. If the pipes in a drain field become clogged, they must be excavated and cleaned or replaced. If the field itself clogs, the leech bed must be cleaned or removed and replaced. Neither of these propositions is cheap.

HOW CAN CLOGS BE AVOIDED?

How can clogs be avoided? Clogs can be avoided by careful attention to what types of waste enter the septic system. Grease, for example, can cause a septic system to become clogged. Bacteria does not do a good job in breaking down grease. Therefore, the grease can enter the slotted drains and leach field with enough bulk to clog them. Paper, other than toilet paper, can also clog up a septic system. If the paper is not broken down before entering the drain field, it can plug up the works.

WHAT ABOUT GARBAGE DISPOSERS, DO THEY HURT A SEPTIC SYSTEM?

What about garbage disposers, do they hurt a septic system? The answers offered to this question vary from yes, to maybe, to no. Many people, including numerous code enforcement offices, believe garbage disposers should not be used in conjunction with septic systems. Other people disagree and believe that disposers have no adverse effect on a septic system.

It is possible for the waste of a disposer to make it into the distribution pipes and drain field. If this happens, the risk for clogging is elevated. Another argument against disposers is the increased load of solids they put on a septic tank. Obviously, the amount of solid waste will depend on the frequency with which the disposer is used.

What is my opinion? My opinion is that disposers increase the risk of septic system failure and should not be used with such systems. However, I know of many houses using disposers with septic systems that are not experiencing any problems. If you check with your local plumbing inspector this question may become a moot point. Many local plumbing codes prevent the use of disposers with septic systems.

PIPING CONSIDERATIONS

There are some additional piping considerations for plumbers to observe. Septic tanks are designed to handle routine sewage. They are not meant to modify chemical discharges and high volumes of water. If, as a plumber, you pipe the discharge from a sump pump into the sanitary plumbing system, which you are not supposed to do, the increased volume of water in the tank could disrupt its normal operation.

Chemical drain openers used in high quantities can also destroy the natural order of a septic tank. Chemicals from photography labs are another risk plumbers should be aware of when piping drainage to a septic system.

GAS CONCENTRATIONS

Gas concentrations in a septic tank can cause problems for plumbers. The gases collected in a septic tank have the potential to explode. If you remove the top of a septic tank with a flame close by, you might be blown up. Also, breathing the gases for an extended period of time can cause health problems.

SEWAGE PUMPS

There are times when sewage pumps must be used to get sewage to a septic system. The pumps are normally installed in a buried box outside of the building being served. The box is often made of concrete. In these cases, the home's sewer pipe goes to the pumping station. From the pumping station, a solid pipe transports the waste to the septic tank.

Sewage pumps must be equipped with alarm systems. The alarms warn the property owner if the pump is not operating and the pump station is filling with sewage. Without the alarm, the sewage could build to a point where it would flow back into the building.

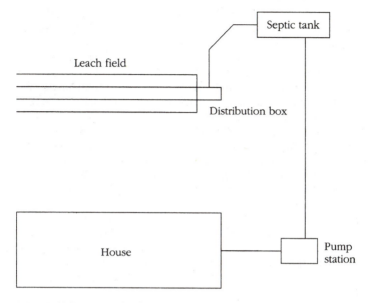

FIGURE 10.10 Example of a pump-station septic system.

Sewage pumps have floats that are lifted as the level of contents in the pump station build. When the float is raised to a certain point, the pump cuts on, emptying the contents of the pump station. The discharge pipe from the pump must be equipped with a check valve. Otherwise, gravity would force waste down the pipe, back into the pump station when the pump cut off. This would result in the pump having to constantly cut on and off, wearing out the pump.

AN OVERFLOWING TOILET

Some homeowners associate an overflowing toilet with a problem in their septic system. It is possible that the septic system is responsible for the toilet backing up, but this is not always the case. A stoppage either in the toilet trap or in the drain pipe can cause a back-up.

If you get a call from a customer who has a toilet flooding their bathroom, there is a quick, simple test you can have the homeowner perform to tell you more about the problem. You know the toilet drain is stopped up, but will the kitchen sink drain properly? Will other toilets in the house drain? If other fixtures drain just fine, the problem is not with the septic tank.

There are some special instructions that you should give your customers prior to having them test other fixtures. First, it is best if they use fixtures that are not in the same bathroom with the plugged-up toilet. Lavatories and bathing units often share the same main drain that a toilet uses. Testing a lavatory that is near a stopped-up toilet can tell you if the toilet is the only fixture affected. It can, in fact, narrow the likelihood of the problem down to the toilet's trap. But, if the stoppage is some way down the drain pipe, it's conceivable that the entire bathroom group will be affected. It is also likely that if the septic tank is the problem, water will back up in a bathtub.

When an entire plumbing system is unable to drain, water will rise to the lowest fixture, which is usually a bathtub or shower. So, if there is no back-up in a bathing unit, there probably isn't a problem with a septic tank. But, back-ups in bathing units can happen even when the major part of a plumbing system is working fine. A stoppage in a main drain could cause the liquids to back up, into a bathing unit.

To determine if a total back-up is being caused, have homeowners fill their kitchen sinks and then release all of the water at once. Get them to do this several times. A volume of water may be needed to expose a problem. Simply running the faucet for a short while might not show a problem with the kitchen drain. If the kitchen sink drains successfully after several attempts, it's highly unlikely that there is a problem with the septic tank. This would mean that you should call your plumber, not your septic installer.

WHOLE-HOUSE BACK-UPS

Whole-house back-ups (where none of the plumbing fixtures drain) indicate either a problem in the building drain, the sewer, or the septic system. There is no way to know where the problem is until some investigative work is done. It's possible that the problem is associated with the septic tank, but you will have to pinpoint the location where trouble is occurring.

For all the plumbing in a house to back up, there must be some obstruction at a point in the drainage or septic system beyond where the last plumbing drain enters the system. Plumbing codes require clean-out plugs along drainage pipes. There should be a clean-out either just inside the foundation wall of a home or just outside the wall. This clean-out location and the access panel of a septic tank are the two places to begin a search for the problem.

If the access cover of the septic system is not buried too deeply, I would start there. But, if extensive digging would be required to expose the cover, I would start with the clean-out at the foundation, hopefully on the outside of the house. Remove the clean-out plug and snake the drain. This will normally clear the stoppage, but you may not know what caused the problem. Habitual stoppages point to a problem in the drainage piping or septic tank.

Removing the inspection cover from the inlet area of a septic tank can show you a lot. For example, you may see that the inlet pipe doesn't have a tee fitting on it and has been jammed into a tank baffle. This could obviously account for some stoppages. Cutting the pipe off and installing the diversion fitting will solve this problem.

Sometimes pipes sink in the ground after they are buried. Pipes sometimes become damaged when a trench is backfilled. If a pipe is broken or depressed during backfilling, there can be drainage problems. When a pipe sinks in uncompacted earth, the grade of the pipe is altered, and stoppages become more likely. You might be able to see some of these problems from the access hole over the inlet opening of a septic tank.

Once you remove the inspection cover of a septic tank, look at the inlet pipe. It should be coming into the tank with a slight downward pitch. If the pipe is pointing upward, it indicates improper grading and a probable caused for stoppages. If the inlet pipe either doesn't exist or is partially pulled out of the tank, there's a very good chance that you have found the cause of your backup. If a pipe is hit with a heavy load of dirt during backfilling, it can be broken off or pulled out of position. This won't happen if the pipe is supported properly before backfilling, but someone may have cheated a little during the installation.

In the case of a new septic system, a total backup is most likely to be the result of some failure in the piping system between the house and the septic tank. If your problem is occurring during very cold weather, it is possible that the drain pipe has retained water in a low spot and that the water has since frozen. I've seen it happen several times in Maine with older homes.

Running a snake from the house to the septic tank will tell you if the problem is in the piping. This is assuming that the snake used is a pretty big one. Little snakes might slip past a blockage that is capable of causing a backup. An electric drain-cleaner with a full-size head is the best tool to use.

THE PROBLEM IS IN THE TANK

There are times, even with new systems, when the problem causing a whole-house back-up is in the septic tank. These occasions are rare, but they do exist. When this is the case, the top of the septic tank must be uncovered. Some tanks, like the one at my house, are only a few inches beneath the surface. Other tanks can be buried several feet below the finished grade.

Once a septic tank is in full operation, it works on a balance basis. The inlet opening of a septic tank is slightly higher than the outlet opening. When water enters a working septic tank, an equal amount of effluent leaves the tank. This maintains the needed balance. But, if the outlet opening is blocked by an obstruction, water can't get out. This will cause a backup.

Strange things sometimes happen on construction sites, so don't rule out any possibilities. It may not seem logical that a relatively new septic tank could be full or clogged, but don't bet on it. I can give you all kinds of things to think about. Suppose a septic installer was using up old scraps of pipe for drops and short pieces, and one of the pieces had a plastic test cap glued into the end of it that was not noticed? This could certainly render the septic system inoperative once the liquid rose to a point where it would be attempting to enter the outlet drain. Could this really happen? I've seen the same type of situation happen with interior plumbing, so it could happen with the piping at a septic tank.

What else could block the outlet of a new septic tank? Maybe a piece of scrap wood found its way into the septic tank during construction and is now blocking the outlet. If the wood floated in the tank and became aligned with the outlet drop, pressure could hold it in place and create a blockage. The point is, almost anything could be happening in the outlet opening, so take a snake and see if it is clear.

If the outlet opening is free of obstructions, and all drainage to the septic tank has been ruled out as a potential problem, you must look further down the line. Expose the distribution box and check it. Run a snake from the tank to the box. If it comes through without a hitch, the problem is somewhere in the leach field. In many cases, a leach field problem will cause the distribution box to flood. So, if you have liquid come rushing of the distribution box, you should be alerted to a probable field problem.

PROBLEMS WITH A LEACH FIELD

Problems with a leach field are uncommon among new installations. Unless the field was poorly designed or installed improperly, there is very little reason why it should fail. However, extremely wet ground conditions, due to heavy or constant rains, could force a field to become saturated. If the field saturates with ground water, it cannot accept the effluent from a septic tank. This, in turn, causes back-ups in houses. When this is the case, the person who created the septic design should be looked to in terms of fault.

Older Fields

Older fields sometimes clog up and fail. Some drain fields become clogged with solids. Financially, this is a devastating discovery. A clogged field has to be dug up and replaced. Much of the crushed stone might be salvageable, but the pipe, the excavation, and whatever new stone is needed can cost thousands of dollars. The reasons for a problem of this nature is either a poor design, bad workmanship, or abuse.

If the septic tank installed for a system is too small, solids are likely to enter the drain field. An undersized tank could be the result of a poor septic design, or it could come about as a family grows and adds onto their home. A tank that is adequate for two people may not be able to keep up with the usage seen when four people are involved. Unfortunately, finding out that a tank is too small often doesn't happen until the damage has already been done.

Why would a small septic tank create problems with a drain field? Septic tanks accept solids and liquids. Ideally, only liquids should leave the septic tank and enter the leach field. Bacterial action occurs in a septic tank to break down solids. If a tank is too small, there is not adequate time for the break down of solids to occur. Increased loads on a small tank can force solids down into the drain field. After this happens for awhile, the solids plug up the drainage areas in the field. This is when digging and replacement is needed.

Is there any such thing as having too much pitch on a drain pipe. Yes, there is. A pipe that is graded with too much pitch can cause several problems. In interior plumbing, a pipe with a fast pitch may allow water to race by without removing all the solids. A properly graded pipe floats the solids in the liquid as drainage occurs. If the water is allowed to rush out, leaving the solids behind, a stoppage will eventually occur.

In terms of a septic tank, a pipe with a fast grade can cause solids to be stirred up and sent down the outlet pipe. When a 4 inch wall of water dumps into a septic tank at a rapid rate, it can create quite a ripple effect. The force of the water might generate enough stir to float solids that should be sinking. If these solids find their way into a leach field, clogging is likely.

We talked a little bit about garbage disposers earlier. When a disposer is used in conjunction with a septic system, there are more solids involved than would exist without a disposer. This, where code calls for a larger septic tank. Due to the increase in solids, a larger tank is needed for satisfactory operation and a reduction in the risk of a clogged field. I remind you again, some plumbing codes prohibit the use of garbage disposers where a septic system is present.

Other causes for field failures can be related to collapsed piping. This is not common with today's modern materials, but it is a fact of life with some old drain fields. Heavy vehicular traffic over a field can compress it and cause the field to fail. This is true even of modern fields. Saturation of a drain field will cause it to fail. This could be the result of seasonal water tables or prolonged use of a field that is giving up the ghost.

Septic tanks should have the solids pumped out of them on a regular basis. For a normal residential system, pumping once every two years should be adequate. Septic professionals can measure sludge levels and determine if pumping is needed. Failure to pump a system routinely can result in a build-up of solids that may invade and clog a leach field.

Normally, septic systems are not considered to be a plumber's problem. Once you establish that a customer's grief is coming from a failed septic system, you should be off the hook. Advise your customers to call septic professionals and go onto your next service call—you've earned your money.

PRIVATE WATER
SUPPLY SYSTEMS

Plumbers in rural locations work with well systems regularly. They install them, and they service them. City-based plumbers have little experience with well systems. Even plumbers in the country don't always have a lot of experience with well systems. In some regions, well drillers control most of the pump business. But, if you work in an area where water wells and pumps are used, you should be prepared for the service calls you may get.

There is good money to be made in well systems. Whether you are installing new systems or fixing problems with existing pumps, the pay can be lucrative. How much do you have to know? It depends on what your plans are. It takes less knowledge to install new well systems than it does to troubleshoot and repair them.

Most pump makers offer booklets to professional plumbers (through plumbing suppliers) that offer troubleshooting tips and techniques. These publications are invaluable when you are in the field with a pump system that you are not familiar with. Ask your supplier for a booklet on all the brands of pumps used in your area. You will probably have to visit several suppliers to obtain literature on all the various types of pumps. If suppliers can't help you, contact the pump makers directly. They should be happy to send you all sorts of information on their products. When you combine specific manufacturer information with the data in this chapter, you will be a formidable force in the installation and repair of well systems. The use of routine checklists can also make your work with pumps easier.

This check list is intended to help in making reliable submersible pump installations. Other data for specific pumps may be needed.

1. Motor Inspection

_____ A. Verify that the model, HP or KW, voltage, phase and hertz on the motor nameplate match the installation requirements. Consider any special corrosion resistance required.

_____ B. Check that the motor lead assembly is tight in the motor and that the motor and lead are not damaged.

_____ C. Test insulation resistance using a 500 or 1000 volt DC megohmmeter, from each lead wire to the motor frame. Resistance should be at least 20 megohms, motor only, no cable.

_____ D. Keep a record of motor model number, HP or KW, voltage, date code and serial number.

2. Pump Inspection

_____ A. Check that the pump rating matches the motor, and that it is not damaged.

_____ B. Verify that the pump shaft turns freely.

3. Pump/Motor Assembly

_____ A. If not yet assembled, check that pump and motor mounting faces are free from dirt and uneven paint thickness.

_____ B. Assemble the pump and motor together so their mounting faces are in contact, then tighten assembly bolts or nuts evenly to manufacturer specifications. If it is visible, check that the pump shaft is raised slightly by assembly to the motor, confirming impeller running clearance.

_____ C. If accessible, check that the pump shaft turns freely.

_____ D. Assemble the pump lead guard over the motor leads. Do not cut or pinch lead wire during assembly or handling of the pump during installation.

4. Power Supply and Controls

_____ A. Verify that the power supply voltage, hertz, and KVA capacity match motor requirements.

_____ B. Use a matching control box with each single phase three wire motor.

_____ C. Check that the electrical installation and controls meet all safety regulations and match the motor requirements, including fuse or circuit breaker size and motor overload protection. Connect all metal plumbing and electrical enclosures to the power supply ground to prevent shock hazard. Comply with National and local codes.

5. Lightning and Surge Protection

_____ A. Use properly rated surge (lightning) arrestors on all submersible pump installations unless the installation is operated directly from an individual generator and/or is not exposed to surges. Motors 5HP and smaller which are marked "Equipped with Lightning Arrestors" contain internal arrestors.

_____ B. Ground all above ground arrestors with copper wire directly to the motor frame, or to metal drop pipe or casing which reaches below the well pumping level. Connecting to a ground rod does not provide good surge protection.

6. Electrical Cable

_____ A. Use cable suitable for use in water, sized to carry the motor current without overheating in water and in air, and complying with local regulations. To maintain adequate voltage at the motor, use lengths no longer than specified in the motor manufacturer's cable charts.

_____ B. Include a ground wire to the pump if required by codes or surge protection, connected to the power supply ground. Always ground any pump operated outside a drilled well.

7. Well Conditions

_____ A. For adequate cooling, motors must have at least the water flow shown on its nameplate. If well conditions and construction do not assure this much water flow will always come from below the motor, use a flow sleeve as shown in the Application, Installation & Maintenance Manual

_____ B. If water temperature exceeds 30 degrees C (86 °F), reduce the motor loading or increase the flow rate to prevent overheating, as specified in the Application, Installation & Maintenance Manual.

8. Pump/Motor Installation

_____ ◉ Splice motor leads to supply cable using electrical grade solder or compression connectors, and carefully insulate each splice with watertight tape or adhesive-lined shrink tubing, as shown in motor or pump installation data.

_____ B. Support the cable to the delivery pipe every 10 feet (3 meters) with straps or tape strong enough to prevent sagging. Use pads between cable and any metal straps.

_____ C. A check valve in the delivery pipe is recommended, even though a pump may be reliable without one. More than one check valve may be required, depending on valve rating and pump setting. Install the lowest check valve below the lowest pumping level of the well, to avoid hydraulic shocks which may damage pipes, valve or motor.

_____ D. Assemble all pipe joints as tightly as practical, to prevent unscrewing from motor torque. Recommended torque is at least 10 pound feet per HP (2 meter-KG per KW).

_____ E. Set the pump far enough below the lowest pumping level to assure the pump inlet will always have at least the Net Positive Suction Head (NPSH) specified by the pump manufacturer, but at least 10 feet (3 meters) from the bottom of the well to allow for sediment build up.

FIGURE 11.1 Submersible pump installation checklist.

____ F. Check insulation resistance from dry motor cable
ends to ground as the pump is installed, using a 500
or 1000 volt DC megohmmeter. Resistance may
drop gradually as more cable enters the water, but
any sudden drop indicates possible cable, splice or
motor lead damage. Resistance should meet motor
manufacturer data.

9. After Installation

____ A. Check all electrical and water line connections and
parts before starting the pump. Make sure water
delivery will not wet any electrical parts, and
recheck that overload protection in three phase
controls meets requirements.

____ B. Start the pump and check motor amps and pump
delivery. If normal, continue to run the pump until
delivery is clear. If three phase pump delivery is
low, it may be running backward because phase
sequence is reversed. Rotation may be reversed
(with power off) by interchanging any two motor
lead connections to the power supply.

____ C. Connect three phase motors for current balance
within 5% of average, using motor manufacturer
instructions. Unbalance over 5% will cause higher
motor temperatures and may cause overload trip,
vibration, and reduced life.

____ D. Make sure that starting, running and stopping cause
no significant vibration or hydraulic shocks.

____ E. After at least 15 minutes running, verify that pump
output, electrical input, pumping level, and other
characteristics are stable and as specified.

Date _____   Filled In By

10. Installation Data

Well Identification _____

Check By _____

Date _____ / _____ / _____

Notes _____

_____

_____

_____

_____

_____

_____

_____

_____

FIGURE 11.1 *(continued)*   Submersible pump installation checklist.

INSTALLER'S NAME _____   OWNER'S NAME _____

ADDRESS_____   ADDRESS _____

CITY _____ STATE_____ ZIP_____   CITY _____ STATE_____ ZIP_____

PHONE (____) _____FAX (____) _____   PHONE (____) _____FAX (____) _____

CONTACT NAME _____   CONTACT NAME _____

WELL NAME/ID_____ DATE INSTALLED_____

MOTOR:

Motor No. _____ Date Code _____ HP _____ Voltage _____ Phase _____

PUMP:

Manufacturer _____ Model No. _____ Curve No. _____ Rating: _____ GPM@_____ft. TDH

NPSH Required: _____ ft. NPSH Available:_____ ft. Actural Pump Delivery_____GPM@ _____ PSI

Operating Cycle: _____ON (Min./Hr.) _____ OFF (Min./Hr.) (Circle Min. or Hr. as appropriate)

YOUR NAME _____ DATE _____/_____/_____

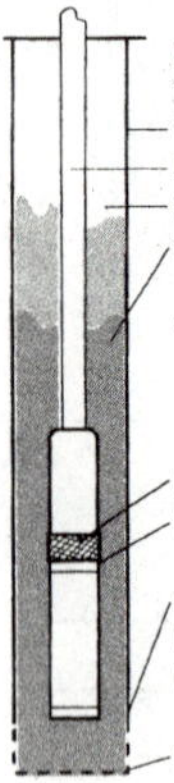

WELL DATA:
Total Dynamic Head _____ft.
Casing Diameter_____in.
Drop Pipe Diameter_____in.
Static Water Level _____ft.
Drawdown (pumping) Water Level_____ft.

Checkvalves at _____&_____&
_____&_____ft.
❑ Solid ❑ Drilled

Pump Inlet Setting _____ ft.
Flow Sleeve: ____No____ Yes, Dia._____in.

Casing Depth_____ft.
❑ Well Screen ❑ Perforated Casing
From_____to_____ft. & _____to_____ft.

Well Depth_____ft.

TOP PLUMBING:
Please sketch the plumbing after the well head (check valves, throttling valves, pressure tank, etc.) and indicate the setting of each device.

Form No. 2207 2/94

FIGURE 11.2 Submersible motor installation record.

POWER SUPPLY:

Cable: Service Entrance to Control _____ ft._____ AWG/MCM ❑ Copper or ❑ Aluminum, ❑ Jacketed or ❑ Individual Conductors

Cable: Control to Motor _____ ft._____ AWG/MCM ❑ Copper or ❑ Aluminum, ❑ Jacketed or ❑ Individual Conductors

Transformers:

KVA _____ #1 _____ #2 _____ #3

Intial Megs (motor & lead)   T1_____T2_____T3_____

Final Megs (motor, lead & cable)   T1_____T2_____T3_____

Incoming Voltage:

No Load: L1-L2_____L2-L3_____L1-L3_____

Full Load: L1-L2_____L2-L3_____L1-L3_____

Running Amps:

HOOKUP 1:

Full Load: L1_____L2_____L3_____

　　%Unbalance_____

HOOKUP 2:

Full Load: L1_____L2_____L3_____

　　%Unbalance_____

HOOKUP 3:

Full Load: L1_____L2_____L3_____

　　%Unbalance_____

Control Panel:

Panel Manufacturer:_____

Short Circuit Device: ❑ Circuit Breaker: Rating_____Setting_____

❑ Fuses:　　Rating_____Type_____

❑ Standard ❑ Delay

Starter Manufacturer:_____ Starter Size_____

Type of Starter:　❑ Full Voltage　　❑ Autotransformer

❑ Other:_____ Full Voltage in _____sec.

Heaters Manufacturer:_____

Number_____Adjustable Set at:_____amps.

Subtrol-Plus: ❑ No ❑ Yes: Registration No._____

If yes, Overload Set?　❑ No　❑ Yes Set at _____amps.

Underload Set?　❑ No　❑ Yes Set at _____amps.

Controls are Grounded to: ❑ Well Head ❑ Motor

❑ Rod　　❑ Power Supply

Ground Wire Size:_____ AWG/MCM

Comments:_____

_____

_____

_____

_____

_____

_____

_____

_____

_____

FIGURE 11.2 *(continued)*   Submersible motor installation record.

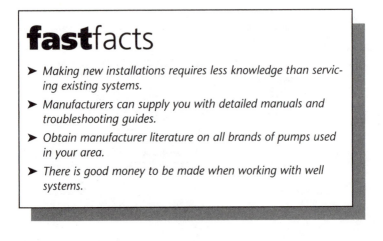

fastfacts

➤ *Making new installations requires less knowledge than servicing existing systems.*

➤ *Manufacturers can supply you with detailed manuals and troubleshooting guides.*

➤ *Obtain manufacturer literature on all brands of pumps used in your area.*

➤ *There is good money to be made when working with well systems.*

SHALLOW WELL JET PUMPS

Shallow well pumps are used with wells where the lowest water level will not be more than 25 feet below the pump. There are many factors that influence the type and size of pump an installation will require. The first consideration is the height that water will need to be pumped. If a pump will have to pump water higher than 25 feet, a shallow well pump is not a viable choice. Shallow well pumps are not intended to lift water more than 25 feet. This limitation is due to the way that a shallow well pump works.

These pumps work on a suction principal. The pump sucks water up the pipe and into the home. With a perfect vacuum at sea level, a shallow well pump may be able to lift water to thirty feet. This maximum lift is not recommended and is rarely achieved. If you will have to lift water higher than 25 feet, investigate other types of pumps. When conditions allow the use of a shallow well jet pump, this section will tell you how to install it. Talk to your pump dealer or refer to manufacturer recommendations for proper sizing of the pump.

A Suction Pump

When you install a suction pump, the single pipe from the well to the pump will usually have a diameter of 1¼ inches. A standard well pipe material is polyethylene, rated for 160 psi. You will want to be sure the suction pipe is not coiled or in any condition that may cause

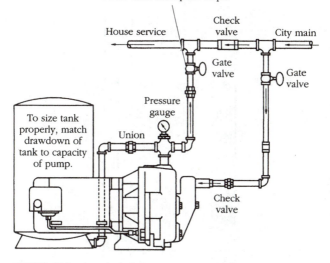

Use flow control or manual valve on discharge to throttle pump, must be sized, or set, to load motor below max. nameplate amps.

FIGURE 11.3 A typical piping arrangement for a jet pump. *(Courtesy Goulds Pumps, Inc.)*

Table 1 number of starts

Motor rating	Maximum starts per 24 hr. day	
	Single phase	Three phase
Up to 3/4 hp	300	300
1 hp thru 5 hp	100	300
7 1/2 hp thru 30 hp	50	100
40 hp and over		100

FIGURE 11.4 Recommended maximum number of times a pump should start in a 24 hour period. *(Courtesy Amtrol, Inc.)*

8100 Series Specifications ▼ Shallow Well

MODEL NO.	HP	VOLTS	IMPELLER MATERIAL	PRES. SWITCH SETTING	SUCTION PIPE SIZE	DISCHARGE SIZE	SHIPPING WEIGHT
8130	1/3	115	Plastic	20-40	1 1/4"	3/4"	46 lbs.
8131	1/3	115	Brass	20-40	1 1/4"	3/4"	48 lbs.
8150	1/2	115/230	Plastic	20-40	1 1/4"	3/4"	48 lbs.
8151	1/2	115/230	Brass	20-40	1 1/4"	3/4"	50 lbs.
8170	3/4	115/230	Plastic	30-50	1 1/4"	3/4"	50 lbs.
8171	3/4	115/230	Brass	30-50	1 1/4"	3/4"	52 lbs.
8110	1	115/230	Plastic	30-50	1 1/4"	3/4"	52 lbs.
8111	1	115/230	Brass	30-50	1 1/4"	3/4"	53 lbs.

FIGURE 11.5   Performance ratings for jet pumps. (Courtesy A. Y. MacDonald Mfg. Co., Dubuque, Iowa)

Multi-Stage Specifications
▼ 1500 Series Horizontal ▼ 1000 Series Vertical

MODEL	MODEL	HP	VOLTS	IMPELLER MATERIAL	PRES. SWITCH SETTING	SUCTION PIPE SIZE	TWIN TYPE DROP PIPE	SHIPPING WEIGHT
1550	1050	1/2	115/230	Brass	30-50	1 1/4"	1" x 1 1/4"	65 lbs.
1575	1075	3/4	115/230	Brass	30-50	1 1/4"	1" x 1 1/4"	71 lbs.
1575SW	1075SW	3/4	115/230	Brass	30-50	1 1/4"	1" x 1 1/4"	66 lbs.
1510	1010	1	115/230	Brass	30-50	1 1/4"	1" x 1 1/4"	74 lbs.
1510SW	1010SW	1	115/230	Brass	30-50	1 1/4"	1" x 1 1/4"	67 lbs.
1515SW	1015SW	1 1/2	115/230	Brass	30-50	1 1/4"	1" x 1 1/4"	72 lbs.

FIGURE 11.6 Performance ratings for multi-stage pumps. (Courtesy A. Y. MacDonald Mfg. Co., Dubuque, Iowa)

357

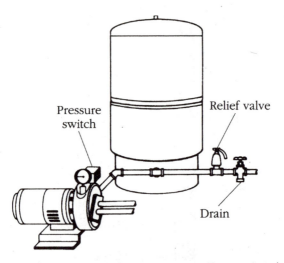

FIGURE 11.7   A typical jet pump set-up. *(Courtesy Amtrol, Inc.)*

it to trap air as it is being installed. If the pipe holds an air pocket, priming the pump can be quite difficult. In most cases a foot valve will be installed on the end of the pipe that is submerged in the well.

Screw a male insert adapter into the foot valve. Place two stainless steel clamps over the well pipe and slide the insert fitting into the pipe. Tighten the clamps to secure the pipe to the insert fitting. When you lower the pipe and foot valve into the well, don't let the foot valve sit on the bottom of the well. If the suction pipe is too close to the bottom of the well, it may suck sand, sediment, or gravel into the foot valve. If this happens, the pipe cannot pull water from the well.

When the pipe reaches the upper portion of the well, it will usually take a 90 degree turn to exit the well casing. This turn will be made with an insert-type elbow. Always use two clamps to hold the pipe to its fittings. When the pipe leaves the well, it should be buried underground. Run the well pipe through a sleeve so that the well casing will not chafe the pipe during use and wear a hole in it. The pipe must be deep enough so that it will not freeze in the winter. This depth will vary from state to state. Your local plumbing inspector will be able to tell you how deep to bury the water supply pipe.

When you place your pipe in a trench, be careful not to lay it on sharp rocks or other objects that might wear a hole in the pipe. Backfill the trench with clean fill dirt. If you dump rocks and clut-

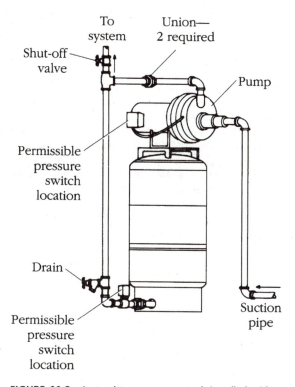

FIGURE 11.8 A stand-type pressure tank installed with a straight-through method not using a tank tee.

tered fill on the pipe, it can be crimped or cut. When you bring the pipe into a building, run it through a sleeve where it comes through or under the foundation. The sleeve should be two pipe sizes larger than the water supply pipe.

Once inside the building, you may have to convert your pipe to copper, CPVC, PB, or one of the other approved materials. When PE pipe is used as a water service, it must be converted to some other type of pipe within five feet of its entry into a building, assuming that the building will have both hot and cold water. If you convert the pipe, the conversion will typically be done with a male insert adapter. The water supply pipe should run directly to the pump. The foot valve acts as a strainer and as a check valve. When you have a foot valve in the well, there is no need for a check valve at the pump.

Your incoming pipe will attach to the pump at the inlet opening with a male adapter. At the outlet opening, you will install a short nipple and a tee fitting. At the top of the tee, you will install reducing bushings and a pressure gauge. From the center outlet of the tee, your pipe will run to another tee fitting. There should be a gate valve installed in this section of pipe, near the pump. At the next tee, the center outlet will be piped to a pressure tank. From the end outlet of the tee, your pipe will run to yet another tee fitting. At this tee, the center outlet will become the main cold water pipe for the house. Another gate valve should be installed in the pipe feeding the water distribution system. On the end outlet of the tee, you will install a pressure-relief valve. All of these tee fittings should be in a close proximity to the pressure tank.

The pump will be equipped with a control box that requires electrical wiring. This job should be done by a licensed electrician. If you are an electrician, you know how to do the job. If you are not an electrician, you shouldn't attempt to wire the controls.

Priming The Pump

The pump will have a removable plug in the top of it to allow the priming of the pump. Remove the plug and pour water into the priming hole. Continue this process until the water is standing in the pump and visible at the hole. Apply pipe dope to the plug and screw it back into the pump. When you turn the pump on, you should have water pressure. If you don't, you must continue the priming process until the pump is pumping water. This can be a time consuming process; don't give up.

Setting The Water Pressure

Once the pump is pumping water and filling the pressure tank, setting the water pressure will become a priority. When the tank is filled, your pressure gauge should give a reading between 40 and 60 pounds of pressure per square inch. The pump's controls will be preset at cut-in and cut-out intervals. These settings regulate when the pump cuts on and off. Typically, a pump will cut on when the tank pressure drops below 20 pounds. The pump will cut off when the tank pressure reaches 40 pounds.

If your customer prefers a higher water pressure, the pressure switch can be altered to deliver higher pressure. You might have the controls set to cut on at 40 pounds and off at 60 pounds. Adjusting these settings is done inside the pressure switch, around electrical wires. There is possible danger from electrocution when making these adjustments. To avoid problems, cut the power off to the pressure switch while you are making the adjustments.

The adjustments are made by turning a nut that sits on top of a spring in the control box. You will see a coiled spring, compressed with a retaining nut. By moving this nut up and down the threaded shaft, you can alter the cut-in and cut-out intervals. Refer to manufacturer's recommendations for precise settings.

DEEP-WELL JET PUMPS

When your water level is more than 25 feet below the pump, you will have to use either a deep-well jet pump or a submersible pump. In today's plumbing applications, submersible pumps are normally used in deep wells. However, deep-well jet pumps will get the job done.

Deep-well jet pumps resemble shallow-well pumps. They are installed above ground and are piped in a similar manner to shallow-well pumps. The noticeable difference is the number of pipes going into the well. A shallow-well pump has only one pipe. Deep-well jet pumps have two pipes. The operating principals of the two types of pumps differ. Shallow-well pumps suck water up from the well. Deep-well jet pumps push water down one pipe and suck water up the other; this is why there are two pipes on deep-well jet pumps.

The only major installation differences between a shallow-well pump and a deep-well pump is the number of pipes used in the installation, and a pressure control valve. Deep-well pumps still use a foot valve. There is a jet-body fitting that is submerged in the well and attached to both pipes and the foot valve. The pressure pipe connects to the jet assembly first. The foot valve hangs below the pressure pipe. There is a molded fitting on the jet body for the suction line to connect to. With this jet body, both pipes are allowed to connect in a natural and efficient manner.

Deep-well jet pumps push pressure down the pressure pipe with water pushed through the jet assembly, it makes it possible for the suction pipe to pull water up from the deep well. From the suction pipe, water is brought into the pump and distributed to the potable water system. When you look at the head of a deep-well jet pump, you will see two openings for your pipes to connect to. The larger opening is for the suction pipe and the smaller opening is for the pressure pipe. The suction pipe will usually have a diameter of 1¼ inches. The pressure pipe will typically have a diameter of one inch.

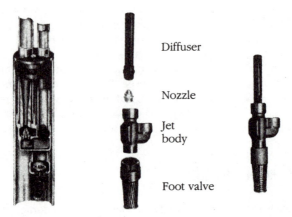

FIGURE 11.9 In-well assembly system for a two-pipe pump. *(Courtesy Goulds Pumps, Inc.)*

The piping from the pump to the pressure tank will need a pressure control valve installed in it. This valve assures a minimum operating pressure for the jet assembly. Shallow-well pumps are not required to have a pressure control valve. Once the pressure control valve is installed, the remainder of the piping is done in the same manner as used for a shallow-well pump.

Pressure tanks for jet pumps come in many sizes and shapes. Tanks are available for a variety of installation methods.

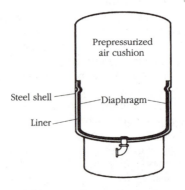

FIGURE 11.10 Diaphragm-type pressure tank. *(Courtesy Goulds Pumps, Inc.)*

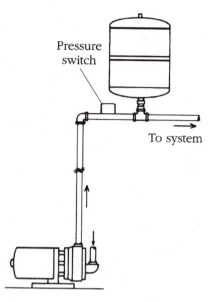

FIGURE 11.11 Small, vertical pressure tank installed above pump. *(Courtesy Amtrol, Inc.)*

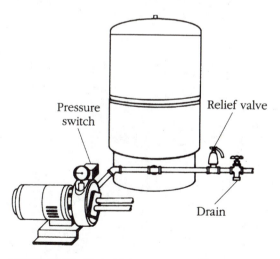

FIGURE 11.12 Typical jet-pump setup. *(Courtesy Amatrol, Inc.)*

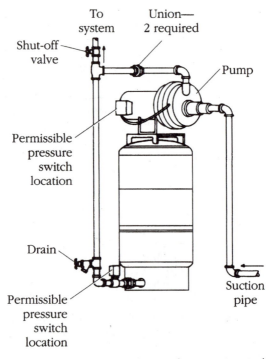

FIGURE 11.13 A jet pump mounted on a pressure tank with a pump bracket. *(Courtesy Amtrol, Inc.)*

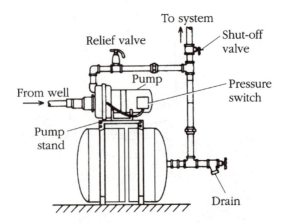

FIGURE 11.14 Bracket-mounted jet pump on a horizontal pressure tank. *(Courtesy Amtrol, Inc.)*

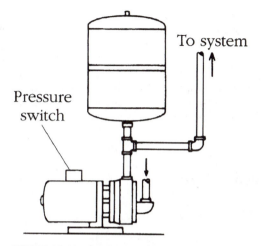

FIGURE 11.15   Small vertical pressure tank installed above pump. *(Courtesy Amtrol, Inc.)*

SUBMERSIBLE PUMPS

Submersible pumps are very different from jet pumps. Jet pumps are installed outside of the well. Submersible pumps are installed in the well, submerged in the water. Jet pumps use suction pipes. Submersible pumps have only one pipe and they push water up the pipe, from the well. Jet pumps use a foot valve, submersible pumps don't. Submersible pumps are much more efficient than jet pumps, and are also easier to install. Under the same conditions, a half-horsepower submersible pump can produce nearly 300 gallons

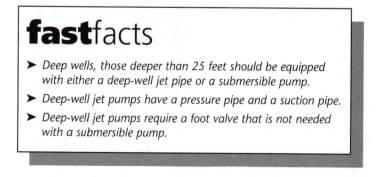

fastfacts

➤ *Deep wells, those deeper than 25 feet should be equipped with either a deep-well jet pipe or a submersible pump.*

➤ *Deep-well jet pumps have a pressure pipe and a suction pipe.*

➤ *Deep-well jet pumps require a foot valve that is not needed with a submersible pump.*

more water than a half-horsepower jet pump. With so many advantages, it is almost foolish to use a jet pump when you could use a submersible pump. The one exception to this is when you are installing a pump for a shallow well. Then a jet pump makes the most sense.

Installing a submersible pump requires different techniques from those used with a jet pump. Since submersible pumps are installed in the well, there must be electrical wires run down the well, to the

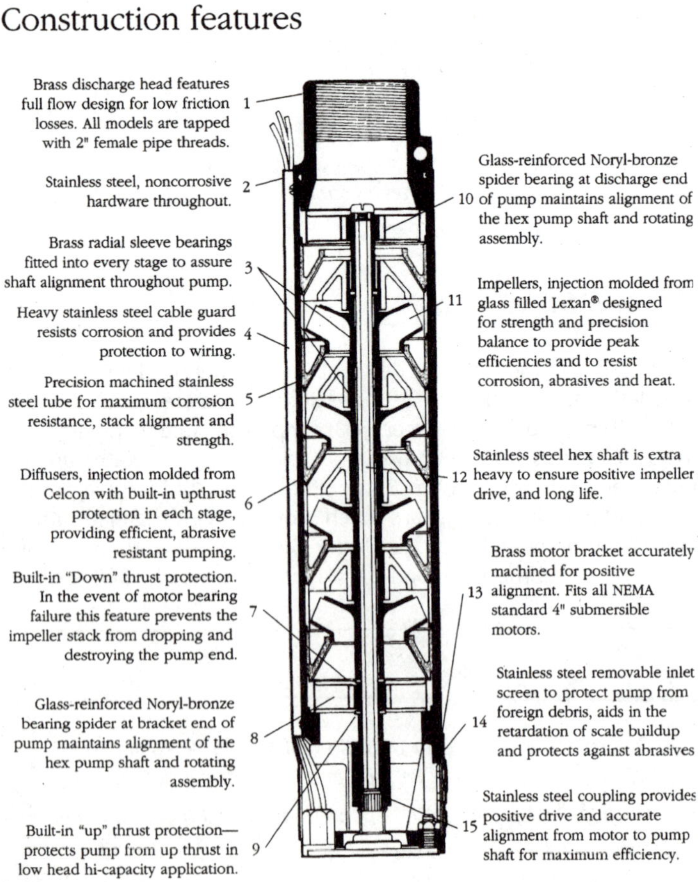

Construction features

Brass discharge head features full flow design for low friction losses. All models are tapped with 2" female pipe threads. 1

Stainless steel, noncorrosive hardware throughout. 2

Brass radial sleeve bearings fitted into every stage to assure shaft alignment throughout pump. 3

Heavy stainless steel cable guard resists corrosion and provides protection to wiring. 4

Precision machined stainless steel tube for maximum corrosion resistance, stack alignment and strength. 5

Diffusers, injection molded from Celcon with built-in upthrust protection in each stage, providing efficient, abrasive resistant pumping. 6

Built-in "Down" thrust protection. In the event of motor bearing failure this feature prevents the impeller stack from dropping and destroying the pump end. 7

Glass-reinforced Noryl-bronze bearing spider at bracket end of pump maintains alignment of the hex pump shaft and rotating assembly. 8

Built-in "up" thrust protection— protects pump from up thrust in low head hi-capacity application. 9

10 Glass-reinforced Noryl-bronze spider bearing at discharge end of pump maintains alignment of the hex pump shaft and rotating assembly.

11 Impellers, injection molded from glass filled Lexan® designed for strength and precision balance to provide peak efficiencies and to resist corrosion, abrasives and heat.

12 Stainless steel hex shaft is extra heavy to ensure positive impeller drive, and long life.

13 Brass motor bracket accurately machined for positive alignment. Fits all NEMA standard 4" submersible motors.

14 Stainless steel removable inlet screen to protect pump from foreign debris, aids in the retardation of scale buildup and protects against abrasives.

15 Stainless steel coupling provides positive drive and accurate alignment from motor to pump shaft for maximum efficiency.

FIGURE 11.16 Cutaway of a submersible pump. *(Courtesy A.Y. McDonald MFG Co. Dubuque, Iowa)*

pump. Before you install a submersible pump, consult a licensed electrician for the wiring needs of the pump. Some plumbers do this wiring themselves, but many jurisdictions require the connections to be made by a licensed electrician.

You will need a hole in the well casing to install a pitless adapter. The pitless adapter provides a watertight seal in the well casing for your well pipe to feed your water service. When you purchase your pitless adapter, it should be packaged with instructions on what size hole you will need in the well casing.

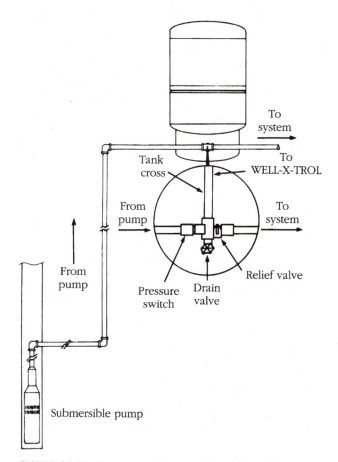

To system

Tank cross

To WELL-X-TROL

From pump

To system

From pump

Pressure switch

Drain valve

Relief valve

Submersible pump

FIGURE 11.17 Pressure tank in use with a submersible pump. *(Courtesy Amtrol, Inc.)*

You can cut a hole in the well casing with a cutting torch or a hole saw. The pitless adapter will attach to the well casing and seal the hole. On the inside of the well casing, you will have a tee fitting on the pitless adapter. This is where your well pipe is attached. This tee fitting is designed to allow you to make all of your pump and pipe connections above ground. After all the connections are made, you lower the pump and pipe into the well and the tee fitting slides into a groove on the pitless adapter.

To make up the pump and pipe connections, you will need to know how deep the well is. The well driller should provide you with the depth and rate of recovery for the well. Once you know the depth, cut a piece of plastic well pipe to the desired length. The pump should hang at least 10 feet above the bottom of the well and at least ten feet below the lowest, expected water level.

Apply pipe dope to a male-insert adapter and screw it into the pump. This fitting is normally made of brass. Slide a torque arrestor over the end of your pipe. Next, slide two stainless steel clamps over the pipe. Place the pipe over the insert adapter and tighten the two clamps. Compress the torque arrestor to a size slightly smaller than the well casing and secure it to the pipe. The torque arrestor absorbs thrust and vibrations from the pump and helps to keep the pump centered in the casing.

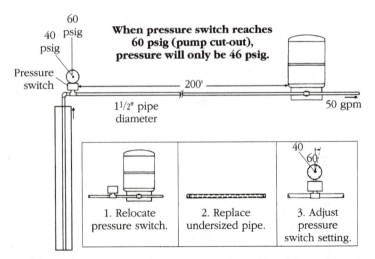

FIGURE 11.18 Sizing suggestions for unusual situations. *(Courtesy Amtrol, Inc.)*

Slide torque stops down the pipe from the end opposite of the pump. Space the torque stops at routine intervals along the pipe to prevent the pipe and wires from scraping against the casing during operation. Secure the electrical wiring to the well pipe at regular intervals to eliminate slack in the wires. Apply pipe dope to a brass, male-insert adapter and screw it into the bottom of the tee fitting for the pitless adapter. Slide two stainless steel clamps over the open end of the pipe and push the pipe onto the insert adapter. After tightening the clamps, you are ready to lower the pump into the well.

Before lowering the pump, it is a good idea to tie a safety rope onto the pump. After the pump is installed, this rope will be tied at the top of the casing to prevent the pump from being lost in the well if the pipe becomes disconnected from the pump. Next, you will need to screw a piece of pipe or an adapter into the top of the tee fitting for the pitless adapter. Most plumbers use a rigid piece of steel pipe for this purpose.

Once you have a pipe extending up from the top of the tee fitting, lower the whole assembly into the well casing. This job is easier if you have someone to help you. Be careful not to scrape the electrical wires

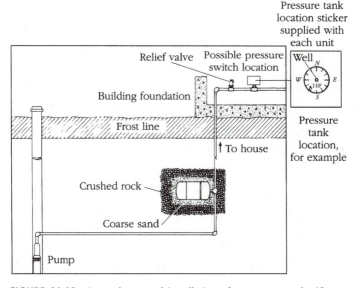

FIGURE 11.19 An underground installation of a pressure tank. *(Courtesy Amtrol, Inc.)*

on the well casing as the pump is lowered. If the insulation on the wires is damaged, the pump may not work. Holding the assembly by the pipe extending from the top of the pitless tee, guide the pitless adapter into the groove of the adapter in the well casing. When the adapter is in the groove, push down to seat it into the mounting bracket. This concludes the well part of the installation.

Attach your water service pipe to the pitless adapter on the outside of the casing. You can do this with a male-insert adapter. Run your pipe to the building in the same way described for jet pumps. Once inside the house, the water pipe should have a union installed in it. The next fitting should be a gate valve, followed by a check valve. From the check valve, your pipe should run to a tank tee.

The tank tee is a device that screws into the pressure tank and allows the installation of all related parts. The switch box, pressure gauge, and boiler drain can all be installed on the tank tee. When your pipe comes to the tank tee, the water is dispersed to the pressure tank, the drain valve, and the water main. When the water main leaves the tank tee, you should install a tee to accommodate a pressure relief valve. After this tee, you can install a gate valve and continue your piping to the water distribution system. All that is left is to test the system. You do not have to prime a submersible pump.

fastfacts

➤ *Submersible pumps are very different from jet pumps.*

➤ *Submersible pumps are installed in wells.*

➤ *Submersible pumps have a single pipe and push water up from the well.*

➤ *Submersible pumps pump more water per minute than a jet pump.*

➤ *Submersible pumps should be installed so that they are at least 10 feet above the bottom of the well.*

➤ *Submersible pumps should be installed so that they will always be at least 10 feet below the static water level.*

PROBLEMS WITH JET PUMPS

Circuit Breakers and Fuses

When your pump won't run, there are six likely causes for the mal-function. The first item to check is the fuse or circuit breaker that controls the pump's electrical circuit. If the fuse is blown or the breaker is tripped, the pump cannot run. When you check your electrical panel and find these conditions, replace the blown fuse or reset the tripped breaker. In future references to fuses or circuit breakers, I will only refer to circuit breakers. If your home has fuses, when I indicate resetting the circuit breaker, you should replace the blown fuse.

Damaged or Loose Wiring

If the circuit breaker is not tripped, inspect all the electrical wiring that affects your pump. There may be a broken or loose wire that is preventing the pump from running. Remember, you are dealing with electricity and must use appropriate caution to avoid electrical shocks.

Pressure Switches

Your pressure switch may be defective. It is also possible that the switch is out of its proper adjustment. To correct this problem, you must adjust or replace the pressure switch. If you remove the cover of your pressure switch, you will see two nuts sitting on top of springs. The nut on the short spring is set at the factory and should not need any adjustment. However, if you want your pump to cut out at a higher pressure, turn this nut clockwise. To have the pump cut out at a lower pressure, turn the nut counterclockwise. The nut on the taller spring controls the cut-in and cut-out cycle of your pump. To make the pump cut in at a higher pressure, turn the nut clockwise. If you want the pump to cut in at a lower pressure, turn the nut counterclockwise.

Stopped-Up Nipples

If the tubing or nipples on the pressure switch become restricted, the pump may not run. If you suspect this to be the problem, take the pipe and fittings apart and clean or replace them. In a new installation, this problem is unlikely, but possible.

Overloaded Motors

If the motor becomes overloaded, the protection contacts will remain open. This problem will solve itself in a short time. The contacts will close automatically after a short time.

Seized Pumps

If your pump becomes mechanically bound, it will not run. To correct this problem, you must first remove the end cap from the pump. Once the end cap is removed, you should be able to turn the motor shaft with your hands. Once the shaft rotates freely, reassemble the pump and test it. Your problem should be resolved.

Voltage Problems

If your pump is not receiving the proper voltage, it will not run.

Defective Motors

Obviously, defective motors can prevent pumps from running. To determine if a pump's motor is bad, you must do extensive electrical troubleshooting. You will have to check the ground, capacitor, switch, overload protector, and winding continuity. If you are experienced in working with electrical systems, you already know how to do this. If you are not, I am afraid to give you instructions on such a dangerous job. One wrong move could deliver a potent electrical shock. If your problem is electrical and beyond your abilities, call in a professional electrician.

WHEN YOUR PUMP RUNS, BUT PRODUCES NO WATER

A Loss of the Pump's Prime

If your pump loses its prime, the pump will run without producing water. With a shallow-well pump, remove the priming plug and pour water into the pump. When the water level stands static at the opening, apply pipe dope to the priming plug and replace it. Start the pump and you should get water pressure. With a deep-well jet pump, trying to prime the pump through the priming hole is a losing proposition.

With a two-pipe system, disconnect the pipes from the pump. Pour water down each pipe until they are both full of water. When the water has filled both pipes and is holding its level, reconnect the pipes to the pump. Prime the pump through the priming hole. Start the pump and you should get water. If you don't, the problem may be in the pressure control valve.

Pressure Control Valves

In a two-pipe system, the pressure control valve may need to be adjusted when your pump runs without producing water. When the pressure control valve is set too high, the air volume control may not function. When the pressure control valve is set too low, the pump may not cut off. To reduce the water pressure, turn the adjusting screw on the pressure reducing valve counterclockwise. To increase pressure, turn the adjusting screw clockwise.

Lack of Water

If your well pipe is not below the water level, the pump can not produce water. With shallow wells, it is not uncommon for the water levels to drop during hot, dry months. If your pump is running, but not producing water, check the water level in your well. You can do this with a roll of string and a weight. Tie the weight on the end of the string and lower it into the well. When the weight hits bottom, withdraw the string. The end of the string should be wet. Measure the distance along the wet string. This will tell you how many feet of water is in the well. This information, along with the records of how long your well pipe is, will tell you if the end of the pipe is submerged in the water.

Clogged Foot Valves

If your foot valve is too close to the bottom of the well, it may become clogged with mud, sand, or sediment. It is not uncommon for shallow wells to fill in with sediment. What starts out as a 30 foot well may become a 25 foot well as the earth settles and shifts. Even if your foot valve was originally set at the appropriate height, it may now be hanging too low in the well. Sometimes, shaking the well pipe will clear the foot valve of its restriction. In other cases, you will have to pull the pipe out of the well and clean or replace the foot valve. If the foot valve is pulling up sediment from the bottom of the well, you will have to shorten the well pipe.

Malfunctioning Foot Valves and Check Valves

Foot valves act as a check valve. If the foot valve or check valve is stuck in the closed position, you will not get any water from your pump. Pull the well line and inspect the check valve or foot valve. If the valve is stuck, replace it.

Defective Air-Volume Control

It is possible that the diaphragm in your air-volume control has a hole in it. By disconnecting the tubing and plugging the connection in the pump, you can determine if you have a hole in the diaphragm of your air-volume control. If plugging the hole corrects your problem, you must replace the air-volume control.

Failing Jet Assemblies

If the jet assembly fails, the pump will not produce water. Inspect the jet assembly for possible obstructions. If an obstruction is found, attempt to clear it. If you cannot clean the assembly, replace it.

Suction Leaks

A common aliment of shallow-well systems is a leak in the suction pipe. To check for this problem, you must pressurize the system and inspect it for leaks. If you find a leak, when it is stopped, you will have water again.

PUMPS THAT WILL NOT CUT OUT

If your pump will build pressure, but will not cut out, you may have a bad pressure switch. Before you replace the switch, there are a few things you can check. Check all the tubing, nipples, and fittings associated with the switch. If they are obstructed, clear them and test the pump. If the problem persists, check the pressure switch to see that it is set properly. You do this by checking the cut-in and cut-out controls discussed earlier. If all else fails, replace the pressure switch.

WHEN YOU DON'T GET ENOUGH PRESSURE

A lack of water pressure can be caused by several problems. If you are trying to lift water too high, the pump may not be able to handle the job. This requires checking the rating of your pump to be sure it is capable of performing under the given conditions. If you have leaks in your piping, the pump will not be able to build adequate pressure. Check all piping to confirm that no leaks are present. If the jet or the foot valve is partially obstructed, your water pressure will suffer. Check these items to be sure they are free to operate. If they are blocked, clear the obstructions and test the pump.

When your air-volume control is on the blink, you may not be able to build suitable pressure. Test the control as described above and replace it if necessary. On older pumps, the impeller hub or guide vane bore may be worn. If this is the case, you will need to call a professional to verify your diagnosis. If either of these parts are worn, they must be replaced.

WHEN YOUR PUMP CYCLES TOO OFTEN

Leaks in your piping can cause your pump to cut on and off more frequently than it should. When the system is pressurized, check all piping for leaks. If any are found, repair them. If the pressure switch is not working correctly, the pump may run at random. Follow the directions given earlier to troubleshoot the pressure switch. If necessary, replace the switch. Your problem could be in the suction lift of the system. If water floods the pump through the suction line, you must control the water flow with a partially closed valve. You also may not be developing enough vacuum on the suction line to get the job done. During your troubleshooting, check out the air-volume control. It is possible this valve is defective. The instructions given earlier will explain how to check the air-volume control.

Pressure Tanks

The most likely cause of pumps that cycle too often is a faulty pressure tank. If the tank has a leak in it, it can not hold pressure. In old tanks, this is a common problem. If you find a leak in the tank, you should replace the tank. If the air pressure in the pressure tank is not right, the pump will cut on frequently. The pump may cut on every time the water to a faucet is turned on. This condition should not be ignored, it could burn the motor of the pump up.

Modern captive air tanks come with a factory pre-charge of air. These tanks use a diaphragm to control the air volume. Older tanks do not have these diaphragms. With older tanks, it is not uncommon for them to become waterlogged. When the tank is waterlogged, it does not contain enough air and does contain too much water.

Checking the Tank's Air Pressure

To check your pressure tank's air pressure, cut off the electrical power to the pump. Drain the tank until the pressure gauge is at zero. A rule of thumb for the proper air charge dictates that the air pressure should be two pounds per square inch less than the cut-in pressure of the pressure switch. For example, if your pressure switch is set to cut in at forty pounds of pressure, the air in the tank should be set at 38 pounds. You can check the air pressure by putting a tire gauge on the air valve of the tank.

If the air pressure is too low, pump air into the tank. You can do this with a bicycle pump or an air compressor. Monitor the pressure as you fill the tank with air. When the air pressure is at the proper setting, turn the electrical power back on to the pump. When the water floods the tank, you should have a normal reading on your pressure gauge. The system should work properly if the air pressure remains at the proper setting. If the pressure tank is old, it may have a small leak that allows the air to escape. It is possible that a new tank is defective. If this procedure does not work and keep the system working, consider replacing the pressure tank.

SUBMERSIBLE PUMPS

Circuit Breakers

The first item to check when your pump wont start is the circuit breaker that controls the pump's electrical circuit. If breaker is tripped, the pump cannot run. When you check your electrical panel and find the breaker tripped, reset the tripped breaker.

Voltage Problems

If the circuit breaker is not tripped, the voltage to the pump must be checked. This inspection can be done quickly with a voltage meter, in experienced hands. If you are not experienced in such matters, call in a professional.

WHEN YOUR PUMP WILL NOT RUN

If your pump starts, but fails to continue running, check the following items.

Fuses

Check all fuses to see that they are not blown. If the improper fuses are installed, the pump may start, but not run.

Voltage

Incorrect voltage can allow the pump to start, but not allow it to continue running.

Seized Pump

If your pump is seized with sand or gravel, pull the pump and clean it. When the pump is seized, you will receive a high amperage reading from your tests with a meter.

Damaged Wiring

If the electrical wiring to the pump is damaged, the pump may not continue to run. This condition can be tested with a meter by reading the resistance. The insulation on the wires may have been damaged when the pump was lowered into the well. A splice connection may have become open or shorted. When the problem is located, the wiring must be repaired or replaced.

Pressure Switch

Your problem may be being caused by a defective pressure switch. Check the points on the pressure switch and replace the switch if the points are bad.

Pipe Obstructions

If the tubing or nipples for the pressure switch are clogged, the pump may not run. Check for obstructions and clear the tubing or nipples if necessary.

Loose Wires

It is possible that you are a victim of loose wires. Check the wiring at the control box and the motor. If the wires are loose, they may allow the pump to start without letting it run.

Control Box

If the pump has never run properly, you may have the wrong control box. Inspect the box and confirm it to be the correct box for your pump. When the control box is exposed to high temperatures, you may experience trouble with your pump. If your control box is located in an area where the temperature exceeds 120 degrees Fahrenheit, the high temperature may be causing your problem.

WHEN YOUR PUMP RUNS, BUT GIVES NO WATER

Clogs

If the protective screening on the pump is obstructed, water can not enter the pump. This will cause the pump to run without providing water. If the impellers are blocked by obstructions, you will experience the same problem. Pull the pump and check the strainer and impellers for clogs.

Bad Motor

If the pump motor is badly worn, it may not produce water, even if it runs. You will have to pull the pump to inspect the motor. If it is worn, replace it.

Electrical Problems

Improper voltage, incorrect connections, or loose connections could be the cause of your problem.

Leaks

If there are leaks in your piping, the water may not be able to get to the house. If you can connect an air compressor to the pipe, you can pump it up with air to test for leaks. Inspect all your piping for leaks and repair or replace the pipe as needed.

Check Valves

Sometimes a check valve will be installed backwards. If you are experiencing problems with a new installation, check to see that the check valve is properly installed. It is also possible that the check valve has stuck in the closed position. Do a visual inspection and replace or reinstall the check valve if necessary.

No Water

Normally, drilled wells maintain a good water level, but there are times when the water level drops below the pump. If the pump is not submerged, it can not pump water. Check the water level in the well with a string and a weight. If the water is low, you may be able to lower the pump, but don't hang the pump too close to the bottom of the well.

When a deep well is drilled, the installer should provide you with the recovery rate of the well. The recovery rate will be stated in gallons per minute (gpm). The minimal recovery rate normally accepted is three gallons per minute. Five gallons per minute is a better rate. If your pump is pumping faster than the well can recover, you may run out of water. Pumps are rated to tell you how many gallons per minute they pump. Compare your pump's rating to the recovery rate of your well and you may find the cause of your trouble.

LOW WATER PRESSURE

If you have leaks in your piping, the pump will not be able to build adequate pressure. Check all piping to confirm that no leaks are present. Check the setting on your pressure switch. An improper setting will affect your water pressure. If the pump is worn or the voltage to the pump is incorrect, you may not have strong water pressure.

WHEN YOUR PUMP CYCLES TOO OFTEN

If your check valve is stuck in the open position, your pump will have to work overtime. Inspect the check valve to be sure it is operating freely. If the valve is bad, replace it. Leaks in the piping may cause your pump to cut on and off frequently. Check all piping for leaks. If any are found, repair them. If the pressure switch is not working correctly, the pump may cycle too often. Follow the directions given earlier to troubleshoot the pressure switch. If necessary, replace the switch.

Pressure Tanks

The most likely cause of this problem is a faulty pressure tank. If the tank has a leak in it, it can not hold pressure. In old tanks, this is a common problem. If you find a leak in the tank, you should replace the tank. If the air pressure in the pressure tank is not right, the pump will cut on frequently. The pump may cut on every time the water to a faucet is turned on. This condition should not be ignored, it could burn the motor of the pump up.

Modern captive air tanks come with a factory pre-charge of air. These tanks use a diaphragm to control the air volume. Older tanks do not have these diaphragms. With older tanks, it is not uncommon for them to become waterlogged. When the tank is waterlogged, it does not contain enough air and does contain too much water.

Checking the Tank's Air Pressure

To check your pressure tank's air pressure, cut off the electrical power to the pump. Drain the tank until the pressure gauge is at zero. A rule of thumb for the proper air charge dictates that the air pressure should be two pounds per square inch less than the cut-in pressure of the pressure switch. For example, if your pressure switch is set to cut in at forty pounds of pressure, the air in the tank should be set at thirty-eight pounds. You can check the air pressure by putting a tire gauge on the air valve of the tank.

If the air pressure is too low, pump air into the tank. You can do this with a bicycle pump or an air compressor. Monitor the pressure as you fill the tank with air. When the air pressure is at the proper setting, turn the electrical power back on to the pump. When the water floods the tank, you should have a normal reading on your pressure gauge. The system should work properly if the air pressure remains at the proper setting. The pressure tank may have a small leak in it, allowing air to escape. If the tank cannot hold air, it cannot build pressure.

TROUBLESHOOTING

Symptoms	Probable cause
Faucet drips from spout	Bad washers or cartridge Bad faucet seats
Faucet leaks at base of spout	Bad O ring
Faucet will not shut off	Bad washers or cartridge Bad faucet seats
Poor water pressure	Partially closed valve Clogged aerator Not enough water pressure Blockage in the faucet Partially frozen pipe
No water	Closed valve Broken pipe Frozen pipe
Drains slowly	Partial obstruction in drain or trap
Will not drain	Blocked drain or trap
Gurgles as it drains	Partial drainage blockage Partial blockage in the vent
Won't hold water	Bad basket strainer Bad putty seal on drain
Spray attachment will not spray	Clogged holes in spray head Kinked spray hose
Spray attachment will not cut off	Bad spray head

FIGURE 12.1 Troubleshooting kitchen sinks.

Symptoms	Probable cause
Faucet drips from spout	Bad washers or cartridge Bad faucet seats
Faucet leaks at base of spout	Bad O ring
Faucet will not shut off	Bad washers or cartridge Bad faucet seats
Poor water pressure	Partially closed valve Clogged aerator Not enough water pressure Blockage in the faucet Partially frozen pipe
No water	Closed valve Broken pipe Frozen pipe
Drains slowly	Partial obstruction in drain or trap
Will not drain	Blocked drain or trap
Gurgles as it drains	Partial drainage blockage Partial blockage in the vent
Won't hold water	Bad basket strainer Bad putty seal on drain

FIGURE 12.2 Troubleshooting laundry tubs.

Symptoms	Probable cause
Faucet drips from spout	Bad washers or cartridge Bad faucet seats
Faucet leaks at base of spout	Bad O ring
Faucet will not shut off	Bad washers or cartridge Bad faucet seats
Poor water pressure	Partially closed valve Clogged aerator Not enough water pressure Blockage in the faucet Partially frozen pipe
No water	Closed valve Broken pipe Frozen pipe
Drains slowly	Hair on pop-up assembly Partial obstruction in drain or trap Pop-up needs to be adjusted
Will not drain	Blocked drain or trap Pop-up is defective
Gurgles as it drains	Partial drainage blockage Partial blockage in the vent
Won't hold water	Pop-up needs adjusting Bad putty seal on drain

FIGURE 12.3 Troubleshooting lavatories.

Symptoms	Probable cause
Won't drain	Clogged drain Clogged strainer Clogged trap
Drains slowly	Hair in strainer Partial drainage blockage
Gurgles as it drains	Partial drainage blockage Partial blockage in the vent
Water drips from shower head	Bad faucet washers/cartridge Bad faucet seats
Faucet will not shut off	Bad washers or cartridge Bad faucet seats
Poor water pressure	Partially closed valve Not enough water pressure Blockage in the faucet Partially frozen pipe
No water	Closed valve Broken pipe Frozen pipe

FIGURE 12.4 Troubleshooting showers.

Symptoms	Probable cause
Will not flush	No water in tank
	Stoppage in drainage system
Flushes poorly	Clogged flush holes
	Flapper or tank ball is not staying open long enough
	Not enough water in tank
	Partial drain blockage
	Defective handle
	Bad connection between handle and flush valve
	Vent is clogged
Water droplets covering tank	Condensation
Tank fills slowly	Defective ballcock
	Obstructed supply pipe
	Low water pressure
	Partially closed valve
	Partially frozen pipe
Makes unusual noises when flushed	Defective ballcock
Water runs constantly	Bad flapper or tank ball
	Bad ballcock
	Float rod needs adjusting
	Float is filled with water
	Ballcock needs adjusting
	Pitted flush valve
	Undiscovered leak
	Cracked overflow tube
Water seeps from base of toilet	Bad wax ring
	Cracked toilet bowl
Water dripping from tank	Condensation
	Bad tank-to-bowl gasket
	Bad tank-to-bowl bolts
	Cracked tank
	Flush-valve nut is loose
No water comes into the tank	Closed valve
	Defective ballcock
	Frozen pipe
	Broken pipe

FIGURE 12.5 Troubleshooting toilets.

Symptoms	Probable cause
Wont't drain	Clogged drain Clogged tub waste Clogged trap
Drains slowly	Hair in tub waste Partial drainage blockage
Won't hold water	Tub waste needs adjusting
Won't release water	Tub waste need adjusting
Gurgles as it drains	Partial drainage blockage Partial blockage in the vent
Water drips from spout	Bad faucet washers/cartridge Bad faucet seats
Water comes out spout and shower at the same time	Bad diverter washer Bad diverter seat Bad diverter
Faucet will not shut off	Bad washers or cartridge Bad faucet seats
Poor water pressure	Partially closed valve Not enough water pressure Blockage in the faucet Partially frozen pipe
No water	Closed valve Broken pipe Frozen pipe

FIGURE 12.6   Troubleshooting bathtubs.

Symptoms	Probable cause
Relief valve leaks slowly	Bad relief valve
Relief valve blows off periodically	High water temperature High pressure in tank Bad relief valve
No hot water	Electrical power is off Elements are bad Defective thermostat Inlet valve is closed
Water not hot enough	An element is bad Bad thermostat Thermostat needs adjusting
Water too hot	Thermostat needs adjusting Controls are defective
Water leaks from tank	Hole in tank Rusted-out fitting in tank

FIGURE 12.7 Troubleshooting electric water heaters.

Symptoms	Probable cause
Relief valve leaks slowly	Bad relief valve
Relief valve blows off periodically	High water temperature High pressure in tank Bad relief valve
No hot water	Out of gas Pilot light is out Bad thermostat Control valve is off Gas valve closed
Water not hot enough	Bad thermostat Thermostat needs adjusting
Water too hot	Thermostat needs adjusting Controls are defective Burner will not shut off
Water leaks from tank	Hole in tank Rusted-out fitting in tank

FIGURE 12.8 Troubleshooting gas water heaters.

Troubleshooting Guide for Electric Water Heaters

Complaint	Cause	Solution
Water leaks (See leakage checkpoints on page 7.)	Improperly sealed hot or cold supply connections, relief valve, or thermostat threads	Tighten threaded connections
	Leakage from other appliances or water lines	Inspect other appliances near water heater
Leaking t & p valve	Thermal expansion in closed water system	Install thermal expansion tank. (do not plug t & p valve)
	Improperly seated valve	Check relief valve for proper operation
Hot water odors (Caution: unauthorized removal of the anode(s) will void the warranty. For further information, contact your dealer.)	High sulfate or mineral content in water supply	Drain and flush heater thoroughly; refill
	Bacteria in water supply	Chlorinate water supply
Not enough or no hot water	Power supply to heater is not on	Turn disconnect switch on or contact electrician
	Thermostat set too low	Refer to temperature regulation
	Heater undersized	Reduce hot water use
	Incoming water is unusually cold water (winter)	Allow more time for heater to reheat
	Leaking hot water from pipes or fixtures	Check and repair leaks
	High-temperature limit switch activated	Contact dealer to determine cause; refer to temperature regulation
Water too hot	Thermostat set too high	Refer to temperature regulation
	High-temperature limit switch activated	Contact dealer to determine cause; see temperature regulation
Water heater sounds	Scale accumulation on elements	Contact dealer to clean or replace elements
	Sediment buildup in tank bottom	Drain & flush thoroughly

SOURCE: A. O. Smith Water Products Co.

FIGURE 12.9 Troubleshooting guide for electric water heaters. *(Courtesy A.O. Smith Water Products Co.)*

Symptoms	Probable cause
Won't start	No electrical power Wrong voltage Bad pressure switch Bad electrical connection
Starts, but shuts off fast	Circuit breaker or fuse is inadequate Wrong voltage Bad control box Bad electrical connections Bad pressure switch Pipe blockage Pump is seized Control box is too hot
Runs, but does not produce water, or produces only a small quantity	Check valve stuck in closed position Check valve installed backward Bad electrical wiring Wrong voltage Pump is sitting above the water in the well Leak in the piping Bad pump or motor Broken pump shaft Clogged strainer Jammed impeller
Low water pressure in pressure tank	Pressure switch needs adjusting Bad pump Leak in piping Wrong voltage
Pump runs too often	Check valve stuck open Pressure tank is waterlogged and needs air injected Pressure switch needs adjusting Leak in piping Wrong-size pressure tank

FIGURE 12.10 Troubleshooting submersible potable-water pumps.

Troubleshooting Suggestions

Causes of trouble	Motor runs continuously	
	Checking procedure	Corrective action
Pressure switch.	Switch contacts may be "welded" in closed position. Pressure switch may be set too high.	Clean contacts, replace switch, or readjust setting.
Low-level well.	Pump may exceed well capacity. Shut off pump, wait for well to recover. Check static and drawdown level from well head.	Throttle pump output or reset pump to lower level. Do not lower if sand may clog pump.
Leak in system.	Check system for leaks.	Replace damaged pipes or repair leaks.
Worn pump.	Symptoms of worn pump are similar to that of drop pipe leak or low water level in well. Reduce pressure switch setting. If pump shuts off, worn parts may be at fault. Sand is usually present in tank.	Pull pump and replace worn impellers, casing, or other close fitting parts.
Loose or broken motor shaft.	No or little water will be delivered if coupling between motor and pump shaft is loose or if a jammed pump has caused the motor shaft to shear off.	Check for damaged shafts if coupling is loose, and replace worn or defective units.
Pump screen blocked.	Restricted flow may indicate a clogged intake screen on pump. Pump may be installed in mud or sand.	Clean screen and reset at less depth. It may be necessary to clean well.
Check valve stuck closed.	No water will be delivered if check valve is in closed position.	Replace if defective.
Control box malfunction.	See pages 296, 297, and 305 for single phase.	Repair or replace.
	Motor runs but overload protector tips	
Incorrect voltage.	Using voltmeter, check the line terminals. Voltage must be within $\pm$ 10% of rated voltage.	Contact power company if voltage is incorrect.
Overheated protectors.	Direct sunlight or other heat source can make control box hot, causing protectors to trip. The box must not be hot to touch.	Shade box, provide ventilation, or move box away from heat source.
Defective control box.	For detailed procedures, see pages 295, 300, and 301.	Repair or replace.
Defective motor or cable.	For detailed procedures, see pages 294, 301, and 304.	Repair or replace.
Worn pump or motor.	Check running current. See pages 294, 305, and 306.	Replace pump and/or motor.

SOURCE: A. Y. McDonald Manufacturing Co.

FIGURE 12.11 Troubleshooting suggestions. *(Courtesy A.Y. McDonald Manufacturing Co., Dubuque, Iowa)*

Preliminary tests—all sizes—single and three phase	
What is to be done	What it means
Measure resistance from any cable to ground (insulation resistance)	1. If the ohm value is normal, the motor windings are not grounded and the cable insulation is not damaged.
	2. If the ohm value is below normal, either the windings are grounded or the cable insulation is damaged. Check the cable at the well seal as the insulation is sometimes damaged by being pinched.
Measure winding resistance (resistance between leads)	1. If all ohm values are normal, the motor windings are neither shorted nor open, and the cable colors are correct.
	2. If any one ohm value is less than normal, the motor is shorted.
	3. If any one ohm value is greater than normal, the winding or the cable is open, or there is a poor cable joint or connection.
	4. If some ohm values are greater than normal and some less on single phase motors, the leads are mixed.

FIGURE 12.12 Troubleshooting motors.

1. What well conditions might possibly limit the capacity of the pump?	The rate of flow from the source of supply, the diameter of a cased deep well, and the pumping level of the water in a cased deep well.
2. How does the diameter of a cased deep well and pumping level of the water affect the capacity?	They limit the size pumping equipment which can be used.
3. If there are no limiting factors, how is capacity determined?	By the maximum number of outlets or faucets likely to be in use at the same time.
4. What is suction?	A partial vacuum, created in the suction chamber of the pump, obtained by removing pressure due to atmosphere, thereby allowing greater pressure outside to force something (air, gas, water) into the container.
5. What is atmospheric pressure?	The atmosphere surrounding the earth presses against the earth and all objects on it, producing what we call atmospheric pressure.
6. How much is the pressure due to atmosphere?	This pressure varies with elevation or altitude. It is greatest at sea level (14.7 pounds per square inch) and gradually decreases as elevation above sea level is increased. The rate is approximately 1 foot per 100 feet of elevation.
7. What is maximum theoretical suction lift?	Since suction lift is actually that height to which atmospheric pressure will force water into a vacuum, theoretically we can use the maximum amount of this pressure 14.7 pounds per square inch at sea level which will raise water 33.9 feet. From this, we obtain the conversion factor of 1 pound per square inch of pressure equals 2.31-feet head.
8. How does friction loss affect suction conditions?	The resistance of the suction pipe walls to the flow of water uses up part of the work which can be done by atmospheric pressure. Therefore, the amount of loss due to friction in the suction pipe must be added to the vertical elevation which must be overcome, and the total of the two must not exceed 25 feet at sea level. This 25 feet must be reduced 1 foot for every 1000-feet elevation above sea level, which corrects for a lessened atmospheric pressure with increased elevation.
9. When and why do we use a deep-well jet pump?	The resistance of the suction pipe walls to below the pump because this is the maximum practical suction lift which can be obtained with a shallow-well pump at sea level.

FIGURE 12.13 Questions and answers about pumps. *(Courtesy A.Y. McDonald Manufacturing Co., Dubuque, Iowa)*

10. What do we mean by water system?	A pump with all necessary accessories, fittings, etc., necessary for its completely automatic operation.
11. What is the purpose of a foot valve?	It is used on the end of a suction pipe to prevent the water in the system from running back into the source of supply when the pump isn't operating.
12. Name the two basic parts of a jet assembly.	Nozzle and diffuser.
13. What is the function of the nozzle?	The nozzle converts the pressure of the driving water into velocity. The velocity thus created causes a vacuum in the jet assembly or suction chamber.
14. What is the purpose of the diffuser?	The diffuser converts the velocity from the nozzle back into pressure.
15. What do we mean by "driving water"?	That water which is supplied under pressure to drive the jet.
16. What is the source of the driving water?	The driving water is continuously recirculated in a closed system.
17. What is the purpose of the centrifugal pump?	The centrifugal pump provides the energy to circulate the driving water. It also boosts the pressure of the discharged capacity.
18. Where is the jet assembly usually located in a shallow-well jet system?	Bolted to the casing of the centrifugal pump.
19. What is the principal factor which determines if a shallow-well jet system can be used?	Total suction lift.
20. When is a deep-well jet system used?	When the total suction sift exceeds that which can be overcome by atmospheric pressure.
21. Can a foot valve be omitted from a deep-well jet system? Why or why not?	No, because there are no valves in the jet assembly, and the foot valve is necessary to hold water in the system when it is primed. Also, when the centrifugal pump isn't running, the foot valve prevents the water from running back into the well.

FIGURE 12.14 Questions and answers about pumps. *(Courtesy A.Y. McDonald Manufacturing Co., Dubuque, Iowa)*

22. What is the function of a check valve in the top of a submersible pump?	To hold the pressure in the line when the pump isn't running.
23. A submersible pump is made up of two basic parts. What are they?	Pump end and motor.
24. Why did the name submersible pump come into being?	Because the whole unit, pump and motor, is designed to be operated under water.
25. Can a submersible pump be installed in a 2-inch well?	No, they require a 4-inch well or larger for most domestic use. Larger pumps with larger capacities require 6-inch wells or larger.
26. A stage in a submersible pump is made up of three parts. What are they?	Impeller, diffuser, and bowl.
27. Does a submersible pump have only one pipe connection?	Yes, the discharge pipe.
28. What are two reasons we should always consider using a submersible first?	It will pump more water at higher pressure with less horsepower. It also has easier installation.
29. The amount of pressure a pump is capable of making is controlled by what?	The diameter of the impeller.
30. What do the width of an impeller and guide vane control?	The amount of water or capacity the pump is capable of pumping.

FIGURE 12.15   Questions and answers about pumps. *(Courtesy A.Y. McDonald Manufacturing Co., Dubuque, Iowa)*

Troubleshooting Suggestions

	Motor does not start	
Cause of trouble	Checking procedure	Corrective action
No power or incorrect voltage.	Using voltmeter, check the line terminals. Voltage must be ± 10% of rated voltage.	Contact power company if voltage is incorrect.
Fuses blown or circuit fuse breakers tripped.	Check fuses for recommended size and check for loose, dirty, or corroded connections in fuse receptacle. Check for tripped circuit breaker.	Replace with proper or reset circuit breaker.
Defective pressure switch.	Check voltage at contact points. Improper contact of switch points can cause voltage less than line voltage.	Replace pressure switch or clean points.
Control box malfunction.	For detailed procedure, see ***	Repair or replace.
Defective wiring.	Check for loose or corroded connections. Check motor ead terminals with voltmeter lfor power.	Correct faulty wiring or connections.
Bound pump.	Locked rotor conditions can result from misalignment between pump and motor or a sand bound pump. Amp readings 3 to 6 times higher than normal will be indicated.	If pump will not start with several trials, it must be pulled and the cause corrected. New installations should always be run without turning off until water clears.
Defective cable or motor.	For detailed procedure, see pp. 291, 294, and 295.	Repair or replace.
	Motor starts too often	
Pressure switch.	Check setting on pressure switch and examine for defects.	Reset limit or replace switch.
Check valve, stuck open.	Damaged or defective check valve will not hold pressure.	Replace if defective.
Waterlogged tank (air supply).	Check air-charging system for proper operation.	Clean or replace.
Leak in system.	Check system for leaks.	Replace damaged pipes or repair leaks.

SOURCE: A. Y. McDonald Manufacturing Co.

FIGURE 12.16 Troubleshooting suggestions. *(Courtesy A.Y. McDonald Manufacturing Co., Dubuque, Iowa)*

Normal ohm and megohm values between all leads and ground		
Insulation resistance varies very little with rating. Motors of all hp, voltage, and phase rating have similar values of insulation resistance.		
Condition of motor and leads	Ohm value	Megohm value
A new motor (without drop cable).	20,000,000 (or more)	20.0 (or more)
A used motor which can be reinstalled in the well.	10,000,000 (or more)	10.0 (or more)
Motor in well. Ohm readings are for drop cable plus motor.		
A new motor in the well.	2,000,000 (or more)	2.0 (or more)
A motor in the well in reasonably good condition.	500,000–2,000,000	0.5–2.0
A motor which may have been damaged by lightning or with damaged leads. Do not pull the pump for this reason.	20,000–500,000	0.02–0.5
A motor which definitely has been damaged or with damaged cable. The pump should be pulled and repairs made to the cable or motor replaced. The motor will not fail for this reason alone, but it will probably not operate for long.	10,000–20,000	0.01–0.02
A motor which has failed or with completely destroyed cable insulation. The pump must be pulled and the cable repaired or the motor replaced.	less than 10,000	0–0.01

FIGURE 12.17   Resistance readings.

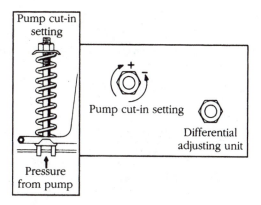

FIGURE 12.18   Fine-tuning instructions for pressure switches.

Meter connections for motor testing

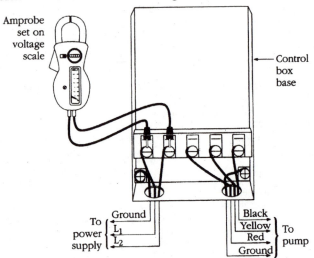

To check voltage

1. Turn power OFF

2. Remove QD cover to break all motor connections.

Caution: L1 and L2 are still connected to the power supply.

3. Turn power ON.

4. Use voltmeter as shown.

Caution: Both voltage and current tests require live circuits with power ON.

FIGURE 12.19 Meter connections for motor testing.

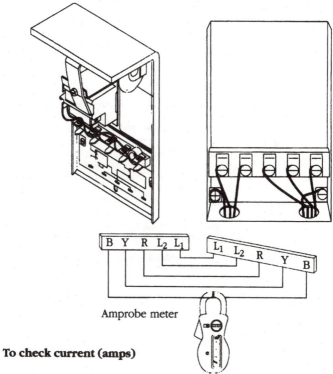

To check current (amps)

1. Turn power OFF

2. Connect test cord as shown.

3. Turn power ON.

4. Use hook-on type ammeter as shown.

FIGURE 12.20 Checking amperage.

Single-phase control boxes

Checking and repairing procedures
(Power off)

   Caution: Turn power off at the power supply panel and discharge capacitors
before using ohmmeter.

   A. General procedures:
         1. Disconnect line power.
         2. Inspect for damaged or burned parts, loose connections, etc.
         3. Check against diagram in control box for misconnections.
         4. Check motor insulation and winding resistance.

   B. Use of ohmmeter:
         1. Ohmmeter such as Simpson Model 372 or 260. Triplet Model 630 or
            666 may be used.
         2. Whenever scales are changed, clip ohmmeter lead together and "zero
            balance" meter.

   C. Ground (insulation resistance) test:
         1. Ohmmeter Setting: Highest scale R × 10K, or R × 100K
         2. Terminal Connections: One ohmmeter lead to "Ground" terminal or
            Q.D. control box lid and touch other lead to the other terminals on the
            terminal board.
         3. Ohmmeter Reading: Pointer should remain at infinity (∞).

Additional tests

Solid state capacitor run
(CRC) control box

   A. Run capacitor
         1. Meter setting: R × 1,000
         2. Connections: Red and Black leads
         3. Correct meter reading: Pointer should swing toward zero, then drift
            back to infinity.

   B. Inductance coil
         1. Meter setting: R × 1
         2. Connections: Orange leads
         3. Correct meter reading: Less than 1 ohm.

   C. Solid state switch
         Step 1 triac test
         1. Meter setting: R × 1,000
         2. Connections: R(Start) terminal and Orange lead on start switch.
         3. Correct meter reading: Should be near infinity after swing.
         Step 2 coil test
         1. Meter setting: R × 1
         2. Connections: Y(Common) and L2.
         3. Correct meter reading: Zero ohms

FIGURE 12.21   Troubleshooting motors.

Single-phase control boxes

Checking and repairing procedures
(Power on)

Caution: Power must be on for these tests. Do not touch any live parts.

A. General procedures:
    1. Establish line power.
    2. Check no load voltage (pump not running).
    3. Check load voltage (pump running).
    4. Check current (amps) in all motor leads.

B. Use of volt/amp meter:
    1. Meter such as Amprobe Model RS300 or equivalent may be used.
    2. Select scale for voltage or amps depending on tests.
    3. When using amp scales, select highest scale to allow for inrush current, then select for midrange reading.

C. Voltage measurements:
    Step 1, no load.
    1. Measure voltage at L1 and L2 of pressure switch or line contractor.
    2. Voltage Reading: Should be ±10% of motor rating.
    Step 2, load.
    1. Measure voltage at load side of pressure switch or line contractor with pump running.
    2. Voltage Reading: Should remain the same except for slight dip on starting.

D. Current (amp) measurements:
    1. Measure current on all motor leads. Use 5 conductor test cord for Q.D. control boxes.
    2. Amp Reading: Current in Red lead should momentarily be high, then drop within one second. This verifies relay or solid state relay operation.

E. Voltage symptoms:
    1. Excessive voltage drop on starting.
    2. Causes: Loose connections, bad contacts or ground faults, or inadequate power supply.

F. Current symptoms:
    1. Relay or switch failures will cause Red lead current to remain high and overload tripping.
    2. Open run capacitor(s) will cause amps to be higher than normal in the Black and Yellow motor leads and lower than normal or zero amps in the Red motor lead.
    3. Relay chatter is caused by low voltage or ground faults.
    4. A bound pump will cause locked rotor amps and overloading tripping.
    5. Low amps may be caused by pump running at shutoff, worn pump or stripped splines.
    6. Failed start capacitor or open switch/relay are indicated if the red lead current is not momentarily high at starting.

FIGURE 12.22 Troubleshooting motors.

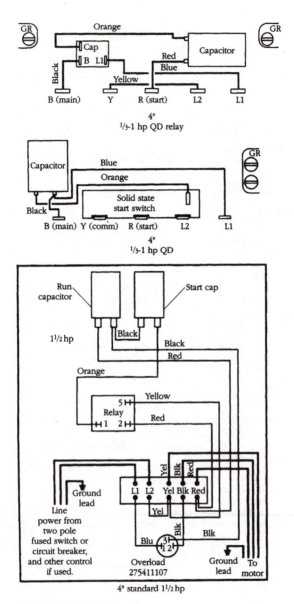

FIGURE 12.23 Wiring diagrams.

Integral horsepower control box parts					
Motor rating hp dia.	Control box (1) model no.	Part no. (2)	Capacitors MFD	Volts	Qty.
5–6″	282 2009 202	275 468 117 S	130–154	330	2
		275 479 103 (5)	15	370	2
	282 2009 203	275 468 117 S	130–154	330	2
		155 327 101 R	30	370	1
5–6″ DLX	282 2009 303	275 468 117 S	130–154	330	2
		155 327 101 R	30	370	1
7½–6″	282 2019 210	275 468 119 S	270–324	330	1
		275 468 117 S	130–154	330	1
		155 327 109 R	45	370	1
	282 2019 202	275 468 117S	130–154	330	3
		275 479 103 R (5)	15	370	3
	282 2019 203	275 468 117 S	130–154	330	3
		155 327 101 R	30	370	1
		155 328 101 R	15	370	1
7½–6″ DLX	282 2019 310	275 468 119 S	270–324	330	1
		275 468 117 S	130–154	330	1
		155 327 109 R	45	370	1
	282 2019 303	275 468 117 S	130–154	330	3
		155 327 101 R	30	370	1
		155 328 101 R	15	370	1
10–6″	282 2029 210	275 468 119 S	270–324	330	2
		155 327 102 R	35	370	2
	282 2029 202	275 468 117 S	130–154	330	4
		275 479 103 R (5)	15	370	5
	282 2029 203	275 468 117 S	130–154	330	4
		155 327 101 R	30	370	2
		155 328 101 R	15	370	1
	282 2029 207	275 468 119 S	270–324	330	2
		155 327 101 R	30	370	2
		155 328 101 R	15	370	1

FIGURE 12.24 Data chart for single-phase motors.

10–6" DLX	282 2029 310	275 468 119 S 155 327 102 R	270–324 35	330 370	2 2
	282 2029 303	275 468 117 S 155 327 101 R 155 328 101 R	130–154 30 15	330 370 370	4 2 1
	282 2029 307	275 468 119 S 155 327 101 R 155 328 101 R	270–324 30 15	330 370 370	2 2 1
15–6" DLX	282 2039 310	275 468 119 S 155 327 109 R	270–324 45	330 370	2 3
	282 2039 303	275 468 119 S 155 327 101 R 155 328 101 R	270–324 30 15	330 370 370	2 4 1

FOOTNOTES:
(1) Lightning arrestor 150 814 902 suitable for all control boxes
(2) S = Start M = Main L = Line R = Run DXL = Deluxe
    control box with line contractor.
(3) Capacitor and overload assembly.
(4) 2 required
(5) These parts may be replaced as follows:

Old	New
275 479 102	155 328 102
275 479 103	155 328 101
275 479 105	155 328 103
275 481 102	155 327 102

FIGURE 12.25 Data chart for single-phase motors.

Integral horsepower control box parts					
Motor rating hp dia.	Control box (1) model no.	Part no.	Capacitors MFD	Volts	Qty.
1½–4″	282 3008 110	275 464 113 S 155 328 102 R	105–126 10	220 370	1 1
	282 3007 202 or 282 3007 102	275 461 107 S 275 479 102 R (5)	105–126 10	220 370	1 1
	282 3007 203 or 282 3007 103	275 461 107 S 155 328 102 R	105–126 10	220 370	1 1
2–4″	282 3018 110	275 464 113 S 155 328 103 R	105–126 20	220 370	1 1
	282 3018 202	275 464 113 S 275 479 105 R (5)	105–126 20	220 370	1 1
	282 3018 203 or 282 3018 103	275 464 113 S 155 328 103 R	105–126 20	220 370	1 1
2–4″ DLX	282 3018 310	275 464 113 S 155 328 103 R	105–126 20	220 370	1 1
	282 3019 103	275 464 113 S 155 328 103 R	105–126 20	220 370	1 1
3–4″	282 3028 110	275 463 111 S 155 327 102 R	208–250 35	220 370	1 1
	282 3028 202	275 463 111 S 275 481 102 R (5)	208–250 35	220 370	1 2
	282 3028 203 or 282 3028 103	275 463 111 S 155 327 102 R	208–250 35	220 370	1 1
3–4″ DLX	282 3028 310	275 463 111 S 155 327 102 R	208–250 35	220 370	1 1
	282 3029 103	275 463 111 S 155 327 102 R	208–250 35	220 370	1 1
5–4″ & 6″	282 1138 110	275 468 118 S 155 327 101 R	216–259 30	330 370	1 2
5–4″	282 1139 202	275 468 118 S 275 479 103 R (5)	216–259 15	330 370	1 4
	282 1139 203 or 282 1139 003	275 468 118 S 155 327 101 R	216–259 30	330 370	1 2
5–4″ & 6″ DLX	282 1138 310 or 282 1139 310	275 468 118 S 155 327 101 R	216–259 30	330 370	1 2
5–4″ DLX	282 1139 303 or 282 1139 103	275 468 118 S 155 327 101 R	216–259 30	330 370	1 2

FOOTNOTES:
(1) Lightning arrestor 150 814 902 suitable for all control boxes
(2) S = Start M = Main L = Line R = Run DXL = Deluxe
control box with line contactor.
(3) Capacitor and overload assembly.
(4) 2 required
(5) These parts may be replaced as follows:

Old	New
275 479 102	155 328 102
275 479 103	155 328 101
275 479 105	155 328 103
275 481 102	155 327 102

FIGURE 12.26 Data chart for single-phase motors.

			QD control box parts		
Hp	Volts	Control box model no.	(1) Solid state SW or QD (blue) relay	Start capacitor	MFD
⅓	115	2801024910	152138905(5)	275464125	159–191
		2801024915	223415905(5)	275464125	159–191
⅓	230	2801034910	152138901(5)	275464126	43–53
		2801034915	223415901(5)	275464126	43–53
½	115	2801044910	152138906(5)	275464201	250–300
		2801044915	223415906(5)	275464201	250–300
½	230	2801054910	152138902(5)	275464105	59–71
		2801054915	223415902(5)	275464105	59–71
½	230	2824055010	152138912	275470115	43–52
		2824055015	223415912(6)	275464105	59–71
¾	230	2801074910	152138903(5)	275464118	86–103
		2801074915	223415903(5)	275464118	86–103
¾	230	2824075010	152138913	275470114	108–130
		2824075015	223415913(6)	275470114	86–103
1	230	2801084910	152138904(5)	275464113	105–126
		2801084915	223415904(5)	275464113	105–126
1	230	2824085010	152138914	275470114	108–130
		2824085015	223415914(6)	275470114	108–130

FOOTNOTES:

(1) Prefixes 152 are solid state switches. Prefixes 223 are QD (Blue) Relays.

(2) Control boxes supplied with solid state relays are designed to operate on normal 230 V systems. For 208 V systems or where line voltage is between 200 V use the next larger cable size, or use boost transformer to raise the voltage to 230 V.

(3) Voltage relay kits 115 V, 305 102 901 and 230 V. 305 102 902 will replace either current voltage or QD Relays, and solid state switches.

(4) QD control boxes produced H85 or later do not contain an overload in the capacitor. On winding thermal overloads were added to three-wire motors rated ⅓-1 hp in A85. If a control box dated H85 or later is applied with a motor dated M84 or earlier, overload protection can be provided by adding an overload kit to the control box.

(5) May be replaced with QD relay kits 305 101 901 thru 906. Use same kit suffix as switch or relay suffix.

(6) Replace with CRC QD Relaying Kits, 223 415 912 with 305 105 901, 223 415 913 with 305 105 902 and 223 415 914 with 305 105 903.

FIGURE 12.27 Data chart for single-phase motors.

chapter

13

WORKSITE SAFETY

Worksite safety doesn't get as much attention as it should. Far too many people are injured on jobs every year. Most of the injuries could be prevented, but they are not. Why is this? People are in a hurry to make a few extra bucks, so they cut corners. This happens with plumbing contractors and piece workers. It even affects hourly plumbers who what to shave 15 minutes off their workday, so that they can head back to the shop early.

Based on my field experience, most accidents occur as a result of negligence. Plumbers try to cut a corner, and they wind up getting hurt. This has proved true with my personal injuries. I've only suffered two serious on the job injuries, and both of them were a direct result of my carelessness. I knew better than to do what I was doing when I was hurt, but I did it anyway. Well, sometimes you don't get a second chance, and the life you affect may not be your own. So, let's look at some sensible safety procedures that you can implement in your daily activity.

VERY DANGEROUS

Plumbing can be a very dangerous trade. The tools of the trade have potential to be killers. Requirements of the job can place you in positions where a lack of concentration could result in serious injury or death. The fact that plumbing can be dangerous is no reason to rule

General Safe Working Habits

1. Wear safety equipment.

2. Observe all safety rules at the particular location.

3. Be aware of any potential dangers in the specific situation.

4. Keep tools in good condition.

FIGURE 13.1   General safe working habits.

out the trade as your profession. Driving can be extremely dangerous, but few people never set foot in an automobile out of fear.

Fear is generally a result of ignorance. When you have a depth of
knowledge and skill, fear begins to subside. As you become more
accomplished at what you do, fear is forgotten. While it is advisable
to learn to work without fear, you should never work without
respect. There is a huge difference between fear and respect.

If, as a plumber, you are afraid to climb up on a roof to flash a
pipe, you are not going to last long in the plumbing trade. However,
if you scurry up the roof recklessly, you could be injured severely, per-
haps even killed. You must respect the position you are putting yourself in. If you are using a ladder to get on the roof, you must respect
the outcome of what a mistake could have. Once you are on the
roof, you must be conscious of footing conditions, and the way that
you negotiate the pitch of the roof. If you pay attention, are properly
trained, and don't get careless, you are not likely to get hurt.

Being afraid of a roof will limit or eliminate your plumbing career.
Treating your trip to and from the roof like a walk into your living
room could be deadly. Respect is the key. If you respect the conse-
quences of your actions, you are aware of what you are doing and
the odds for a safe trip improve.

Many young plumbers are fearless in the beginning. They think
nothing of darting around on a roof or jumping down in a sewer
trench. As their careers progress, they usually hear about or see on-
the-job accidents. Someone gets buried in the cave-in of a trench.
Somebody falls off a roof. A metal ladder being set up hits a power
line. A careless plumber steps into a flooded basement and is elec-

trocuted because of submerged equipment. The list of possible job-related injuries is a long one.

Millions of people are hurt every year in job-related accidents. Most of these people were not following solid safety procedures. Sure, some of them were victims of unavoidable accidents, but most were hurt by their own hand, in one way or another. You don't have to be one of these statistics.

In over 25 years of plumbing, I have only been hurt seriously on the job twice. Both times were my fault. I got careless. In one of the instances, I let loose clothing and a powerful drill work together to chew up my arm. In the other incident, I tried to save myself the trouble of repositioning my stepladder while drilling holes in floor joists. My desire to save a minute cost me torn stomach muscles and months of pain from a twisting drill.

My accidents were not mistakes, they were stupidity. Mistakes are made through ignorance. I wasn't ignorant of what could happen to me. I knew the risk I was taking, and I knew the proper way to perform my job. Even with my knowledge, I slipped up and got hurt. Luckily, both of my injuries healed, and I didn't pay a lifelong price for my stupidity.

During my long plumbing career I have seen a lot of people get hurt. Most of these people have been helpers and apprentices. Of all the on the job accidents I have witnessed, every one of them could have been avoided. Many of the incidents were not extremely serious, but a few were.

As a plumber, you will be doing some dangerous work. You will be drilling holes, running threading machines, snaking drains, installing roof flashings, and a lot of other potentially dangerous jobs. Hopefully, your employer will provide you with quality tools and equipment. If you have the right tool for the job, you are off to a good start in staying safe.

Safety training is another factor you should seek from your employer. Some plumbing contractors fail to tell their employees how to do their jobs safely. It is easy for someone, like an experienced plumber, who knows a job inside and out to forget to inform an inexperienced person of potential danger.

For example, a plumber might tell you to break up the concrete around a pipe to allow the installation of a closet flange and never consider telling you to wear safety glasses. The plumber will assume you know the concrete is going to fly up in your face as it is chiseled up. However, as a rookie, you might not know about the reaction concrete has when hit with a cold chisel. One swing of the hammer could cause extreme damage to your eyesight.

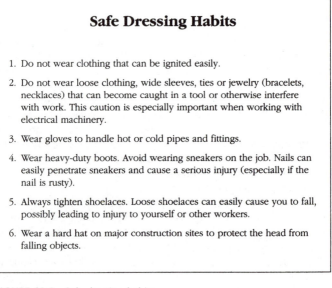

Safe Dressing Habits

1. Do not wear clothing that can be ignited easily.

2. Do not wear loose clothing, wide sleeves, ties or jewelry (bracelets, necklaces) that can become caught in a tool or otherwise interfere with work. This caution is especially important when working with electrical machinery.

3. Wear gloves to handle hot or cold pipes and fittings.

4. Wear heavy-duty boots. Avoid wearing sneakers on the job. Nails can easily penetrate sneakers and cause a serious injury (especially if the nail is rusty).

5. Always tighten shoelaces. Loose shoelaces can easily cause you to fall, possibly leading to injury to yourself or other workers.

6. Wear a hard hat on major construction sites to protect the head from falling objects.

FIGURE 13.2   Safe dressing habits.

Simple jobs, like the one in the example, are all it takes to ruin a career. You might be really on your toes when asked to scoot across an I-beam, but how much thought are you going to give to tightening the bolts on a toilet? The risk of falling off the I-beam is obvious. Having chips of china, from a broken toilet where the nuts were turned one time too many, flying into your eyes is not so obvious. Either way, you can have a work-stopping injury.

Safety is a serious issue. Some job sites are very strict in the safety requirements maintained. But a lot of jobs have no written rules of safety. If you are working on a commercial job, supervisors are likely to make sure you abide by the rules of the Occupational Safety and Health Administration (OSHA). Failure to comply with OSHA regulations can result in stiff financial penalties. However, if you are working in residential plumbing, you may never set foot on a job where OSHA regulations are observed.

In all cases, you are responsible for your own safety. Your employer and OSHA can help you to remain safe, but in the end, it is up to you. You are the one who has to know what to do and how to do it. And not only do you have to take responsibility for your own actions, you also have to watch out for the actions of others. It

---

Safe Operation of Grinders

1. Read the operating instructions before starting to use the grinder.

2. Do not wear any loose clothing or jewelry.

3. Wear safety glasses or goggles.

4. Do not wear gloves while using the machine.

5. Shut the machine off promptly when you are finished using it.

FIGURE 13.3 Safe operation of grinders.

is not unlikely that you could be injured by someone else's careless-ness. Now that you have had the primer course, let's get down to the specifics of job-related safety.

As we move into specifics, you will find the suggestions in this chapter broken down into various categories. Each category will deal with specific safety issues related to the category. For example, in the section on tool safety, you will learn procedures for working safely with tools. As you move from section to section, you may notice some overlapping of safety tips. For example, in the section on general safety, you will see that it is wise to work without wearing jewelry. Then you will find jewelry mentioned again in the tool section. The duplication is done to pinpoint definite safety risks and procedures. We will start into the various sections with general safety.

GENERAL SAFETY

General safety covers a lot of territory. It starts from the time you get into the company vehicle and carries you right through to the end of the day. Much of the general safety recommendations involve the use of common sense. Now, let's get started.

Vehicles

Many plumbers are given company trucks for their use in getting to and from jobs. You will probably spend a lot of time loading and unloading company trucks. And, of course, you will spend time either riding in or driving them. All of these areas can threaten your safety.

If you will be driving the truck, take the time to get used to how it handles. Loaded plumbing trucks don't drive like the family car. Remember to check the vehicle's fluids, tires, lights, and related equipment. Many plumbing trucks are old and have seen better days. Failure to check the vehicle's equipment could result in unwanted headaches. Also, remember to use the seat belts—they do save lives.

Apprentices are normally charged with the duty of unloading the truck at the job site. There are a lot of ways to get hurt in doing this job. Many plumbing trucks use roof racks to haul pipe and ladders. If you are unloading these items, make sure they will not come into contact with low-hanging electrical wires. Copper pipe and aluminum ladders make very good electrical conductors, and they will carry the power surge through you on the way to the ground. If you are unloading heavy items, don't put your body in awkward positions. Learn the proper ways for lifting, and never lift objects inappropriately. If the weather is wet, be careful climbing on the truck. Step bumpers get slippery, and a fall can impale you on an object or bang up your knee.

When it is time to load the truck, observe the same safety precautions you did in unloading. In addition to these considerations, always make sure your load is packed evenly and well secured. Be especially careful of any load you attach to the roof rack, and always double check the cargo doors on trucks with utility bodies.

It will not only be embarrassing to lose your load going down the road, it could be deadly. I have seen a one-piece fiberglass tub\shower unit fly out of the back of a pick-up truck as the truck was rolling up an interstate highway. As a young helper, I lost a load of pipe in the middle of a busy intersection. In that same year, the cargo doors on the utility body of my truck flew open as I came off a ramp, onto a major highway. Tools scattered across two lanes of traffic. These types of accidents don't have to happen. It's your job to make sure they don't.

CLOTHING

Clothing is responsible for a lot of on-the-job injuries. Sometimes it is the lack of clothing that causes the accidents, and there are many times when too much clothing creates the problem. Generally, it is wise not to wear loose fitting clothes. Shirttails should be tucked in, and short-sleeve shirts are safer than long-sleeved shirts when operating some types of equipment.

Caps can save you from minor inconveniences, like getting glue in your hair, and hard hats provide some protection from potentially damaging accidents, like having a steel fitting dropped on your head. If you have long hair, keep it up and under a hat.

Good footwear is essential in the trade. Normally a strong pair of hunting-style boots will be best. The thick soles provide some protection from nails and other sharp objects you may step on. Boots with steel toes can make a big difference in your physical well being. If you are going to be climbing, wear foot gear with a flexible sole that grips well. Gloves can keep your hands warm and clean, but they can also contribute to serious accidents. Wear gloves sparingly, depending upon the job you are doing.

JEWELRY

On the whole, jewelry should not be worn in the workplace. Rings can inflict deep cuts in your fingers. They can also work with machinery to amputate fingers. Chains and bracelets are equally dangerous, probably more so.

EYE AND EAR PROTECTION

Eye and ear protection is often overlooked. An inexpensive pair of safety glasses can prevent you from spending the rest of your life blind. Ear protection reduces the effect of loud noises, such as jackhammers and drills. You may not notice much benefit now, but in later years you will be glad you wore it. If you don't want to lose your hearing, wear ear protection when subjected to loud noises.

PADS

Kneepads not only make a plumber's job more comfortable, they help to protect the knees. Some plumbers spend a lot of time on their knees, and pads should be worn to ensure that they can continue to work for many years.

The embarrassment factor plays a significant role in job-related injuries. People, especially young people, feel the need to fit in and to make a name for themselves. Plumbing is sort of a macho trade. There is no secret that plumbers often fancy themselves as strong human specimens. Many plumbers are strong. The work can be hard and in doing it, becoming strong is a side benefit. But you can't allow safety to be pushed aside for the purpose of making a macho statement.

All too many people believe that working without safety glasses, ear protection, and so forth makes them tough. That's just not true. It may make them appear dumb, and it may get them hurt, but it does not make them look tough. If anything, it makes them look stupid or inexperienced.

Don't fall into the trap so many young plumbers do. Never let people goad you into bad safety practices. Some plumbers are going to laugh at your kneepads. Let them laugh, you will still have good knees when they are hobbling around on canes. I'm dead serious about this issue. There is nothing sissy about safety. Wear your gear in confidence, and don't let the few jokesters get to you.

TOOL SAFETY

Tool safety is a big issue in plumbing. Anyone in the plumbing trade will work with numerous tools. All of these tools are potentially dan-

---

Safe Use of Hand Tools

1. Use the right tool for the job.

2. Read any instructions that come with the tool unless you are thoroughly familiar with its use.

3. Wipe and clean all tools after each use. If any other cleaning is necessary, do it periodically.

4. Keep tools in good condition. Chisels should be kept sharp and any mushroomed heads kept ground smooth; saw blades should be kept sharp; pipe wrenches should be kept free of debris and the teeth kept clean; etc.

5. Do not carry small tools in your pocket, especially when working on a ladder or scaffolding. If you should fall, the tools might penetrate your body and cause serious injury.

FIGURE 13.4 Safe use of hand tools.

gerous, but some of them are especially hazardous. This section is broken down by the various tools used on the job. You cannot afford to start working without the basics in tool safety. The more you can absorb on tool safety, the better off you will be.

The best starting point is reading all literature available from the manufacturers of your tools. The people that make the tools provide some good safety suggestions with them. Read and follow the manufacturers' recommendations.

The next step in working safely with your tools is to ask questions. If you don't understand how a tool operates, ask someone to explain it to you. Don't experiment on your own, the price you pay could be much too high.

Common sense is irreplaceable in the safe operation of tools. If you see an electrical cord with cut insulation, you should have enough common sense to avoid using it. In addition to this type of simple observation, you will learn some interesting facts about tool safety. Now, let me tell you what I've learned about tool safety over the years.

Safe Use of Electric Tools

1. Always use a three-prong plug with an electric tool.

2. Read all instructions concerning the use of the tool (unless you are thoroughly familiar with its use).

3. Make sure that all electrical equipment is properly grounded. Ground fault circuit interrupters (GFCI) are required by OSHA regulations in many situations.

4. Use proper-sized extension cords. (Undersized wires can burn out a motor, cause damage to the equipment, and present a hazardous situation.

5. Never run an extension cord through water or through any area where it can be cut, kinked, or run over by machinery.

6. Always hook up an extension cord to the equipment and then plug it into the main electrical outlet—not vice versa.

7. Coil up and store extension cords in a dry area.

FIGURE 13.5 Safe use of electric tools.

There are some basic principals to apply to all of your work with tools. We will start with the basics, and then we will move on to specific tools. Here are the basics:

- *Keep body parts away from moving parts*
- *Don't work with poor lighting conditions*
- *Be careful of wet areas when working with electrical tools*
- *If special clothing is recommended for working with your tools, wear it*
- *Use tools only for their intended purposes*
- *Get to know your tools well*
- *Keep your tools in good condition*

Now, let's take a close look at the tools you are likely to use. Plumbers use a wide variety of hand tools and electrical tools. They also use specialty tools. So let's see how you can use all these tools without injury.

Torches

Torches are often used by plumbers on a daily basis. Some plumbers use propane torches, and others use acetylene torches. In either case, the containers of fuel for these torches can be very dangerous. The flames produced from torches can also do a lot of damage.

To Prevent Fires

1. Always keep fire extinguishers handy, and be sure that the extinguisher is full and that you know how to use it quickly.

2. Be sure to disconnect and bleed all hoses and regulators used in welding, brazing, soldering, etc.

3. Store cylinders of acetylene, propane, oxygen, and similar substances in an upright position in a well-vented area.

4. Operate all air acetylene, welding, soldering, and related equipment according to the manufacturer's directions.

5. Do not use propane torches or other similar equipment near material that can easily catch fire.

6. Be careful at all times. Be prepared for the worst, and be ready to act.

FIGURE 13.6 Preventing fires.

fastfacts

➤ *Don't allow torch equipment fall on a hard surface.*
➤ *Check torch connections carefully.*
➤ *Pay attention to where the flame of your torch is pointed.*
➤ *Concrete can explode when heated.*
➤ *Use a heat shield when using a torch near potentially flammable materials.*
➤ *Turn off your torch's fuel tank when not in use.*
➤ *Use a striker to light your touch.*

When working with torches and tanks of fuel, you must be very careful. Don't allow your torch equipment to fall on a hard surface, the valves may break. Check all your connections closely, a leak allowing fuel to fill an area could result in an explosion when you light the torch.

Always pay attention to where your flame is pointed. Carelessness with the flame could start unwanted fires. This is especially true when working near insulation and other flammable substances. If the flame is directed at concrete, the concrete may explode. Since moisture is retained in concrete, intense heat can cause the moisture to force the concrete to explode.

It's not done often enough, but you should always have a fire extinguisher close by when working with a torch. If you have to work close to flammable substances, use a heat shield on the torch. When your flame is close to wood or insulation, try to remove the insulation or wet the flammable substance, before applying the flame. When you are done using the torch, make sure the fuel tank is turned off and the hose is drained of all fuel. Use a striker to light your torch. The use of a match or cigarette lighter puts your hand too close to the source of the flame.

Lead Pots and Ladles

Lead pots and ladles offer their own style of potential danger. Plumbers today don't use molten lead as much as they used to, but hot lead is still used. It is used to make joints with cast-iron pipe and fittings.

Rules for Working Safely in Ditches or Trenches

1. Be careful of underground utilities when digging.

2. Do not allow people to stand on the top edge of a ditch while workers are in the ditch.

3. Shore all trenches deeper than 4 feet.

4. When digging a trench, be sure to throw the dirt away from the ditch walls (2 feet or more).

5. Be careful to see that no water gets into the trench. Be especially careful in areas with a high water table. Water in a trench can easily undermine the trench walls and lead to a cave-in.

6. Never work in a trench alone.

7. Always have someone nearby—someone who can help you and locate additional help.

8. Always keep a ladder nearby so you can exit the trench quickly if need be.

9. Be watchful at all times. Be aware of any potentially dangerous situations. Remember, even heavy truck traffic nearby can cause a cave-in.

FIGURE 13.7 Working safely in ditches.

When working with a small quantity of lead, many plumber heat the lead in a ladle. They melt the lead with their torch and pour it straight from the ladle. When larger quantities of lead are needed, lead pots are used. The pots are filled with lead and set over a flame. All of this type of work is dangerous.

Never put wet materials in hot lead. If the ladle is wet or cold when it is dipped into the pot of molten lead, the hot lead can explode. Don't add wet lead to the melting pot if it contains molten lead. Before you pour the hot lead in a waiting joint, make sure the joint is not wet. As another word of caution, don't leave a working lead pot where rain dripping off a roof can fall into it.

Obviously, molten lead is hot, so you shouldn't touch it. Be very careful to avoid tipping the pot of hot lead over. I remember one accident when a pot of hot lead was knocked over on a plumber's foot. Let me just say the scene was terrifying.

Drills and Bits

Drills have been my worst enemy in plumbing. The two serious injuries I have received were both related to my work with a right angle drill. The drills used most by plumbers are not the little pistol-grip, hand-held types of drills most people think of. The day-to-day drilling done by plumbers involves the use of large, powerful right-angle drills. These drills have enormous power when they get in a bind. Hitting a nail or a knot in the wood being drilled can do a lot of damage. You can break fingers, lose teeth, suffer head injuries, and a lot more. As with all electrical tools, you should always check the electrical cord before using your drill. If the cord is not in good shape, don't use the drill.

Always know what you are drilling into. If you are doing new-construction work it is fairly easy to look before you drill. However, drilling in a remodeling job can be much more difficult. You cannot always see what you are getting into. If you are unfortunate enough to drill into a hot wire, you can get a considerable electrical shock.

The bits you use in a drill are part of the safe operation of the tool. If your drill bits are dull, sharpen them. Dull bits are much more dangerous than sharp ones. When you are using a standard plumber's bit to drill through thin wood, like plywood, be careful. Once the worm driver of the bit penetrates the plywood fully, the big teeth on the bit can bite and jump, causing you to lose control of the drill. If you will be drilling metal, be aware that the metal shavings will be sharp and hot.

Power Saws

Plumbers don't use power saws as much as carpenters, but they do use them. The most common type of power saw used by plumbers is the reciprocating saw. These saws are used to cut pipe, plywood, floor joists, and a whole lot more. In addition to reciprocating saws, plumbers use circular saws and chop saws. All of the saws have the potential for serious injury.

Reciprocating saws are reasonably safe. Most models are insulated to help avoid electrical shocks if a hot wire is cut. The blade is typically a safe distance from the user, and the saws are pretty easy to hold and control. However, the brittle blades do break. This could result in an eye injury. Circular saws are used by plumbers occasionally. The blades on these saws can bind and cause the saws to kick back. Chop saws are sometimes used to cut pipe. If you keep your body parts out of the way and wear eye protection, chop saws are not unusually dangerous.

Hand-Held Drain Cleaners

Hand-held drain cleaners don't get a lot of use from plumbers that do new-construction work, but they are a frequently-used tool of service plumbers. Most of these drain cleaners resemble, to some extent, a straight, hand-held drill. There are models that sit on stands, but most small snakes are hand-held. These small-diameter snakes are not nearly as dangerous as their big brothers, but they do deserve respect. These units carry all the normal hazards of an electric tool, but there is more.

The cables used for small drain cleaning jobs are usually very flexible. They are basically springs. The heads attached to the cables take on different shapes and looks. When you look at these thin cables, they don't look dangerous, but they can be. When the cables are being fed down a drain and turning, they can hit hard stoppages and go out of control. The cable can twist and kink. If your finger, hand, or arm is caught in the cable, injury can be the result. To avoid this, don't allow excessive cable to exist between the drain pipe and the machine.

Large Sewer Machines

Large sewer machines are much more dangerous than small ones. These machines have tremendous power and their cables are capable of removing fingers. Broken bones, severe cuts, and assorted other injuries are also possible with big snakes.

One of the most common conflicts with large sewer machines is with the cables. When the cutting heads hit roots or similar hard items in the pipe, the cable goes wild. The twisting, thrashing cable can do a lot of damage. Again, limiting excess cable is one of the best protections possible. Special sleeves are also available to contain unruly cables.

Most big machines can be operated by one person, but it is wise to have someone standing by in case help is needed. Loose clothing results in many drain machine accidents. The use of special mitts can help reduce the risk of hand injuries. Electrical shocks are also possible when doing drain cleaning.

Power Pipe Threaders

Power pipe threaders are very nice to have if you are doing much work with threaded pipe. However, these threading machines can grind body parts as well as they thread pipe. Electric threaders are very dangerous in the hands of untrained people. It is critical to keep fingers and clothing away from the power mechanisms. The metal shavings produced by pipe threaders can be very sharp, and burrs left on the threaded pipe can slash your skin. The cutting oil used to keep the dies from getting too hot can make the floor around the machine slippery.

Air-Powered Tools

Air-powered tools are not used often by plumbers. Jackhammers are probably the most used air-powered tools for plumbers. When using tools with air hoses, check all connections carefully. If you experience a blow-out, the hose can spiral wildly out of control.

Powder-Actuated Tools

Powder-actuated tools are used by plumbers to secure objects to hard surfaces, like concrete. If the user is properly trained, these tools are not too dangerous. However, good training, eye protection, and ear protection are all necessary. Misfires and chipping hard surfaces are the most common problem with these tools.

Ladders

Ladders are used frequently by plumbers, both stepladders and extension ladders. Many ladder accidents are possible. You must always be aware of what is around you when handling a ladder. If you brush against a live electrical wire with a ladder you are carrying, your life could be over. Ladders often fall over when the people using them are not careful. Reaching too far from a ladder can be all it takes to take a fall.

When you set up a ladder, or rolling scaffolds make sure it is set up properly. The ladder should be on firm footing, and all safety braces and clamps should be in place. When using an extension ladder, many plumbers use a rope to tie rungs together where the sections overlap. The rope provides an extra guard against the ladder's safety clamps failing and the ladder collapsing. When using an extension ladder, be sure to secure both the base and the top. I had an unusual accident on a ladder that I would like to share with you.

Safety on Rolling Scaffolds

1. Do not lay tools or other materials on the floor of the scaffold. They can easily move and you could trip over them, or they might fall, hitting someone on the ground.

2. Do not move a scaffold while you are on it.

3. Always lock the wheels when the scaffold is positioned and you are using it.

4. Always keep the scaffold level to maintain a steady platform on which to work.

5. Take no shortcuts. Be watchful at all times and be prepared for any emergencies.

FIGURE 13.8 Safety on rolling scaffolds.

I was on a tall extension ladder, working near the top of a commercial building. The top of my ladder was resting on the edge of the flat roof. There was metal flashing surrounding the edge of the roof, and the top of the ladder was leaning against the flashing. There was a picket fence behind me and electrical wires entering the building to my right. The entrance wires were a good ways away, so I was in no immediate danger. As I worked on the ladder, a huge gust of wind blew around the building. I don't know where it came from—it hadn't been very windy when I went up the ladder.

The wind hit me and pushed me and the ladder sideways. The top of the ladder slid easily along the metal flashing, and I couldn't grab anything to stop me. I knew the ladder was going to go down, and I didn't have much time to make a decision. If I pushed off of the ladder, I would probably be impaled on the fence. If I rode the ladder down, it might hit the electrical wires and fry me. I waited until the last minute and jumped off of the ladder.

I landed on the wet ground with a thud, but I missed the fence. The ladder hit the wires and sparks flew. Fortunately, I wasn't hurt and electricians were available to take care of the electrical problem. This was a case where I wasn't really negligent, but I could have been killed. If I had secured the top of the ladder, my accident wouldn't have happened.

Working Safely on a Ladder

1. Use a solid and level footing to set up the ladder.

2. Use a ladder in good condition; do not use one that needs repair.

3. Be sure step ladders are opened fully and locked.

4. When using an extension ladder, place it at least ¼ of its length away from the base of the building.

5. Tie an extension ladder to the building or other support to prevent it from falling or blowing down in high winds.

6. Extend a ladder at least 3 feet over the roof line.

7. Keep both hands free when climbing a ladder.

8. Do not carry tools in your pocket when climbing a ladder. (If you fall, the tools could cut into you and cause serious injury.)

9. Use the ladder the way it should be used. For example, do not allow two people on a ladder designed for use by one person.

10. Keep the ladder and all its steps clean—free of grease, oil, mud, etc.—in order to avoid a fall and possible injury.

FIGURE 13.9   Safety on a ladder.

Screwdrivers and Chisels

Eye injuries and puncture wounds are common when working with screwdrivers and chisels. When the tools are use properly and safety glasses are worn, few accidents occur.

The key to avoiding injury with most hand tools is simply to use the right tool for the job. If you use a wrench as a hammer or a screwdriver as a chisel, you are asking for trouble.

There are, of course, other types of tools and safety hazards found in the plumbing trade. However, this list covers the ones that result in the most injuries. In all cases, observe proper safety procedures and utilize safety gear, such as eye and ear protection.

CO-WORKER SAFETY

Co-worker safety is the last segment of this chapter. I am including it because workers are frequently injured by the actions of co-workers. This section is meant to protect you from others and to make you aware of how your actions might affect your co-workers.

Most plumbers find themselves working around other people. This is especially true on construction jobs. When working around other people, you must be aware of their actions, as well as your own. If you are walking out of a house to get something off the truck and a roll of roofing paper gets away from a roofer, you could get an instant headache.

If you don't pay attention to what is going on around you, it is possible to wind up in all sorts of trouble. Cranes lose their loads some times, and such a load landing on you is likely to be fatal. Equipment operators don't always see the plumber kneeling down for a piece of pipe. It's not hard to have a close encounter with heavy equipment. While we are on the subject of equipment, let me bore you with another war story.

One day I was in a sewer ditch, connecting the sewer from a new house to the sewer main. The section of ditch that I was working in was only about four feet deep. There was a large pile of dirt near the edge of the trench; it had been created when the ditch was dug. The dirt wasn't laid back like it should have been; it was piled up.

As I worked in the ditch, a backhoe came by. The operator had no idea I was in the ditch. When he swung the backhoe around to make a turn, the small scorpion-type bucket on the back of the equipment hit the dirt pile.

I had stood up when I heard the hoe approaching, and it was a good thing I had. When the equipment hit the pile of dirt, part of the mound caved in on me. I tried to run, but it caught both of my legs and the weight drove me to the ground. I was buried from just below my waist. My head was okay, and my arms were free. I was still holding my shovel.

I yelled, but nobody heard me. I must admit, I was a little panicked. I tried to get up and couldn't. After a while, I was able to move enough dirt with the shovel to crawl out from under the dirt. I was lucky. If I had been on my knees making the connection or grading the pipe, I might have been smothered. As it was, I came out of the ditch no worse for the wear. But, boy, was I mad at the careless backhoe operator. I won't go into the details of the little confrontation I had with him.

That accident is a prime example of how other workers can hurt you and never know they did it. You have to watch out for yourself at all times. As you gain field experience, you will develop a second nature for impending co-worker problems. You will learn to sense when something is wrong or is about to go wrong. But you have to stay alive and healthy long enough to get that experience.

Always be aware of what is going on over your head. Avoid working under other people and hazardous overhead conditions. Let people know where you are, so you won't get stranded on a roof or in an attic when your ladder is moved or falls over.

You must also remember that your actions could harm co-workers. If you are on a roof to flash a pipe and your hammer gets away from you, somebody could get hurt. Open communication between workers is one of the best ways to avoid injuries. If everyone knows where everyone else is working, injuries are less likely. Primarily, think and think some more. There is no substitute for common sense. Try to avoid working alone, and remain alert at all times.

14

FIRST AID

veryone should invest some time in learning the basics of first aid. You never know when having skills in first aid treatments may save your life. Plumbers live what can be a dangerous life. On the job injuries are not uncommon. Most injuries are fairly minor, but they often require treatment. Do you know the right way to get a sliver of copper out of your hand? If your helper suffers from an electrical shock when a drill cord goes bad, do you know what to do? Well, many plumbers don't possess good first aid skills.

Before we get too far into this chapter, there are a few points are want to make. First of all, I'm not a medical doctor or any type of trained medical-care person. I've taken first aid classes, but I'm certainly not an authority on medical issues. The suggestions that I will give you in this chapter are for informational purposes only. This book is not a substitute for first aid training offered by qualified professionals.

My intent here is to make you aware of some basic first aid procedures that can make life on the job much easier. But, I want you to understand that I'm not advising you to use my advice to administer first aid. Hopefully, this chapter will show you the types of advantages you can gain from taking first aid classes. Before you attempt first aid on anyone, including yourself, you should attend a structured, approved first aid class. I'm going to give you information that is as accurate as I can make it, but don't assume that my words are enough. Take a little time to seek professional training in the art of first aid. You may never use what you learn, but the one time it is needed, you will be glad you made the effort to learn what to do. With this said, let's jump right into some tips on first aid.

OPEN WOUNDS

Open wounds are a common problem for plumbers. Many tools and materials used by plumbers can create open wounds. What should you do if you or one of your workers is cut?

- *Stop the bleeding as soon as possible*
- *Disinfect and protect the wound from contamination*
- *You may have to take steps to avoid shock symptoms*
- *Once the patient is stable, seek medical attention for severe cuts*

When a bad cut is encountered, the victim may slip into shock. A loss of consciousness could result from a loss of blood. Death from extreme bleeding is also a risk. As a first aid provider, you must act quickly to reduce the risk of serious complications.

Bleeding

To stop bleeding, direct pressure is normally a good tactic. This may be as crude as clamping your hand over the wound, but a cleaner compression is desirable. Ideally, a sterile material should be placed over the wound and secured, normally with tape (even if it's duct tape). Whenever possible, wear rubber gloves to protect yourself from possible disease transfer if you are working on someone else. Thick gauze used as a pressure material can absorb blood and begin to allow the clotting process to start.

Bad wounds may bleed right through the compress material. If this happens, don't remove the blood-soaked material. Add a new layer of material over it. Keep pressure on the wound. If you are not prepared with a first aid kit, you could substitute gauze and tape with strips cut from clothing that can be tied in place over the wound.

When you are dealing with a bleeding wound, it is usually best to elevate it. If you suspect a fractured or broken bone in the area of the wound, elevation may not be practical. When we talk about elevating a wound, it simply means to raise the wound above the level of the victim's heart. This helps the blood flow to slow, due to gravity.

Super Serious

Super serious bleeding might not stop even after a compression bandage is applied and the wound is elevated. When this is the case, you must resort to putting pressure on the main artery which is producing the blood. Constricting an artery is not an alternative for the steps that we have discussed previously.

Putting pressure on an artery is serious business. First, you must be able to locate the artery, and you should not keep the artery constricted any longer than necessary. You may have to apply pressure for awhile, release it, and then apply it again. It's important that you do not restrict the flow of blood in arteries for long periods of time. I hesitate to go into too much detail on this process, as I feel it is a method that you should be taught in a controlled, classroom situation. However, I will hit the high spots. But remember, these words are not a substitute for professional training from qualified instructors.

Open arm wounds are controlled with the brachial artery. The location of this artery is in the area between the biceps and triceps, on the inside of the arm. It's about halfway between the armpit and the elbow. Pressure is created with the flat parts of your fingertips. Basically, you are holding the victim's wrist with one hand and closing off the artery with your other hand. Pressure exerted by your fingers pushes the artery against the arm bone and restricts blood flow. Again, don't attempt this type of first aid until you have been trained properly in the execution of the procedure.

Severe leg wounds may require the constriction of the femoral artery. This artery is located in the pelvic region. Normally, bleeding victims are placed on their backs for this procedure. The heel of a hand is placed on the artery to restrict blood flow. In some cases, fingertips are used to apply pressure. I'm uncomfortable with going into great detail on these procedures, because I don't want you to rely solely on what I'm telling you. It's enough that you understand that knowing when and where to apply pressure to arteries can save lives and that you should seek professional training in these techniques.

Tourniquets

Tourniquets get a lot of attention in movies, but they can do as much harm as they do good if not used properly. A tourniquet should only be used in a life-threatening situation. When a tourniquet is applied, there is a risk of losing the limb to which the restriction is applied. This is obviously a serious decision and one that must be made only when all other means of stopping blood loss have been exhausted.

Unfortunately, plumbers might run into a situation where a tourniquet is the only answer. For example, if a worker allowed a power saw to get out of control, a hand might be severed or some other type of life-threatening injury could occur. This would be cause for the use of a tourniquet. Let me give you a basic overview of what's involved when a tourniquet is used.

Tourniquets should be at least two inches wide. A tourniquet should be placed at a point that is above a wound, between the bleeding and the victim's heart. However, the binding should not encroach directly on the wound area. Tourniquets can be fashioned out of many materials. If you are using strips of cloth, wrap the cloth around the limb that is wounded and tie a knot in the material. Use a stick, screwdriver, or whatever else you can lay your hands on to tighten the binding.

Once you have made a commitment to apply a tourniquet, the wrapping should be removed only by a physician. It's a good idea to note the time that a tourniquet is applied, as this will help doctors later in assessing their options. As an extension of the tourniquet treatment, you will most likely have to treat the patient for shock.

Infection

Infection is always a concern with open wounds. When a wound is serious enough to require a compression bandage, don't attempt to clean the cut. Keep pressure on the wound to stop bleeding. In cases of sever wounds, be on the look out for shock symptoms, and be prepared to treat them. Your primary concern with a serious open wound is to stop the bleeding and gain professional medical help as soon as possible.

Lesser cuts, which are more common than deep ones, should be cleaned. Regular soap and water can be used to clean a wound before applying a bandage. Remember, we are talking about minor cuts and scrapes at this point. Flush the wound generously with clean water. A piece of sterile gauze can be used to pat the wound dry. Then a clean, dry bandage can be applied to protect the wound while in transport to a medical facility.

SPLINTERS AND SUCH

Splinters and such foreign objects often invade the skin of plumbers. Getting these items out cleanly is best done by a doctor, but there

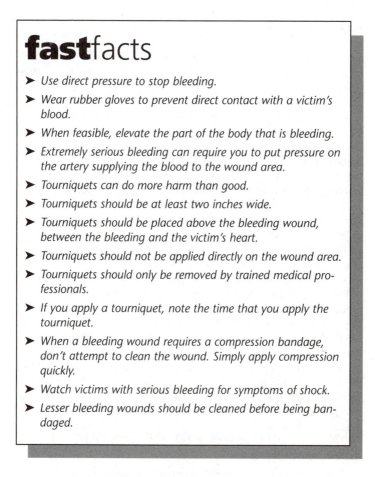

fastfacts

➤ *Use direct pressure to stop bleeding.*

➤ *Wear rubber gloves to prevent direct contact with a victim's blood.*

➤ *When feasible, elevate the part of the body that is bleeding.*

➤ *Extremely serious bleeding can require you to put pressure on the artery supplying the blood to the wound area.*

➤ *Tourniquets can do more harm than good.*

➤ *Tourniquets should be at least two inches wide.*

➤ *Tourniquets should be placed above the bleeding wound, between the bleeding and the victim's heart.*

➤ *Tourniquets should not be applied directly on the wound area.*

➤ *Tourniquets should only be removed by trained medical professionals.*

➤ *If you apply a tourniquet, note the time that you apply the tourniquet.*

➤ *When a bleeding wound requires a compression bandage, don't attempt to clean the wound. Simply apply compression quickly.*

➤ *Watch victims with serious bleeding for symptoms of shock.*

➤ *Lesser bleeding wounds should be cleaned before being bandaged.*

are some on the job methods that you might want to try. A magnifying glass and a pair of tweezers work well together when removing embedded objects, such as splinters and slivers of copper tubing. Ideally, tweezers being used should be sterilized either over an open flame, such as the flame of your torch, or in boiling water.

Splinters and slivers that are submerged beneath the skin can often be lifted out with the tip of a sterilized needle. The use of a needle in conjunction with a pair of tweezers is very effective in the removal of most simple splinters. If you are dealing with something that has gone extremely deep into tissue, it is best to leave the object alone until a doctor can remove it.

EYE INJURIES

Eye injuries are very common on construction and remodeling jobs. Most of these injuries could be avoided if proper eye protection was worn, but far too many workers don't wear safety glasses and goggles. This sets the stage for eye irritations and injuries.

Before you attempt to help someone who is suffering from an eye injury, you should wash your hands thoroughly. I know this is not always possible on construction sites, but cleaning your hands is advantageous. In the meantime, keep the victim from rubbing the injured eye. Rubbing can make matters much worse.

Never attempt to remove a foreign object from someone's eye with the use of a rigid device, such as a toothpick. Cotton swabs that have been wetted can serve well as a magnet to remove some types of invasion objects. If the person you are helping has something embedded in an eye, get the person to a doctor as soon as possible. Don't attempt to remove the object yourself.

When you are investigating the cause of an eye injury, you should pull down the lower lid of the eye to determine if you can see the object causing trouble. A floating object, such as a piece of sawdust trapped between an eye and an eyelid can be removed with a tissue, a damp cotton swab, or even a clean handkerchief. Don't allow dry cotton material to come into contact with an eye. If looking under the lower lid doesn't a source of discomfort, check under the lower lid. Clean water can be used to flush out many eye contaminants without much risk of damage to the eye. Objects that cannot be removed easily should be left alone until a physician can take over.

- *Wash your hands, if possible, before treating eye injuries*
- *Don't rub an eye wound*
- *Don't attempt to remove embedded items from an eye*
- *Clean water can be used to flush out some eye irritants*

SCALP INJURIES

Scalp injuries can be misleading. What looks like a serious wound can be a fairly minor cut. On the other hand, what appears to be only a cut can involve a fractured skull. If you or someone around you sustains a scalp injury, such as having a hammer fall on your head from an overhead worker, take it seriously. Don't attempt to clean the wound. Expect profuse bleeding.

If you don't suspect a skull fracture, raise the victim's head and shoulders to reduce bleeding. Try not to bend the neck. Put a sterile bandage over the wound, but don't apply excessive pressure. If there is a bone fracture, pressure could worsen the situation. Secure the bandage with gauze or some other material that you can wrap around it. Seek medical attention immediately.

FACIAL INJURIES

Facial injuries can occur on plumbing jobs. I've seen helpers let their right-angle drills get away from them with the result being hard knocks to the face. On one occasion, I remember a tooth being lost, and split lips and tongues that have been bitten are common when a drill goes on a rampage.

Extremely bad facial injuries can cause a blockage of the victim's air passages. This, of course, is a very serious condition. It's critical that air passages be open at all times. If the person's mouth contains broken teeth or dentures, remove them. Be careful not to jar the individual's spine if you have reason to believe there may be injury to the back or neck. Conscious victims should be positioned, when possible, so that secretions from the mouth and nose will drain out. Shock is a potential concern in severe facial injuries. For most on the job injuries, plumbers should be treated for comfort and sent for medical attention.

NOSE BLEEDS

Nose bleeds are not usually difficult to treat. Typically, pressure applied to the side of the nose where bleeding is occurring will stop the flow of blood. Applying cold compresses can also help. If external pressure is not stopping the bleeding, use a small, clean pad of gauze to create a dam on the inside of the nose. Then, apply pressure on the outside of the nose. This will almost always work. If it doesn't, get to a doctor.

BACK INJURIES

There is really only one thing that you need to know about back injuries. Don't move the injured party. Call for professional help and see that the victim remains still until help arrives. Moving someone who has suffered a back injury can be very risky. Don't do it unless there is a life-threatening cause for your action, such as a person trapped in a fire or some other type of deadly situation.

LEGS AND FEET

Legs and feet sometimes become injured on job sites. The worst case of this type that I can remember was when a plumber knocked a pot of molten lead over on his foot. It sends shivers up my spine just to recall that incident. Anyway, when someone suffers a minor foot or leg injury, you should clean and cover the wound. Bandages should be supportive without being constrictive. The appendage should be elevated above the victim's heart level when possible. Prohibit the person from walking. Remove boots and socks so that you can keep an eye on the person's toes. If the toes begin to swell or turn blue, loosen the supportive bandages.

Blisters

Blisters may not seem like much of an emergency, but they can sure take the steam out of a helper or plumber. In most cases, blisters can be covered with a heavy gauze pad to reduce pain. It is generally recommended to leave blisters unbroken. When a blister breaks, the area should be cleaned and treated as an open wound. Some blisters tend to be more serious than others. For example, blisters in the palm of a hand or on the sole of a foot should be look at by a doctor.

HAND INJURIES

Hand injuries are common in the plumbing trade. Little cuts are the most frequent complaint. Getting flux in a cut is an eye-opening experience, so even the smallest break in the skin should be covered. Serious hand injuries should be elevated. This tends to reduce swelling. You should not try to clean really bad hand injuries. Use a pressure bandage to control bleeding. If the cut is on the palm of a hand, a roll of gauze can be squeezed by the victim to slow the flow of blood. Pressure should stop the bleeding, but if it doesn't, seek medical assistance. As with all injuries, use common sense on whether or not professional attention is needed after first aid is applied.

SHOCK

Shock is a condition that can be life threatening even when the injury responsible for a person going into shock is not otherwise fatal. We are talking about traumatic shock, not electrical shock. Many factors can lead to a person going into shock. A serious injury is a common cause, but many other causes exist. There are certain signs of shock which you can look for.

If a person's skin turns pale or blue and is cold to the touch, it's a likely sign of shock. Skin that becomes moist and clammy can indicate shock is present. A general weakness is also a sign of shock. When a person is going into shock, the individual's pulse is likely to exceed 100 beats per minute. Breathing is usually increased, but it may be shallow, deep, or irregular. Chest injuries usually result in shallow breathing. Victims who have lost blood may be thrashing about as they enter into shock. Vomiting and nausea can also signal shock.

As a person slips into deeper shock, the individual may become unresponsive. Look at the eyes, they may be widely dilated. Blood pressure can drop, and in time, the victim will lose consciousness. Body temperature will fall, and death will be likely if treatment is not rendered.

There are three main goals when treating someone for shock. Get the person's blood circulating well. Make sure an adequate supply of oxygen is available to the individual, and maintain the person's body temperature. When you have to treat a person for shock, you should keep the victim lying down. Cover the individual so that the loss of body heat will be minimal. Get medical help as soon as possible. The reason it's best to keep a person lying down is that the individual's blood should circulate better. Remember, if you suspect back or neck injuries, don't move the person.

People who are unconscious should be placed on one side so that fluids will run out of the mouth and nose. It's also important to make sure that air passages are open. A person with a head injury may be laid out flat or propped up, but the head should not be lower than the rest of the body. It is sometimes advantageous to elevate a person's feet when they are in shock. However is there is any difficulty in breathing or if pain increases when the feet are raised, lower them.

Body temperature is a big concern with shock patients. You want to overcome or avoid chilling. However, don't attempt to add additional heat to the surface of the person's body with artificial means. This can be damaging. Use only blankets, clothes, and other similar items to regain and maintain body temperature.

Avoid the temptation to offer the victim fluids, unless medical care is not going to be available for a long time. Avoid fluids completely if the person is unconscious or is subject to vomiting. Under most job-site conditions, fluids should not be administered.

Checklist of Shock Symptoms

✔ Skin that is pale, blue, or cold to the touch

✔ Skin that is moist and clammy

✔ General weakness

✔ Pulse rate in excess of 100 beats per minute

✔ Increased breathing

✔ Shallow breathing

✔ Thrashing

✔ Vomiting and nausea

✔ Unresponsive action

✔ Widely dilated eyes

✔ A drop in blood pressure

BURNS

Burns are not real common among plumbers, but they can occur in the workplace. There are three types of burns that you may have to deal with. First-degree burns are the least serious. These burns typically come from overexposure to the sun, which construction workers often suffer from, quick contact with a hot object, like the tip of a torch, and scalding water, which could be the case when working with a boiler or water heater.

Second-degree burns are more serious. They can come from a deep sunburn or from contact with hot liquids and flames. A person who is affected by a second-degree burn may have a red or mottled appearance, blisters, and a wet appearance of the skin within the burn area. This wet look is due to a loss of plasma through the damaged layers of skin.

Third-degree burns are the most serious. They can be caused by contact with open flames, hot objects, or immersion in very hot water. Electrical injuries can also result in third-degree burns. This type of burn can look similar to a second-degree burn, but the difference will be the loss of all layers of skin.

Treatment

Treatment for most job-related burns can be administered on the job site and will not require hospitalization. First-degree burns should be washed with or submerged in cold water. A dry dressing can be applied if necessary. These burns are not too serious. Eliminating pain is the primary goal with first-degree burns.

Second-degree burns should be immersed in cold (but not ice) water. The soaking should continue for at least one hour and up to two hours. After soaking, the wound should be layered with clean cloths that have been dipped in ice water and wrung out. Then the wound should be dried by blotting, not rubbing. A dry, sterile gauze should then be applied. Don't break open any blisters. It is also not advisable to use ointments and sprays on severe burns. Burned arms and legs should be elevated, and medical attention should be acquired.

Bad burns, the third-degree type, need quick medical attention. First, don't remove a burn victim's clothing, skin might come off with it. A thick, sterile dressing can be applied to the burn area. Personally, I would avoid this if possible. A dressing might stick to the mutilated skin and cause additional skin loss when the dressing is removed. When hands are burned, keep them elevated above the victim's heart. The same goes for feet and legs. You should not soak a third-degree burn in cold water, it could induce more shock symptoms. Don't use ointments, sprays, or other types of treatments. Get the burn victim to competent medical care as soon as possible.

HEAT RELATED PROBLEMS

Heat related problems can include heat stroke and heat exhaustion. Cramps are also possible when working in hot weather. There are people who don't consider heat stroke to be serious. They are wrong. Heat stroke can be life threatening. People affected by heat stroke can develop body temperatures in excess of 106 degrees F. Their skin is likely to be hot, red, and dry. You might think sweating would take place, but it doesn't. Pulse is rapid and strong, and victims can sink into an unconscious state.

If you are dealing with heat stroke, you need to lower the person's body temperature quickly. There is a risk, however, of cooling the body too quickly once the victim's temperature is below 102 degrees F. You can lower body temperature with rubbing alcohol, cold packs, cold water on clothes or in a bathtub of cold water. Avoid the use of ice in the cooling process. Fans and air conditioned space can be used to achieve your cooling goals. Get the body temperature down to at least 102 degrees and then go for medical help.

Cramps

Cramps are not uncommon among workers during hot spells. A simple massage can be all it takes to cure this problem. Salt-water solutions are another way to control cramps. Mix one teaspoonful of salt per glass of water and have the victim drink half a glass about every 15 minutes.

Exhaustion

Heat exhaustion is more common than heat stroke. A person affected by heat exhaustion is likely to maintain a fairly normal body temperature. But, the person's skin may be pale and clammy. Sweating may be very noticeable, and the individual will probably complain of being tired and weak. Headaches, cramps, and nausea may accompany the symptoms. In some cases, fainting might occur.

The salt-water treatment described for cramps will normally work with heat exhaustion. Victims should lie down and elevate their feet about a foot off the floor or bed. Clothing should be loosened, and cool, wet cloths can be used to add comfort. If vomiting occurs, get the person to a hospital for intravenous fluids.

We could continue talking about first aid for a much longer time. However, the help I can give you here for medical procedures is limited. You owe it to yourself, your family, and the people you work with to learn first aid techniques. This can be done best by attending formal classes in your area. Most towns and cities offer first aid classes on a regular basis. I strongly suggest that you enroll in one. Until you have some hands-on experience in a classroom and gain the depth of knowledge needed, you are not prepared for emergencies. Don't get caught short. Prepare now for the emergency that might never happen.

chapter 2

REFERENCE AND CONVERSION TABLES AND DATA

This chapter contains a wealth of useful information. It is all presented in a visual format. You can use the material in this chapter to do everything from metric conversions to finding the weight of cast iron pipe. Many plumbing questions are answered in the illustrations found in this chapter.

COMMONLY USED ABBREVIATIONS

ABS	acrylonitrile butadiene styrene
AGA	American Gas Association
AWWA	American Water Works Association
BOCA	Building Officials Conference of America
B&S	bell-and-spigot (cast-iron) pipe
BT	bathtub
C-to-C	center-to-center
CI	cast iron
CISP	cast-iron soil pipe
CISPI	Cast Iron Soil Pipe Institute
CO	clean-out
CPVC	chlorinated polyvinyl chloride
CW	cold water
DF	drinking fountain
DWG	drawing
DWV	drainage, waste, and vent system
EWC	electric water cooler
FG	finish grade
FPT	female pipe thread
FS	federal specifications
FTG	fitting
FU	fixture-unit
GALV	galvanized
GPD	gallons per day
GPM	gallons per minute
HWH	hot-water heater
ID	inside diameter
IPS	iron pipe size
KS	kitchen sink
LAV	lavatory
LT	laundry tray
MAX	maximum
MCA	Mechanical Contractors Association
MGD	million gallons per day
MI	malleable iron
MIN (min)	minute or minimum

FIGURE 15.1 Commonly used abbreviations.

Abbreviation	Meaning
A or a	Area
A.W.G.	American wire gauge
bbl.	Barrels
B or b	Breadth
bhp	Brake horse power
B.M.	Board measure
Btu	British thermal units
B.W.G.	Birmingham wire gauge
B & S	Brown and Sharpe wire gauge (American wire gauge)
C of g	Center gravity
cond.	Condensing
cu.	Cubic
cyl.	Cylinder
D or d	Depth or diameter
evap.	Evaporation
F	Coefficient of friction; Fahrenheit
F or f	Force or factor of safety
ft. lbs.	Foot pounds
gals.	Gallons
H or h	Height or head of water
HP	Horsepower
IHP	Indiated horsepower
L or l	Length
lbs.	Pounds
lbs. per sq. in.	Pounds per square inch
o.d.	Outside diameter (pipes)
oz.	Ounces
pt.	Pint
P or p	Pressure or load
psi	Pounds per square inch
R or r	Radius
rpm	Revolutions per minute
sq. ft.	Square foot
sq. in.	Square inch

FIGURE 15.2 Abbreviations.

A or a	Area, acre
AWG	American Wire Gauge
B or b	Breadth
bbl	Barrels
bhp	Brake horsepower
BM	Board measure
Btu	British thermal units
BWG	Birmingham Wire Gauge
B & S	Brown and Sharpe Wire Gauge (American Wire Gauge)
C of g	Center of gravity
cond	Condensing
cu	Cubic
cyl	Cylinder
D or d	Depth, diameter
dr	Dram
evap	Evaporation
F	Coefficient of friction; Fahrenheit
F or f	Force, factor of safety
ft (or ')	Foot
ft lb	Foot pound
fur	Furlong
gal	Gallon
gi	Gill
ha	Hectare
H or h	Height, head of water
HP	horsepower
IHP	Indicated horsepower
in (or ")	Inch
L or l	Length
lb	Pound
lb/sq in.	Pounds per square inch
mi	Mile
o.d.	Outside diameter (pipes)
oz	Ounces
pt	Pint
P or p	Pressure, load
psi	Pounds per square inch
R or r	Radius
rpm	Revolutions per minute
sq ft	Square foot
sq in.	Square inch
sq yd	Square yard
T or t	Thickness, temperature
temp	Temperature
V or v	Velocity
vol	Volume
W or w	Weight
W. I.	Wrought iron

FIGURE 15.3 Abbreviations.

Material	Chemical symbol
Aluminum	AL
Antimony	Sb
Brass	..
Bronze	..
Chromium	Cr
Copper	Cu
Gold	Au
Iron (cast)	Fe
Iron (wrought)	Fe
Lead	Pb
Manganese	Mn
Mercury	Hg
Molybdenum	Mo
Monel	..
Platinum	Pt
Steel (mild)	Fe
Steel (stainless)	..
Tin	Sn
Titanium	Ti
Zinc	Zn

FIGURE 15.4 Symbols of various materials.

AVAILABLE LENGTHS OF COPPER PLUMBING TUBE

Drawn (hard copper) (feet)		*Annealed (soft copper) (feet)*	
Type K Tube			
Straight Lengths:		Straight Lengths:	
Up to 8-in. diameter	20	Up to 8-in. diameter	20
10-in. diameter	18	10-in. diameter	18
12-in. diameter	12	12-in. diameter	12
		Coils:	
		Up to 1-in. diameter	60
			100
		1¼-in. diameter	60
			100
		2-in. diameter	40
			45
Type L Tube			
Straight Lengths:		Straight Lengths:	
Up to 10-in. diameter	20	Up to 10-in. diameter	20
12-in. diameter	18	12-inch diameter	18
		Coils:	
		Up to 1-in. diameter	60
		100	
		1¼- and 1½-in. diameter	60
		100	
		2-in. diameter	40
		45	
DWV Tube			
Straight Lengths:		Not available	
All diameters	20		
Type M Tube			
Straight Lengths:		Not available	
All diameters	20		

FIGURE 15.5 Available lengths of copper plumbing tube.

WEIGHT OF CAST-IRON PIPE

	Diameter (inches)	Service Weight (lb)	Extra Heavy Weight (lb)
Double Hub, *5-ft Lengths*	2	21	26
	3	31	47
	4	42	63
	5	54	78
	6	68	100
	8	105	157
	10	150	225
Double Hub, *30-ft Length*	2	11	14
	3	17	26
	4	23	33
Single Hub, *5-ft Lengths*	2	20	25
	3	30	45
	4	40	60
	5	52	75
	6	65	95
	8	100	150
	10	145	215
Single Hub, *10-ft Lengths*	2	38	43
	3	56	83
	4	75	108
	5	98	133
	6	124	160
	8	185	265
	10	270	400
No-Hub Pipe, *10-ft Lengths*	1½	27	
	2	38	
	3	54	
	4	74	
	5	95	
	6	118	
	8	180	

FIGURE 15.6 Weight of cast iron pipe.

COMMON SEPTIC TANK CAPACITIES

Single-Family Dwellings; Number of Bedrooms	Multiple Dwelling Units or Apartments; One Bedroom Each	Other Uses; Maximum Fixture-Units Served	Minimum Septic Tank Capacity in Gallons
1–3		20	1000
4	2	25	1200
5–6	3	33	1500
7–8	4	45	2000
	5	55	2250
	6	60	2500
	7	70	2750
	8	80	3000
	9	90	3250
	10	100	3500

FIGURE 15.7 Common septic tank capacities.

FACTS ABOUT WATER

1 ft^3 of water contains 7½ gal, 1728 $in.^3$, and weighs 62½ lb.

1 gal of water weighs 8⅓ lb and contains 231 $in.^3$

Water expands 1/23 of its volume when heated from 40° to 212°.

The height of a column of water, equal to a pressure of 1 $lb/in.^2$, is 2.31 ft.

To find the pressure in $lb/in.^2$ of a column of water, multiply the height of the column in feet by 0.434.

The average pressure of the atmosphere is estimated at 14.7 $lb/in.^2$ so that with a perfect vacuum it will sustain a column of water 34 ft high.

The friction of water in pipes varies as the square of the velocity.

To evaporate 1 ft^3 of water requires the consumption of 7½ lb of ordinary coal or about 1 lb of coal to 1 gal of water.

1 $in.^3$ of water evaporated at atmospheric pressure is converted into approximately 1 ft^3 of steam.

FIGURE 15.8 Facts about water.

RATES OF WATER FLOW

Fixture	*Flow Rate (gpm)*[a]
Ordinary basin faucet	2.0
Self-closing basin faucet	2.5
Sink faucet, 3/8 in.	4.5
Sink faucet, 1/2 in.	4.5
Bathtub faucet	6.0
Laundry tub cock, 1/2 in.	5.0
Shower	5.0
Ballcock for water closet	3.0
Flushometer valve for water closet	15-35
Flushometer valve for urinal	15.0
Drinking fountain	.75
Sillcock (wall hydrant)	5.0

[a]Figures do not represent the use of water-conservation devices.

FIGURE 15.9   Rates of water flow

CONSERVING WATER[a]

Activity	*Normal Use (gallons)*	*Conservative Use (gallons)*
Shower	25 (water running)	4 (wet down, soap up, rinse off)
Tub bath	36 (full)	10–12 (minimal water level)
Dishwashing	50 (tap running)	5 (wash and rinse in sink)
Toilet flushing	5–7 (depends on tank size)	1½–3 (water-saver toilets or tank displacement bottles)
Automatic dishwasher	16 (full cycle)	7 (short cycle)
Washing machine	60 (full cycle, top water level)	27 (short cycle, minimal water level)
Washing hands	2 (tap running)	1 (full basin)
Brushing teeth	1 (tap running)	1/2 (wet and rinse briefly)

[a]Amounts based on 2½–3½ gpm.

FIGURE 15.10   Conserving water.

WEIGHTS OF VARIOUS MATERIALS

Material	Pounds per Cubic Inch	Pounds per Cubic Foot
Aluminum	0.093	160
Antimony	0.2422	418
Brass	0.303	524
Bronze	0.320	552
Chromium	0.2348	406
Copper	0.323	558
Gold	0.6975	1205
Iron (cast)	0.260	450
Iron (wrought)	0.2834	490
Lead	0.4105	710
Manganese	0.2679	463
Mercury	0.491	849
Molybdenum	0.309	534
Monel	0.318	550
Platinum	0.818	1413
Steel (mild)	0.2816	490
Steel (stainless)	0.277	484
Tin	0.265	459
Titanium	0.1278	221
Zinc	0.258	446

FIGURE 15.11 Weights of various materials.

MELTING POINTS OF COMMERCIAL METALS

Metal	Degrees Fahrenheit
Aluminum	1200
Antimony	1150
Bismuth	500
Brass	1700/1850
Copper	1940
Cadmium	610
Iron (cast)	2300
Iron (wrought)	2900
Lead	620
Mercury	139
Steel	2500
Tin	446
Zinc, cast	785

FIGURE 15.12 Melting points of commercial metals.

Pipe size (in inches)	PSI	Length of pipe is 50 feet
¾	20	16
¾	40	24
¾	60	29
¾	80	34
1	20	31
1	40	44
1	60	55
1	80	65
1¼	20	84
1¼	40	121
1¼	60	151
1¼	80	177
1½	20	94
1½	40	137
1½	60	170
1½	80	200

FIGURE 15.13   Discharge of pipes in gallons per minute.

Pipe size (in inches)	PSI	Length of pipe is 100 feet
¾	20	11
¾	40	16
¾	60	20
¾	80	24
1	20	21
1	40	31
1	60	38
1	80	44
1¼	20	58
1¼	40	84
1¼	60	104
1¼	80	121
1½	20	65
1½	40	94
1½	60	117
1½	80	137

FIGURE 15.14   Discharge of pipes in gallons per minute.

Pipe diameter (in inches)	Approximate capacity (in U.S. gallons) per foot of pipe
¾	.0230
1	.0408
1¼	.0638
1½	.0918
2	.1632
3	.3673
4	.6528
6	1.469
8	2.611
10	4.018

FIGURE 15.15   Pipe capacities.

Pipe material	Coefficient	
	in/in/°F	(°C)
Metallic pipe		
Carbon steel	0.000005	(14.0)
Stainless steel	0.000115	(69)
Cast iron	0.0000056	(1.0)
Copper	0.000010	(1.8)
Aluminum	0.0000980	(1.7)
Brass (yellow)	0.000001	(1.8)
Brass (red)	0.000009	(1.4)
Plastic pipe		
ABS	0.00005	(8)
PVC	0.000060	(33)
PB	0.000150	(72)
PE	0.000080	(14.4)
CPVC	0.000035	(6.3)
Styrene	0.000060	(33)
PVDF	0.000085	(14.5)
PP	0.000065	(77)
Saran	0.000038	(6.5)
CAB	0.000080	(14.4)
FRP (average)	0.000011	(1.9)
PVDF	0.000096	(15.1)
CAB	0.000085	(14.5)
HDPE	0.00011	(68)
Glass		
Borosilicate	0.0000018	(0.33)

FIGURE 15.16   Thermal expansion of piping materials.

Length (ft)	Temperature Change (°F)						
	40	50	60	70	80	90	100
20	0.278	0.348	0.418	0.487	0.557	0.626	0.696
40	0.557	0.696	0.835	0.974	1.114	1.235	1.392
60	0.835	1.044	1.253	1.462	1.670	1.879	2.088
80	1.134	1.392	1.670	1.879	2.227	2.506	2.784
100	1.192	1.740	2.088	2.436	2.784	3.132	3.480

FIGURE 15.17   Thermal expansion of PVC-DWV pipe.

Pipe size (in inches)	Maximum outside diameter	Threads per inch
¼	1.375	8
1	1.375	8
1¼	1.6718	9
1½	1.990	9
2	2.5156	8
3	3.6239	6
4	5.0109	4
5	6.260	4
6	7.025	4

FIGURE 15.18   Threads per inch for national standard threads.

Pipe size (in inches)	Maximum outside diameter	Threads per inch
¼	1.0353	14
1	1.295	11.5
1¼	1.6399	11.5
1½	1.8788	11.5
2	2.5156	8
3	3.470	8
4	4.470	8

FIGURE 15.19   Threads per inch for American standard straight pipe.

IPS, in	Weight per foot, lb	Length in feet containing 1 ft³ of water	Gallons in 1 linear ft
¼	0.42		0.005
⅜	0.57	754	0.0099
½	0.85	473	0.016
¾	1.13	270	0.027
1	1.67	166	0.05
1¼	2.27	96	0.07
1½	2.71	70	0.1
2	3.65	42	0.17
2½	5.8	30	0.24
3	7.5	20	0.38
4	10.8	11	0.66
5	14.6	7	1.03
6	19.0	5	1.5
8	25.5	3	2.6
10	40.5	1.8	4.1
12	53.5	1.2	5.9

FIGURE 15.20 Weight of steel pipe and contained water.

Nominal rod diameter, in	Root area of thread, in²	Maximum safe load at rod temperature of 650°F, lb
¼	0.027	240
⁵⁄₁₆	0.046	410
⅜	0.068	610
½	0.126	1,130
⅝	0.202	1,810
¾	0.302	2,710
⅞	0.419	3,770
1	0.552	4,960
1⅛	0.693	6,230
1¼	0.889	8,000
1⅜	1.053	9,470
1½	1.293	11,630
1⅝	1.515	13,630
1¾	1.744	15,690
1⅞	2.048	18,430
2	2.292	20,690
2¼	3.021	27,200
2½	3.716	33,500
2¾	4.619	41,600
3	5.621	50,600
3¼	6.720	60,500
3½	7.918	71,260

FIGURE 15.21 Load rating of threaded rods.

Pipe size, in	Rod size, in
2 and smaller	⅜
2½ to 3½	½
4 and 5	⅝
6	¾
8 to 12	⅞
14 and 16	1
18	1⅛
20	1¼
24	1½

FIGURE 15.22 Recommended rod size for individual pipes.

Pipe	Weight factor*
Aluminum	0.35
Brass	1.12
Cast iron	0.91
Copper	1.14
Stainless steel	1.0
Carbon steel	1.0
Wrought iron	0.98

*Average plastic pipe weights one-fifth as much as carbon steel pipe.

FIGURE 15.23 Relative weight factor for metal pipe.

To change	To	Multiply by
Inches	Millimeters	25.4
Feet	Meters	.3048
Miles	Kilometers	1.6093
Square inches	Square centimeters	6.4515
Square feet	Square meters	.09290
Acres	Hectares	.4047
Acres	Square kilometers	.00405
Cubic inches	Cubic centimeters	16.3872
Cubic feet	Cubic meters	.02832
Cubic yards	Cubic meters	.76452
Cubic inches	Liters	.01639
U.S. gallons	Liters	3.7854
Ounces (avoirdupois)	Grams	28.35
Pounds	Kilograms	.4536
Lbs. per sq. in. (P.S.I.)	Kg.'s per sq. cm.	.0703
Lbs. per cu. ft.	Kg.'s per cu. meter	16.0189
Tons (2000 lbs.)	Metric tons (1000 kg.)	.9072
Horsepower	Kilowatts	.746

FIGURE 15.24 English to metric conversion.

Quantity	Equals
100 sq. millimeters	1 sq. centimeter
100 sq. centimeters	1 sq. decimeter
100 sq. decimeters	1 sq. meter

FIGURE 15.25   Metric square measure.

Unit	Equals
1 cubic meter	35.314 cubic feet 1.308 cubic yards 264.2 U.S. gallons (231 cubic inches)
1 cubic decimeter	61.0230 cubic inches .0353 cubic feet
1 cubic centimeter	.061 cubuic inch
1 liter	1 cubic decimeter 61.0230 cubic inches 0.0353 cubic foot 1.0567 quarts (U.S.) 0.2642 gallon (U.S.) 2.2020 lb. of water at 62°F.
1 cubic yard	.7645 cubic meter
1 cubic foot	.02832 cubic meter 28.317 cubic decimeters 28.317 liters
1 cubic inch	16.383 cubic centimeters
1 gallon (British)	4.543 liters
1 gallon (U.S.)	3.785 liters
1 gram	15.432 grains
1 kilogram	2.2046 pounds
1 metric ton	.9842 ton of 2240 pounds 19.68 cwts. 2204.6 pounds
1 grain	.0648 gram
1 ounce avoirdupois	28.35 grams
1 pound	.4536 kilograms
1 ton of 2240 lb.	1.1016 metric tons 1016 kilograms

FIGURE 15.26   Measures of volume and capacity.

To change	To	Multiply by
Atmospheres	Pounds per square inch	14.696
Atmospheres	Inches of mercury	29.92
Atmospheres	Feet of water	34
Long tons	Pounds	2240
Short tons	Pounds	2000
Short tons	Long tons	0.89295

FIGURE 15.27 Measurement conversion factors.

To find	Multiply	By
Microns	Mils	25.4
Centimeters	Inches	2.54
Meters	Feet	0.3048
Meters	Yards	0.19144
Kilometers	Miles	1.609344
Grams	Ounces	28.349523
Kilograms	Pounds	0.4539237
Liters	Gallons (U.S.)	3.7854118
Liters	Gallons (Imperial)	4.546090
Milliliters (cc)	Fluid ounces	29.573530
Milliliters (cc)	Cubic inches	16.387064
Square centimeters	Square inches	6.4516
Square meters	Square feet	0.09290304
Square meters	Square yards	0.83612736
Cubic meters	Cubic feet	2.8316847×10^{-2}
Cubic meters	Cubic yards	0.76455486
Joules	BTU	1054.3504
Joules	Foot-pounds	1.35582
Kilowatts	BTU per minute	0.01757251
Kilowatts	Foot-pounds per minute	2.2597×10^{-5}
Kilowatts	Horsepower	0.7457
Radians	Degrees	0.017453293
Watts	BTU per minute	17.5725

FIGURE 15.28 Conversion factors in converting from customary units to metric units.

To change	To	Multiply by
Inches	Feet	0.0833
Inches	Millimeters	25.4
Feet	Inches	12
Feet	Yards	0.3333
Yards	Feet	3
Square inches	Square feet	0.00694
Square feet	Square inches	144
Square feet	Square yards	0.11111
Square yards	Square feet	9
Cubic inches	Cubic feet	0.00058
Cubic feet	Cubic inches	1728
Cubic feet	Cubic yards	0.03703
Gallons	Cubic inches	231
Gallons	Cubic feet	0.1337
Gallons	Pounds of water	8.33
Pounds of water	Gallons	0.12004
Ounces	Pounds	0.0625
Pounds	Ounces	16
Inches of water	Pounds per square inch	0.0361
Inches of water	Inches of mercury	0.0735
Inches of water	Ounces per square inch	0.578
Inches of water	Pounds per square foot	5.2
Inches of mercury	Inches of water	13.6
Inches of mercury	Feet of water	1.1333
Inches of mercury	Pounds per square inch	0.4914
Ounces per square inch	Inches of mercury	0.127
Ounces per square inch	Inches of water	1.733
Pounds per square inch	Inches of water	27.72
Pounds per square inch	Feet of water	2.310
Pounds per square inch	Inches of mercury	2.04
Pounds per square inch	Atmospheres	0.0681
Feet of water	Pounds per square inch	0.434
Feet of water	Pounds per square foot	62.5
Feet of water	Inches of mercury	0.8824

FIGURE 15.29 Measurement conversion factors.

Pipe size	Projected flow rate (gallons per minute)
½ inch	2 to 5
¾ inch	5 to 10
1 inch	10 to 20
1¼ inch	20 to 30
1½ inch	30 to 40

FIGURE 15.30 Projected flow rates for various pipe sizes.

Material	Weight in pounds per cubic inch	Weight in pounds per cubic foot
Aluminum	.093	160
Antimony	.2422	418
Brass	.303	524
Bronze	.320	552
Chromium	.2348	406
Copper	.323	558
Gold	.6975	1205
Iron (cast)	.260	450
Iron (wrought)	.2834	490
Lead	.4105	710
Manganese	.2679	463
Mercury	.491	849
Molybdenum	.309	534
Monel	.318	550
Platinum	.818	1413
Steel (mild)	.2816	490
Steel (stainless)	.277	484
Tin	.265	459
Titanium	.1278	221
Zinc	.258	446

FIGURE 15.31 Weights of various materials.

GPM	Liters/Minute
1	3.75
2	6.50
3	11.25
4	15.00
5	18.75
6	22.50
7	26.25
8	30.00
9	33.75
10	37.50

FIGURE 15.32 Flow rate conversion from gallons per minute to approximate liters per minute.

Pounds per square inch	Feet head
1	2.31
2	4.62
3	6.93
4	9.24
5	11.54
6	13.85
7	16.16
8	18.47
9	20.78
10	23.09
15	34.63
20	46.18
25	57.72
30	69.27
40	92.36
50	115.45
60	138.54
70	161.63
80	184.72
90	207.81
100	230.90
110	253.98
120	277.07
130	300.16
140	323.25
150	346.34
160	369.43
170	392.52
180	415.61
200	461.78
250	577.24
300	692.69
350	808.13
400	922.58
500	1154.48
600	1385.39
700	1616.30
800	1847.20
900	2078.10
1000	2309.00

FIGURE 15.33 Water pressure in pounds with equivalent feet heads.

Feet head	Pounds per square inch	Feet head	Pounds per square inch
1	0.43	50	21.65
2	0.87	60	25.99
3	1.30	70	30.32
4	1.73	80	34.65
5	2.17	90	38.98
6	2.60	100	43.34
7	3.03	110	47.64
8	3.46	120	51.97
9	3.90	130	56.30
10	4.33	140	60.63
15	6.50	150	64.96
20	8.66	160	69.29
25	10.83	170	73.63
30	12.99	180	77.96
40	17.32	200	86.62

FIGURE 15.34 Water feet head to pounds per square inch.

Vacuum in inches of mercury	Boiling point
29	76.62
28	99.93
27	114.22
26	124.77
25	133.22
24	140.31
23	146.45

FIGURE 15.35 Boiling points of water at various pressures.

To change	to	Multiply by
Inches	Feet	0.0833
Inches	Millimeters	25.4
Feet	Inches	12
Feet	Yards	0.3333
Yards	Feet	3
Square inches	Square feet	0.00694
Square feet	Square inches	144
Square feet	Square yards	0.11111
Square yards	Square feet	9
Cubic inches	Cubic feet	0.00058
Cubic feet	Cubic inches	1728
Cubic feet	Cubic yards	0.03703
Gallons	Cubic inches	231
Gallons	Cubic feet	0.1337
Gallons	Pounds of water	8.33
Pounds of water	Gallons	0.12004
Ounces	Pounds	0.0625
Pounds	Ounces	16
Inches of water	Pounds per square inch	0.0361
Inches of water	Inches of mercury	0.0735
Inches of water	Ounces per square inch	0.578
Inches of water	Pounds per square foot	5.2
Inches of mercury	Inches of water	13.6
Inches of mercury	Feet of water	1.1333
Inches of mercury	Pounds per square inch	0.4914
Ounces per square inch	Inches of mercury	0.127
Ounces per square inch	Inches of water	1.733
Pounds per square inch	Inches of water	27.72
Pounds per square inch	Feet of water	2.310
Pounds per square inch	Inches of mercury	2.04
Pounds per square inch	Atmospheres	0.0681
Feet of water	Pounds per square inch	0.434
Feet of water	Pounds per square foot	62.5
Feet of water	Inches of mercury	0.8824
Atmospheres	Pounds per square inch	14.696
Atmospheres	Inches of mercury	29.92
Atmospheres	Feet of water	34
Long tons	Pounds	2240
Short tons	Pounds	2000
Short tons	Long tons	0.89295

FIGURE 15.36 Measurement conversion factors.

To change	to	Multiply by
Inches	Feet	0.0833
Inches	Millimeters	25.4
Feet	Inches	12
Feet	Yards	0.3333
Yards	Feet	3
Square inches	Square feet	0.00694
Square feet	Square inches	144
Square feet	Square yards	0.11111
Square yards	Square feet	9
Cubic inches	Cubic feet	0.00058
Cubic feet	Cubic inches	1728
Cubic feet	Cubic yards	0.03703
Cubic yards	Cubic feet	27
Cubic inches	Gallons	0.00433
Cubic feet	Gallons	7.48
Gallons	Cubic inches	231
Gallons	Cubic feet	0.1337
Gallons	Pounds of water	8.33
Pounds of water	Gallons	0.12004
Ounces	Pounds	0.0625
Pounds	Ounces	16
Inches of water	Pounds per square inch	0.0361
Inches of water	Inches of mercury	0.0735
Inches of water	Ounces per square inch	0.578
Inches of water	Pounds per square foot	5.2
Inches of mercury	Inches of water	13.6
Inches of mercury	Feet of water	1.1333
Inches of mercury	Feet of water	0.4914
Ounces per square inch	Pounds per square inch	0.127
Ounces per square inch	Inches of mercury	1.733
Pounds per square inch	Inches of water	27.72
Pounds per square inch	Feet of water	2.310
Pounds per square inch	Inches of mercury	2.04
Pounds per square inch	Atmospheres	0.0681
Feet of water	Pounds per square inch	0.434
Feet of water	Pounds per square foot	62.5
Feet of water	Inches of mercury	0.8824
Atmospheres	Pounds per square inch	14.696
Atmospheres	Inches of mercury	29.92
Atmospheres	Feet of water	34
Long tons	Pounds	2240
Short tons	Pounds	2000
Short tons	Long tons	0.89295

FIGURE 15.37 Useful multipliers.

Outside design temperature = average of lowest recorded temperature in each
month from October to March

Inside design temperature = 70°F or as specified by owner

A degree day is one day multiplied by the number of Fahrenheit degrees the mean
temperature is below 65°F. The number of degree days in a year is a good guideline
for designing heating and insulation systems.

FIGURE 15.38 Design temperatures.

Temperature (°F)	Steel	Cast iron	Brass and copper
0	0	0	0
20	0.15	0.10	0.25
40	0.30	0.25	0.45
60	0.45	0.40	0.65
80	0.60	0.55	0.90
100	0.75	0.70	1.15
120	0.90	0.85	1.40
140	1.10	1.00	1.65
160	1.25	1.15	1.90
180	1.45	1.30	2.15
200	1.60	1.50	2.40
220	1.80	1.65	2.65
240	2.00	1.80	2.90
260	2.15	1.95	3.15
280	2.35	2.15	3.45
300	2.50	2.35	3.75
320	2.70	2.50	4.05
340	2.90	2.70	4.35
360	3.05	2.90	4.65
380	3.25	3.10	4.95
400	3.45	3.30	5.25
420	3.70	3.50	5.60
440	3.95	3.75	5.95
460	4.20	4.00	6.30
480	4.45	4.25	6.65
500	4.70	4.45	7.05

FIGURE 15.39 Steam pipe expansion (inches increase per 100 inches).

Unit	Symbol
Length	
Millimeter	mm
Centimeter	cm
Meter	m
Kilometer	km
Area	
Square millimeter	mm^2
Square centimeter	cm^2
Square decimeter	dm^2
Square meter	m^2
Square kilometer	km^2
Volume	
Cubic centimeter	cm^3
Cubic decimeter	dm^3
Cubic meter	m^3
Mass	
Milligram	mg
Gram	g
Kilogram	kg
Tonne	t
Temperature	
Degrees Celsius	°C
Kelvin	K
Time	
Second	s
Plane angle	
Radius	rad
Force	
Newton	N
Energy, work, quantity of heat	
Joule	J
Kilojoule	kJ
Megajoule	MJ
Power, heat flow rate	
Watt	W
Kilowatt	kW
Pressure	
Pascal	Pa
Kilopascal	kPa
Megapascal	MPa
Velocity, speed	
Meter per second	m/s
Kilometer per hour	km/h

FIGURE 15.40 Metric abbreviations.

Appliance	Size (inches)
Clothes washer	2
Bathtub with or without shower	1½
Bidet	1½
Dental unit or cuspidor	1¼
Drinking fountain	1¼
Dishwasher, domestic	1½
Dishwasher, commercial	2
Floor drain	2, 3, or 4
Lavatory	1¼
Laundry tray	1½
Shower stall, domestic	2
Sinks:	
Combination, sink and tray (with disposal unit)	1½
Combination, sink and tray (with one trap)	1½
Domestic, with or without disposal unit	1½
Surgeon's	1½
Laboratory	1½
Flushrim or bedpan washer	3
Service sink	2 or 3
Pot or scullery sink	2
Soda fountain	1½
Commercial, flat rim, bar, or counter sink	1½
Wash sinks circular or multiple	1½
Urinals:	
Pedestal	3
Wall-hung	1½ or 2
Trough (per 6-ft section)	1½
Stall	2
Water closet	3

FIGURE 15.41 Common trap sizes.

1 ft^3 at 50°F weighs 62.41 lb
1 gal at 50°F weighs 8.34 lb
1 ft^3 of ice weighs 57.2 lb
Water is at its greatest density at 39.2°F
1 ft^3 at 39.2°F weighs 62.43 lb

FIGURE 15.42 Water weight.

1 ft^3	62.4 lbs
1 ft^3	7.48 gal
1 gal	8.33 lbs
1 gal	0.1337 ft^3

FIGURE 15.43 Water volume to weight.

Category	Estimated water usage per day
Barber shop	100 gal per chair
Beauty shop	125 gal per chair
Boarding school, elementary	75 gal per student
Boarding school, secondary	100 gal per student
Clubs, civic	3 gal per person
Clubs, country	25 gal per person
College, day students	25 gal per student
College, junior	100 gal per student
College, senior	100 gal per student
Dentist's office	750 gal per chair
Department store	40 gal per employee
Drugstore	500 gal per store
Drugstore with fountain	2000 gal per store
Elementary school	16 gal per student
Hospital	400 gal per patient
Industrial plant	30 gal per employee + process water
Junior and senior high school	25 gal per student
Laundry	2000–20,000 gal
Launderette	1000 gal per unit
Meat market	5 gal per 100 ft^2 of floor area
Motel or hotel	125 gal per room
Nursing home	150 gal per patient
Office building	25 gal per employee
Physician's office	200 gal per examining room
Prison	60 gal per inmate
Restaurant	20–120 gal per seat
Rooming house	100 gal per tenant
Service station	600–1500 gal per stall
Summer camp	60 gal per person
Theater	3 gal per seat

FIGURE 15.44 Estimating guidelines for daily water usage.

Unit	Pounds per square inch	Feet of water	Meters of water	Inches of mercury	Atmospheres
1 pound per square inch	1.0	2.31	0.704	2.04	0.0681
1 foot of water	0.433	1.0	0.305	0.882	0.02947
1 meter of water	1.421	3.28	1.00	2.89	0.0967
1 inch of mercury	0.491	1.134	0.3456	1.00	0.0334
1 atmosphere (sea level)	14.70	33.93	10.34	29.92	1.0000

FIGURE 15.45 Conversion of water values.

To convert	Multiply by	To obtain
	A	
acres	4.35×10^4	sq. ft.
acres	4.047×10^3	sq. meters
acre-feet	4.356×10^4	cu. feet
acre-feet	3.259×10^5	gallons
atmospheres	2.992×10^1	in. of mercury (at 0°C.)
atmospheres	1.0333	kgs./sq. cm.
atmospheres	1.0333×10^4	kgs./sq. meter
atmospheres	1.47×10^1	pounds/sq. in.
	B	
barrels (u.s., liquid)	3.15×10^1	gallons
barrels (oil)	4.2×10^1	gallons (oil)
bars	9.869×10^{-1}	atmospheres
btu	7.7816×10^2	foot-pounds
btu	3.927×10^{-4}	horsepower-hours
btu	2.52×10^{-1}	kilogram-calories
btu	2.928×10^{-4}	kilowatt-hours
btu/hr.	2.162×10^{-1}	ft. pounds/sec.
btu/hr.	3.929×10^{-4}	horsepower
btu/hr.	2.931×10^{-1}	watts
btu/min.	1.296×10^1	ft.-pounds/sec.
btu/min.	1.757×10^{-2}	kilowatts
	C	
centigrade (degrees)	(°C × 9/5) + 32	fahrenheit (degrees)
centigrade (degrees)	°C + 273.18	kelvin (degrees)
centigrams	$1. \times 10^{-2}$	grams
centimeters	3.281×10^{-2}	feet
centimeters	3.937×10^{-1}	inches
centimeters	$1. \times 10^{-5}$	kilometers
centimeters	$1. \times 10^{-2}$	meters
centimeters	$1. \times 10^1$	millimeters
centimeters	3.937×10^2	mils
centimeters of mercury	1.316×10^{-2}	atmospheres
centimeters of mercury	4.461×10^{-1}	ft. of water
centimeters of mercury	1.934×10^{-1}	pounds/sq. in.
centimeters/sec.	1.969	feet/min.
centimeters/sec.	3.281×10^{-2}	feet/sec.
centimeters/sec.	6.0×10^{-1}	meters/min.
centimeters/sec./sec.	3.281×10^{-2}	ft./sec./sec.
cubic centimeters	3.531×10^{-5}	cubic ft.
cubic centimeters	6.102×10^{-2}	cubic in.
cubic centimeters	1.0×10^{-6}	cubic meters
cubic centimeters	2.642×10^{-4}	gallons (u.s. liquid)
cubic centimeters	2.113×10^{-3}	pints (u.s. liquid)
cubic centimeters	1.057×10^{-3}	quarts (u.s. liquid)

FIGURE 15.46 Measurement conversions.

To convert	Multiply by	To obtain
cubic feet	2.8320×10^4	cu. cms.
cubic feet	1.728×10^3	cu. inches
cubic feet	2.832×10^{-2}	cu. meters
cubic feet	7.48052	gallons (u.s. liquid)
cubic feet	5.984×10^1	pints (u.s. liquid)
cubic feet	2.992×10^1	quarts (u.s. liquid)
cubic feet/min.	4.72×10^1	cu. cms./sec.
cubic feet/min.	1.247×10^{-1}	gallons/sec.
cubic feet/min.	4.720×10^{-1}	liters/sec.
cubic feet/min.	6.243×10^1	pounds water/min.
cubic feet/sec.	6.46317×10^{-1}	million gals./day
cubic feet/sec.	4.48831×10^2	gallons/min.
cubic inches	5.787×10^{-4}	cu. ft.
cubic inches	1.639×10^{-5}	cu. meters
cubic inches	2.143×10^{-5}	cu. yards
cubic inches	4.329×10^{-3}	gallons

D

degrees (angle)	1.745×10^{-2}	radians
degrees (angle)	3.6×10^3	seconds
degrees/sec.	2.778×10^{-3}	revolutions/sec.
dynes/sq. cm.	4.015×10^{-4}	in. of water (at 4°C.)
dynes	1.020×10^{-6}	kilograms
dynes	2.248×10^{-6}	pounds

F

fathoms	1.8288	meters
fathoms	6.0	feet
feet	3.048×10^1	centimeters
feet	3.048×10^{-1}	meters
feet of water	2.95×10^{-2}	atmospheres
feet of water	3.048×10^{-2}	kgs./sq. cm.
feet of water	6.243×10^1	pounds/sq. ft.
feet/min.	5.080×10^{-1}	cms./sec.
feet/min.	1.667×10^{-2}	feet/sec.
feet/min.	3.048×10^{-1}	meters/min.
feet/min.	1.136×10^{-2}	miles/hr.
feet/sec.	1.829×10^1	meters/min.
feet/100 feet	1.0	per cent grade
foot-pounds	1.286×10^{-3}	btu
foot-pounds	1.356×10^7	ergs
foot-pounds	3.766×10^{-7}	kilowatt-hrs.
foot-pounds/min.	1.286×10^{-3}	btu/min.
foot-pounds/min.	3.030×10^{-5}	horsepower
foot-pounds/min.	3.241×10^{-4}	kg.-calories/min.
foot-pounds/sec.	4.6263	btu/hr.
foot-pounds/sec.	7.717×10^{-2}	btu/min.
foot-pounds/sec.	1.818×10^{-3}	horsepower
foot-pounds/sec.	1.356×10^{-3}	kilowatts
furlongs	1.25×10^{-1}	miles (u.s.)

FIGURE 15.46 *(continued)* Measurement conversions.

To convert	Multiply by	To obtain
	G	
gallons	3.785×10^3	cu. cms.
gallons	1.337×10^{-1}	cu. feet
gallons	2.31×10^2	cu. inches
gallons	3.785×10^{-3}	cu. meters
gallons	4.951×10^{-3}	cu. yards
gallons	3.785	liters
gallons (liq. br. imp.)	1.20095	gallons (u.s. liquid)
gallons (u.s.)	$8,3267 \times 10^{-1}$	gallons (imp.)
gallons of water	8.337	pounds of water
gallons/min.	2.228×10^{-3}	cu. feet/sec.
gallons/min.	6.308×10^{-2}	liters/sec.
gallons/min.	8.0208	cu. feet/hr.
	H	
horsepower	4.244×10^1	btu/min.
horsepower	3.3×10^4	foot-lbs./min.
horsepower	5.50×10^2	foot-lbs./sec.
horsepower (metric)	9.863×10^{-1}	horsepower
horsepower	1.014	horsepower (metric)
horsepower	7.457×10^{-1}	kilowatts
horsepower	7.457×10^2	watts
horsepower (boiler)	3.352×10^4	btu/hr.
horsepower (boiler)	9.803	kilowatts
horsepower-hours	2.547×10^3	btu
horsepower-hours	1.98×10^6	foot-lbs.
horsepower-hours	6.4119×10^5	gram-calories
hours	5.952×10^{-3}	weeks
	I	
inches	2.540	centimeters
inches	2.540×10^{-2}	meters
inches	1.578×10^{-5}	miles
inches	2.54×10^1	millimeters
inches	1.0×10^3	mils
inches	2.778×10^{-2}	yards
inches of mercury	3.342×10^{-2}	atmospheres
inches of mercury	1.133	feet of water
inches of mercury	3.453×10^{-2}	kgs./sq. cm.
inches of mercury	3.453×10^2	kgs./sq. meter
inches of mercury	7.073×10^1	pounds/sq. ft.
inches of mercury	4.912×10^{-1}	pounds/sq. in.
in. of water (at 4°C.)	7.355×10^{-2}	inches of mercury
in. of water (at 4°C.)	2.54×10^{-3}	kgs./sq. cm.
in. of water (at 4°C.)	5.204	pounds/sq. ft.
in. of water (at 4°C.)	3.613×10^{-2}	pounds/sq. in.

FIGURE 15.46 *(continued)* Measurement conversions.

To convert	Multiply by	To obtain
	J	
joules	9.486×10^{-4}	btu
joules/cm.	1.0×10^{7}	dynes
joules/cm.	1.0×10^{2}	joules/meter (newtons)
joules/cm.	2.248×10^{1}	pounds
	K	
kilograms	9.80665×10^{5}	dynes
kilograms	1.0×10^{3}	grams
kilograms	2.2046	pounds
kilograms	9.842×10^{-4}	tons (long)
kilograms	1.102×10^{-3}	tons (short)
kilograms/sq. cm.	9.678×10^{-1}	atmospheres
kilograms/sq. cm.	3.281×10^{1}	feet of water
kilograms/sq. cm.	2.896×10^{1}	inches of mercury
kilograms/sq. cm.	1.422×10^{1}	pounds/sq. in.
kilometers	1.0×10^{5}	centimeters
kilometers	3.281×10^{3}	feet
kilometers	3.937×10^{4}	inches
kilometers	1.0×10^{3}	meters
kilometers	6.214×10^{-1}	miles (statute)
kilometers	5.396×10^{-1}	miles (nautical)
kilometers	1.0×10^{6}	millimeters
kilowatts	5.692×10^{1}	btu/min.
kilowatts	4.426×10^{4}	foot-lbs./min.
kilowatts	7.376×10^{2}	foot-lbs./sec.
kilowatts	1.341	horsepower
kilowatts	1.434×10^{1}	kg.-calories/min.
kilowatts	1.0×10^{3}	watts
kilowatt-hrs.	3.413×10^{3}	btu
kilowatt-hrs.	2.655×10^{6}	foot-lbs.
kilowatt-hrs.	8.5985×10^{3}	gram calories
kilowatt-hrs.	1.341	horsepower-hours
kilowatt-hrs.	3.6×10^{6}	joules
kilowatt-hrs.	8.605×10^{2}	kg.-calories
kilowatt-hrs.	8.5985×10^{3}	kg.-meters
kilowatt-hrs.	2.275×10^{1}	pounds of water raised from 62° to 212°F.
	L	
links (engineers)	1.2×10^{1}	inches
links (surveyors)	7.92	inches
liters	1.0×10^{3}	cu. cm.
liters	6.102×10^{1}	cu. inches
liters	1.0×10^{-3}	cu. meters
liters	2.642×10^{-1}	gallons (u.s. liquid)
liters	2.113	pints (u.s. liquid)
liters	1.057	quarts (u.s. liquid)

FIGURE 15.46 *(continued)* Measurement conversions.

To convert	Multiply by	To obtain
M		
meters	1.0×10^2	centimeters
meters	3.281	feet
meters	3.937×10^1	inches
meters	1.0×10^{-3}	kilometers
meters	5.396×10^{-4}	miles (nautical)
meters	6.214×10^{-4}	miles (statute)
meters	1.0×10^3	millimeters
meters/min.	1.667	cms./sec.
meters/min.	3.281	feet/min.
meters/min.	5.468×10^{-2}	feet/sec.
meters/min.	6.0×10^{-2}	kms./hr.
meters/min.	3.238×10^{-2}	knots
meters/min.	3.728×10^{-2}	miles/hr.
meters/sec.	1.968×10^2	feet/min.
meters/sec.	3.281	feet/sec.
meters/sec.	3.6	kilometers/hr.
meters/sec.	6.0×10^{-2}	kilometers/min.
meters/sec.	2.237	miles/hr.
meters/sec.	3.728×10^{-2}	miles/min.
miles (neutical)	6.076×10^3	feet
miles (statute)	5.280×10^3	feet
miles/hr.	8.8×10^1	ft./min.
millimeters	1.0×10^{-1}	centimeters
millimeters	3.281×10^{-3}	feet
millimeters	3.937×10^{-2}	inches
millimeters	1.0×10^{-1}	meters
minutes (time)	9.9206×10^{-5}	weeks
O		
ounces	2.8349×10^1	grams
ounces	6.25×10^{-2}	pounds
ounces (fluid)	1.805	cu. inches
ounces (fluid)	2.957×10^{-2}	liters
P		
parts/million	5.84×10^{-2}	grains/u.s. gal.
parts/million	7.016×10^{-2}	grains/imp. gal.
parts/million	8.345	pounds/million gal.
pints (liquid)	4.732×10^2	cubic cms.
pints (liquid)	1.671×10^{-2}	cubic ft.
pints (liquid)	2.887×10^1	cubic inches
pints (liqui)	4.732×10^{-4}	cubic meters
pints (liquid)	1.25×10^{-1}	gallons
pints (liquid)	4.732×10^{-1}	liters
pints (liquid)	5.0×10^{-1}	quarts (liquid)

FIGURE 15.46 *(continued)* Measurement conversions.

To convert	Multiply by	To obtain
pounds	2.56×10^2	drams
pounds	4.448×10^5	dynes
pounds	7.0×10^1	grains
pounds	4.5359×10^2	grams
pounds	4.536×10^{-1}	kilograms
pounds	1.6×10^1	ounces
pounds	3.217×10^1	pounds
pounds	1.21528	pounds (troy)
pounds of water	1.602×10^{-2}	cu. ft.
pounds of water	2.768×10^1	cu. inches
pounds of water	1.198×10^{-1}	gallons
pounds of water/min.	2.670×10^{-4}	cu. ft./sec.
pound-feet	1.356×10^7	cm.-dynes
pound-feet	1.3825×10^4	cm.-grams
pound-feet	1.383×10^{-1}	meter-kgs.
pounds/cu. ft.	1.602×10^{-2}	grams/cu. cm.
pounds/cu. ft.	5.787×10^{-4}	pounds/cu. inches
pounds/sq. in.	6.804×10^{-2}	atmospheres
pounds/sq. in.	2.307	feet of water
pounds/sq. in.	2.036	inches of mercury
pounds/sq. in.	7.031×10^2	kgs./sq. meter
pounds/sq. in.	1.44×10^2	pounds/sq. ft.
	Q	
quarts (dry)	6.72×10^1	cu. inches
quarts (liquid)	9.464×10^2	cu. cms.
quarts (liquid)	3.342×10^{-2}	cu. ft.
quarts (liquid)	5.775×10^1	cu. inches
quarts (liquid)	2.5×10^{-1}	gallons
	R	
revolutions	3.60×10^2	degrees
revolutions	4.0	quadrants
rods (surveyors' meas.)	5.5	yards
rods	1.65×10^1	feet
rods	1.98×10^2	inches
rods	3.125×10^{-3}	miles

FIGURE 15.46 *(continued)* Measurement conversions.

To convert	Multiply by	To obtain
	S	
slugs	3.217×10^1	pounds
square centimeters	1.076×10^{-3}	sq. feet
square centimeters	1.550×10^{-1}	sq. inches
square centimeters	1.0×10^{-4}	sq. meters
square centimeters	1.0×10^2	sq. millimeters
square feet	2.296×10^{-5}	acres
square feet	9.29×10^2	sq. cms.
square feet	1.44×10^2	sq. inches
square feet	9.29×10^{-2}	sq. meters
square feet	3.587×10^{-3}	sq. miles
square inches	6.944×10^{-3}	sq. ft.
square inches	6.452×10^2	sq. millimeters
square miles	6.40×10^2	acres
square miles	2.788×10^7	sq. ft.
square yards	2.066×10^{-4}	acres
square yards	8.361×10^3	sq. cms.
square yards	9.0	sq. ft.
square yards	1.296×10^3	sq. inches
	T	
temperature (°C.) +273	1.0	absolute temperature (°K.)
temperature (°C.) +17.78	1.8	temperature (°F.)
temperature (°F.) +460	1.0	absolute temperature (°R.)
temperature (°F.) −32	⁵⁄₉	temperature (°C.)
tons (long)	2.24×10^2	pounds
tons (long)	1.12	tons (short)
tons (metric)	2.205×10^6	pounds
tons (short)	2.0×10^3	pounds
	W	
watts	3.4129	btu/hr.
watts	5.688×10^{-2}	btu/min.
watts	4.427×10^1	ft.-lbs/min.
watts	7.378×10^{-1}	ft.-lbs./sec
watts	1.341×10^{-3}	horsepower
watts	1.36×10^{-3}	horsepower (metric)
watts	1.0×10^{-3}	kilowatts
watt-hours	3.413	btu
watt-hours	2.656×10^3	foot-lbs.
watt-hours	1.341×10^{-3}	horsepower-hours
watt (international)	1.000165	watt (absolute)
weeks	1.68×10^2	hours
weeks	1.008×10^4	minutes
weeks	6.048×10^5	seconds

Source: *Pump Handbook* by I. J. Karassik et al. Copyright 1976, McGraw-Hill, Inc.

FIGURE 15.46 *(continued)*   Measurement conversions.

Nom. pipe size, in	Relative humidity, %														
	20			50			70			80			90		
	THK*	HG†	ST‡	THK	HG	ST	THK	HG	ST	THK	HG	ST	THK	HG	ST
0.50				0.5	4	64	0.5	4	64	0.5	4	64	1.5	2	68
0.75				0.5	4	64	0.5	4	64	0.5	4	64	1.5	3	67
1.00				0.5	6	63	0.5	6	63	1.0	4	66	1.5	3	67
1.25				0.5	6	63	0.5	6	63	1.0	5	65	1.5	3	67
1.50				0.5	8	62	0.5	8	62	1.0	5	66	1.5	4	67
2.00				0.5	8	63	0.5	8	63	1.0	6	66	1.5	4	67
2.50				0.5	10	63	0.5	10	63	1.0	6	66	1.5	5	67
3.00				0.5	12	62	0.5	12	62	1.0	8	65	1.5	6	67
3.50	Condensation			0.5	14	61	0.5	14	61	1.0	7	66	1.5	6	67
4.00	control not			0.5	15	62	0.5	15	62	1.0	9	65	1.5	7	67
5.00	required for this			0.5	16	63	0.5	16	63	1.0	11	65	2.0	7	67
6.00	condition			0.5	22	61	0.5	22	61	1.0	13	65	2.0	8	67
8.00				1.0	16	65	1.0	16	65	1.0	16	65	2.0	10	67
10.00				1.0	20	65	1.0	20	65	1.0	20	65	2.0	11	67
12.00				1.0	22	65	1.0	22	65	1.0	22	65	2.0	13	67

*THK—Insulation thickness, inches.
†HG—Heat gain/lineal foot (pipe) 28 ft (flat).
‡ST—Surface temperature.

FIGURE 15.47 Insulation thickness to prevent condensation, 34 degree F. service temperature and 70 degree F ambient temperature.

Nom. pipe size, in	Relative humidity, %														
	20			50			70			80			90		
	THK*	HG†	ST‡	THK	HG	ST	THK	HG	ST	THK	HG	ST	THK	HG	ST
0.50				0.5	2	66	0.5	2	66	0.5	2	66	1.0	2	68
0.75				0.5	2	67	0.5	2	67	0.5	2	67	0.5	2	67
1.00				0.5	3	66	0.5	3	66	0.5	3	66	1.0	2	68
1.25				0.5	3	66	0.5	3	66	0.5	3	66	1.0	3	67
1.50				0.5	4	65	0.5	4	65	0.5	4	65	1.0	3	67
2.00				0.5	5	66	0.5	5	66	0.5	5	66	1.0	3	67
2.50				0.5	5	65	0.5	5	65	0.5	5	65	1.0	4	67
3.00				0.5	7	65	0.5	7	65	0.5	7	65	1.0	4	67
3.50	Condensation			0.5	8	65	0.5	8	65	0.5	8	65	1.0	4	68
4.00	control not			0.5	8	65	0.5	8	65	0.5	8	65	1.0	5	67
5.00	required for this			0.5	10	65	0.5	10	65	0.5	10	65	1.0	6	67
6.00	condition			0.5	12	65	0.5	12	65	0.5	12	65	1.0	7	67
8.00				1.0	9	67	1.0	9	67	1.0	9	67	1.0	9	67
10.00				1.0	11	67	1.0	11	67	1.0	11	67	1.0	11	67
12.00				1.0	12	67	1.0	12	67	1.0	12	67	1.0	12	67

*THK—Insulation thickness, inches.
†HG—Heat gain/lineal foot (pipe) 28 ft (flat).
‡ST—Surface temperature.

FIGURE 15.48 Insulation thickness to prevent condensation, 50 degree F. service temperature and 70 degree F ambient temperature.

Fixture	Pressure, psi
Basin faucet	8
Basin faucet, self-closing	12
Sink faucet, ⅜ in (0.95 cm)	10
Sink faucet, ½ in (1.3 cm)	5
Dishwasher	15–25
Bathtub faucet	5
Laundry tub cock, ¼ in (0.64 cm)	5
Shower	12
Water closet ball cock	15
Water closet flush valve	15–20
Urinal flush valve	15
Garden hose, 50 ft (15 m), and sill cock	30
Water closet, blowout type	25
Urinal, blowout type	25
Water closet, low-silhouette tank type	30–40
Water closet, pressure tank	20–30

FIGURE 15.49 Minimum acceptable operating pressures for various fixtures.

Use	Minimum temperature, °F
Lavatory:	
Hand washing	105
Shaving	115
Showers and tubs	110
Commercial and institutional laundry	180
Residential dishwashing and laundry	140
Commercial spray-type dishwashing as required by National Sanitation Foundation:	
Single or multiple tank hood or rack type:	
Wash	150
Final rinse	180 to 195
Single-tank conveyor type:	
Wash	160
Final rinse	180 to 195
Single-tank rack or door type:	
Single-temperature wash and rinse	165
Chemical sanitizing glasswasher:	
Wash	140
Rinse	75

FIGURE 15.50 Minimum hot water temperature for plumbing fixtures and equipment.

| Pipe | Fittings | | Maximum working pressure |
	Schedule	Sizes	
160 psi (SDR 26) (1102.4 kPa)	40	½″ thru 8″ incl. (12.7 mm–203.2 mm)	160 psi–1102.4 kPa
	80	½″ thru 8″ incl. (12.7 mm–203.2 mm)	160 psi–1102.4 kPa
200 psi (SDR 21) (1378 kPa)	40	½″ thru 4″ incl. (12.7 mm–101.6 mm)	200 psi–1378 kPa
	80	½″ thru 8″ incl. (12.7 mm–203.2 mm)	200 psi–1378 kPa
250 psi (SDR 17) (1722.5 kPa)	40	½″ thru 3″ incl. (12.7 mm–76.2 mm)	250 psi–1722.5 kPa
	80	½″ thru 8″ incl. (12.7 mm–101.6 mm)	250 psi–1722.5 kPa
315 psi (SDR 13.5) (2170.4 kPa)	40	½″ thru 1½″ incl. (12.7 mm–38.1 mm)	315 psi–2170.4 kPa
	80	½″ thru 4″ incl. (12.7 mm–101.6 mm)	315 psi–2170.4 kPa
Schedule 40	40 80	½″ thru 1½″ incl. (12.7 mm–38,1 mm)	320 psi–2204.8 kPa
	40 80	2″ thru 4″ incl. (50.8 mm–101.6 mm)	220 psi–1515.8 kPa
	40	5″ thru 8″ incl.	160 psi–1102.4 kPa
Schedule 80	40	½″ thru 1½″ incl. (12.7 mm–38.1 mm)	320 psi–2204.8 kPa
	40	2″ thru 4″ incl. (50.8 mm–101.6 mm)	220 psi–1515.8 kPa
	40	5″ thru 8″ incl.	160 psi–1102.4 kPa
	80	½″ thru 4″ incl. (12.7 mm–101.6 mm)	320 psi–2204.8 kPa
	80	5″ thru 8″ incl. (127 mm–203.2 mm)	250 psi–1722.5 kPa

FIGURE 15.51 Maximum working pressures.

INDEX